Advanced ICTs for Disaster Management and Threat Detection:
Collaborative and Distributed Frameworks

Eleana Asimakopoulou
Loughborough University, UK

Nik Bessis
University of Bedfordshire, UK

INFORMATION SCIENCE REFERENCE

Hershey · New York

Director of Editorial Content:	Kristin Klinger
Director of Book Publications:	Julia Mosemann
Acquisitions Editor:	Lindsay Johnston
Development Editor:	Christine Bufton
Publishing Assistant:	Jamie Snavely
Typesetter:	Keith Glazewski
Production Editor:	Jamie Snavely
Cover Design:	Lisa Tosheff
Printed at:	Yurchak Printing Inc.

Published in the United States of America by
Information Science Reference (an imprint of IGI Global)
701 E. Chocolate Avenue
Hershey PA 17033
Tel: 717-533-8845
Fax: 717-533-8661
E-mail: cust@igi-global.com
Web site: http://www.igi-global.com

Library of Congress Cataloging-in-Publication Data

Advanced ICTs for disaster management and threat detection : collaborative and
distributed frameworks / Eleana Asimakopoulou and Nik Bessis, editors.
    p. cm.
 Includes bibliographical references and index.
 Summary: "This book offers state-of-the-art information and references for work undertaken in the challenging area of utilizing cutting-edge distributed and collaborative ICT to advance disaster management as a discipline to cope with current and future unforeseen threats"--Provided by publisher. ISBN 978-1-61520-987-3 (hbk.) -- ISBN 978-1-61520-988-0 (ebook) 1. Emergency management--Information technology. 2. Emergency management--Communication systems. I. Asimakopoulou, Eleana, 1976- II. Bessis, Nik, 1967- HV551.2.A33 2010
  384.5--dc22
                                    2009053384

British Cataloguing in Publication Data
A Cataloguing in Publication record for this book is available from the British Library.

Table of Contents

Section 1
Current Approaches in Disaster Management

Chapter 1
A Systemic Approach to Managing Natural Disasters... 1
Jaime Santos-Reyes, SEPI-ESIME, IPN, Mexico
Alan N. Beard, Heriot-Watt University, Scotland

Chapter 2
ICT Approaches in Disaster Management: Public Awareness, Education and Training,
Community Resilience in India 22
Sunitha Kuppuswamy, Anna University Chennai, India

Chapter 3
Multimedia Educational Application for Risk Reduction.. 35
Ana Iztúriz, Universidad Pedagógica Experimental Libertador, Venezuela
Yolanda Barrientos, Universidad Pedagógica Experimental Libertador, Venezuela
María A. González, Universidad Pedagógica Experimental Libertador, Venezuela
Larry Rivas, Universidad Pedagógica Experimental Libertador, Venezuela
Matilde V. de Bezada, Universidad Pedagógica Experimental Libertador, Venezuela
Simón Ruíz, Universidad Pedagógica Experimental Libertador, Venezuela

Chapter 4
Natural Hazards: Changing Media Environments and the Efficient Use of ICT for Disaster
Communication.. 46
Helena Zemp, University Zurich, Switzerland

Section 2
Advanced Collaborative Technologies for Disaster Management

Section 3
Next Generation Approaches and Distributed Frameworks for Disaster Management

Detailed Table of Contents

Section 1
Current Approaches in Disaster Management

Chapter 1

Jaime Santos-Reyes, SEPI-ESIME, IPN, Mexico
Alan N. Beard, Heriot-Watt University, Scotland

The objective of this chapter is to present a Systemic Disaster Management System (SDMS) model. The SDMS model is intended to provide a sufficient structure for effective disaster management. It may be argued that it has a fundamentally preventive potentiality in that if all the subsystems (i.e., systems 1-5) and channels of communication are present and working effectively, the probability of failure should be less than otherwise. Moreover, the model is capable of being applied proactively in the case of the design of a new 'disaster management system' as well as reactively. In the latter case, a past disaster may be examined using the model as a 'template' for comparison. In this way, lessons may be learned from past disasters. It may also be employed as a 'template' to examine an existing 'disaster management system'. It is hoped that this approach will lead to more effective management of natural disasters.

Chapter 2

Sunitha Kuppuswamy, Anna University Chennai, India

We live in a world where there are potential sources of disasters with a potential to cause a loss all around us. We could be living close to a coastline that is prone to cyclones or mountainous region vulnerable to earthquakes. On the other hand, we might be living in a place where there may be frequent communal tension. Whatever is the area we live in; we need to be aware of our vulnerability to hazardous events. Better awareness about various hazards, proper education, training and preparedness will enhance the community resilience. Information and communication technologies in the form of audio and video (through community radio, village information centres, video awareness programmes through DVDs, etc) play a vital role in creating public awareness, giving education and training to vulnerable communities. This chapter aims to discuss the various ICT initiatives taken in the coastal districts of Tamilnadu, one of the seven coastal states of India.

Ana Iztúriz, Universidad Pedagógica Experimental Libertador, Venezuela
Yolanda Barrientos, Universidad Pedagógica Experimental Libertador, Venezuela
María A. González, Universidad Pedagógica Experimental Libertador, Venezuela
Larry Rivas, Universidad Pedagógica Experimental Libertador, Venezuela
Matilde V. de Bezada, Universidad Pedagógica Experimental Libertador, Venezuela
Simón Ruíz, Universidad Pedagógica Experimental Libertador, Venezuela

The XXI century has brought changes and innovations in educational technology that should be incorporated in the curriculum at all educational levels. Therefore, it is necessary to apply modern technologies in learning processes. This work was based on basic cognitive learning theories and principles, and used software, such as Freehand, Adobe Photoshop, Jigsaw Puzzle Creator, Brainsbraker and Jigsaw Puzzle Lite, to create a multimedia version of the SALTARIESGOS boardgame. The game was applied and validated with students from elementary school of 2nd, 4th, 5th and 6th grades. The results we obtained can be considered satisfactory based on the student's opinion, and the pedagogical strategy was as an effective tool for achieving our goals. The educational contribution of this interactive version of the game will promote and sensitize school community members at different urban and rural areas, in order to enhance the preventive culture in disaster risk reduction.

Helena Zemp, University Zurich, Switzerland

The growing importance of mass media in the 'information society', combined with society's increased dependence on electronic modes of information is important to the perception, regulation and management of risk at a local, national and international level. However, media organisations have their own logic and goals that are not necessarily compatible with the logic and goals of disaster planning and assistance agencies. Using a detailed study of the media coverage of floods in Switzerland from 1910 to 2005, we will illustrate the salient features of disaster reporting and how these relate to issues of risk perception and risk prevention behaviour in the public sphere. The findings are used to discuss the traditional media's shortcomings for the goal of risk reduction, the public's information seeking behaviour, and the opportunities and limitations arising from the emergence of digital, internet-based information and communication technologies (ICT) for disaster communication.

Chapter 5

The United Arab Emirates (UAE) has more exposure to natural hazards than has been previously recognized. In the last 20 years the UAE has been subject to earthquakes, landslides, floods and tropical storms. This chapter examines the structure and procedures for management of natural disasters in the UAE, in particular issues of governance, accountability and communication within states that are part of a federal system. The study involved interviews with officials at both federal and emirate levels and case studies are presented of the impact of recent natural hazard events. Two emirates were selected for more detailed examination, Fujairah the most hazard prone and a rural emirate and Dubai which is a highly urbanized emirate which has undergone rapid development. There is now increasing awareness of natural hazards in the UAR and progress is being made at regional and federal levels. There needs to be a clear delineation between regional and federal roles and an understanding of the need for effective channels of information to relevant agencies.

From risk identification to emergency response and recovery, information plays vital role and effective use of information is instrumental to reduce the impact of disasters. With the advancement of information and communication technology in the last few decades, lack of information is no more a major issue for disaster risk reduction. The major issue, rather, is managing the information, translating it into a comprehensive knowledge for decision making and disseminating it to the communities at risk for action. The advancement of technology and reach of communication tools at grass-root level have created an opportunity to increase effectiveness of disaster risk management with the optimum use of disaster informatics. This chapter presents an overview of disaster informatics, conceptual framework for information management for disaster risk reduction, review on existing approaches of information dissemination through internet and on use of combined potential of internet with tools which are widely available at grassroots.

Section 2
Advanced Collaborative Technologies for Disaster Management

This chapter will emphasize that efficient integration of various Information and Communication Technology (ICT) in disaster management process can help mitigation of impacts of disasters on people and the environment, minimizing the failures and maximizing the collaboration. It summarizes the nature of information flow and management processes during disasters and the potential of recent ICT at three stages of disaster management. The requirements and problems faced during their deployment at different stages of disaster management process are stated. The solutions for common constraints are discussed as well as the critical factors that should be considered in efficient deployment of ICT in the disaster management process.

The deployment of Early Warning Systems (EWS) and Alerting Technologies (AT) is one of the best measures for improved disaster prevention and mitigation. With the evolution of Information and Communication Technologies (ICT) we face new opportunities as well as new challenges for improving classical warning processes. This chapter concentrates on the main aspects of existing early warning systems and alerting technologies. Beginning with the definition and classifications in this field, we describe general approaches, representative systems, and interoperability aspects of EWS. Furthermore, we introduce a list of criteria for evaluating and comparing existing systems. It is worth noting that the deployment of an operational EWS is a complex challenge and remains a young field of research. This is due to many reasons, ranging from the political to the technical. The most critical issues regarding efficient alerting are described in this chapter, along with areas for future research.

The Medical Information System (MedISys) is a fully automatic 24/7 public health surveillance system monitoring human and animal infectious diseases and chemical, biological, radiological and nuclear (CBRN) threats in open-source media. In this article, we explain the technology behind MedISys, describing the processing chain from the definition of news sources, scraping and grabbing articles from the internet, text mining, event extraction with the Pattern-based Understanding and Learning System (PULS, developed by the University of Helsinki), news clustering and alerting, to the display of results. The web interface and service applications are shown from a user's perspective. Users can display world maps in which event locations are highlighted as well as statistics on the reporting about diseases, countries and combinations thereof and can apply filters for language, disease or location or filters with orthogonal categories, e.g. outbreaks, via their browser. Specific entities such as persons, organizations and locations are identified automatically.

Chapter 10

Miranda Dandoulaki, National Centre of Public Administration and Local Government, Greece
Matina Halkia, European Commission, Joint Research Centre, Italy

Social media technologies such as blogs, social networking sites, microblogs, instant messaging, wikis, widgets, social bookmarking, image/video sharing, virtual worlds, and internet forums, have been identified to have played a role in crises. This chapter examines how social media technologies interact with formal and informal crises communication and information management. We first review the background and history of social media (Web 2.0) in crisis contexts. We then focus on the use of social media in the recent Gaza humanitarian crisis (12.2008-1.2009) in an effort to detect signs of a paradigm shift in crisis information management. Finally, we point to directions in the future development of collaborative intelligence systems for crisis management.

Chapter 11

Kumaresh Rajan, The State University of New York at Buffalo, USA
Rui Chen, Ball State University, USA
Hejamadi Raghav Rao, The State University of New York at Buffalo, USA
JinKyu Lee, Oklahoma State University, USA

The principles of Web 2.0 such as transparency, security, community, usability, and availability are well suited to help effectively manage the effects of a disaster. Many Web 2.0 technologies rely on social collaboration, and as a result these technologies are built with robust communication channels. Utilizing this existing framework will help to create software systems that can efficiently manage disasters. This chapter will examine differing Web 2.0 innovations through the use of Activity Theory, and the benefits and drawbacks of each technology will be analyzed. From this analysis, recommendations and conclusions will be presented to the reader.

Anne M. Hewitt, Seton Hall University, USA
Susan S. Spencer, SetonWorldWide, USA
Danielle Mirliss, Seton Hall University, USA
Anne Riad Twal, Seton Hall University, USA

The maturation of incident and disaster management training has led to opportunities for the inclusion of multi-modal learning frameworks. Virtual reality technology, specifically multi-user virtual environments (MUVEs) such as virtual worlds (VW), offers the potential, through carefully crafted applications, for increasing collaboration, leadership, and decision making skills of diverse adult learners. This chapter presents a review of ICT appropriate learning theories and a synopsis of the educational benefits and practices. A case study, offered as part of a Master of Healthcare Administration (MHA) course for health care managers, demonstrates the application of a virtual world training scenario hosted in Second Life® and using a Play2Train simulation. Students report a strong positive reaction to virtual learning and demonstrate improved crisis communication skills and decision making competencies. Additional research is recommended to demonstrate the utility of virtual world learning as compared to standard training options such as tabletop exercises.

Section 3
Next Generation Approaches and Distributed Frameworks for Disaster Management

José G. Hernández R., Universidad Metropolitana, Venezuela
María J. García G., Minimax Consultores C.A., Venezuela

Immediately after the catastrophes that affected Venezuela at the end of 1999, especially the flood of the State of Vargas, a group of investigators of a consultancy company and of a private university of Caracas Venezuela, started working in decisions support systems (DSS) that could be useful in the moment of a catastrophe, helping to minimize the impact of its three principal stages: Pre-catastrophe, Impact and Post-catastrophe. Clearly, for the development of these DSS, it was indispensable to construct mathematical models to support them. The objective of this chapter is to disclose this experience by presenting some of these mathematical models and its conversion in DSS that supports decision making in the case of catastrophes.

Chapter 14

*Tina Comes, Institute for Industrial Production (IIP), Karlsruhe Institute of Technology (KIT),
Germany*

*Michael Hiete, Institute for Industrial Production (IIP), Karlsruhe Institute of Technology (KIT),
Germany*

Niek Wijngaards, Thales Research & Technology Netherlands/D-CIS Lab, The Netherlands

*Frank Schultmann, Institute for Industrial Production (IIP), Karlsruhe Institute of Technology
(KIT), Germany*

Multi-criteria decision analysis (MCDA) is a technique for decision making among mul¬ti¬ple alternatives for action providing transparent and coherent decision support for com¬plex situa-tions with
conflicting objectives. Managing longer term decisions for en¬viron¬mental incidents is an application
domain in which MCDA has proved useful. Yet a diffi¬culty in applying MCDA is when uncertainties
abound. Contrarily, scenario-based rea¬soning is a method allowing for the assessment of multiple possible future developments of the situation. In this way, the use of scenarios is a transparent and easily
understand¬able way to integrate uncertainties into the reasoning process. We pro¬pose a mechanism to
integrate scenarios. Our theoretical framework can be operationalised by decision sup¬port systems relying on both automated systems and human experts. These facilitate the assessment of conse¬quences
within a scenario, and may propose new scenarios. We il¬lustrate this mechanism taking the decision
making in emergency management after a train crash with potential release of chlorine as an example.

Chapter 15

Pierre Kuonen, University of Applied Sciences of Western Switzerland, Switzerland

Mathias Bavay, WSL Institute for Snow and Avalanche Research SLF, Switzerland

Michael Lehning, WSL Institute for Snow and Avalanche Research SLF, Switzerland

In the developed world, an ever better and finer understanding of the processes leading to natural hazards is expected. This is in part achieved using the invaluable tool of numerical modeling that offers
the possibility of applying scenarios to a given situation. This in turn leads to a dramatic increase in
the complexity of the processes that the scientific community wants to simulate. A numerical model
is becoming more and more like a galaxy of various sub-process models, each with their own numerical characteristics. The traditional approach to High Performance Computing (HPC) can hardly face
this challenge without rethinking its paradigms. A possible evolution would be to move away from the
Single Program, Multi Data (SPMD) approach and towards an approach that leverages the well known
Object Oriented approach. This evolution is at the foundation of the POP parallel programming model
that is presented here, as well as its C++ implementation, POP-C++.

Odysseas Sekkas, University of Athens, Greece
Dimitrios V. Manatakis, University of Athens, Greece
Elias S. Manolakos, University of Athens, Greece
Stathes Hadjiefthymiades, University of Athens, Greece

The SCIER platform is an integrated system of networked sensors and distributed computing facilities, aiming to detect and monitor a hazard predict its evolution and assist the authorities in crisis management for hazards occurring at Wildlife Urban Interface (WUI) areas. The goal of SCIER is to make the vulnerable WUI zone safer for the citizens and protect their lives and property from environmental risks. To achieve its objective, SCIER adopts and combines technologies such as: (1) wireless sensor networks for the detection and monitoring of disastrous natural hazards, (2) advanced sensor data fusion and management for accurately monitoring the dynamics of multiple interrelated risks, (3) environmental risk models for simulating and predicting the evolution of hazardous phenomena using advanced computing (e.g., Grid-computing). In this chapter we focus on the key software components of the SCIER architecture, namely the sensor data fusion component and the predictive modeling and simulation component.

Eleana Asimakopoulou, University of Bedfordshire, UK
Nik Bessis, University of Bedfordshire, UK
Ravikanth Varaganti, University of Bedfordshire, UK
Peter Norrington, University of Bedfordshire, UK

Much work is under way in disaster reduction and emergency management towards the utilization of information and communication technologies (ICT) and the design of relevant services associated with risk management towards sustainable development and livelihood. Recent forest fires occurred in Southern Europe, caused environmental destruction and a number of fatalities. The effective and efficient production of forest fire evacuation plans requires decisions based on integrated data from heterogeneous and distributed sources that change over time very quickly. Recent ICT advances suggest the need for further work in the advanced evacuation systems area. We are particularly interested of how to automatically inform potential victims about the most relevant evacuation routes in the most-timely fashion so they can escape a forest fire safely. With this in mind, this chapter describes the concepts, architecture and implementation of the Personalized Forest Fire Evacuation Data Grid Push Service using data push and next generation grid technologies.

Foreword

During the past decades, the international public opinion is commuted and concerned, with an increasing frequency, about disasters triggered by natural phenomena and technological accidents. Consecutive floods and extreme weather phenomena, climatic changes, earthquakes, tsunamis, landslides and other phenomena manifested around the planet with intense geotectonic effects, give the impression of a gradually accelerating procedure of changes on the earths' surface.

At the same time, technological disasters occur with an increasing frequency during the last years, triggered by human activities and inappropriate management of technological elements, such as transportation accidents, waste and toxic substances contaminations, explosions, biological pollution, urban and forest fires etc.

Taking into account that in the immediate future, the above mentioned disasters will represent the primary concern of countries and administrative authorities, many scientists have been orientated towards the systematic research of the phenomena with impressive results. However, at the same time it is observed that the "know-how" is accumulating in research centers rather than being distributed to the statutory authorities or to the broader social groups. As a result, the response is not as effective as it should be and the consequences unfortunately are increasing.

The content of the present book aims to cover the gaps that exist in a series of disaster management topics. The information techniques for the new pioneer practices of natural and technological disaster management through novel management models and actions have the primary role.

For the information of the authorities and groups of population, the use of modern internet technologies that can transfer directly new scientific knowledge and information on potential natural hazards that may trigger disasters and emergencies, is determinant.

An important sector that is presented in this book, is that of the distant education ability of broader groups of population that can be trained in a variety of subjects easily, quickly and with low cost.

I am certain that this book investigates an interesting field of natural and technological disaster management and contributes to the reduction of the consequences caused by disasters in a global scale.

Efthymios Lekkas
National and Kapodistrian University of Athens, Greece

Efthimios Lekkas *is a professor of Dynamic, Tectonic and Applied Geology at the National and Kapodistrian University of Athens. His main research field is the management of Natural and Technological Disasters and their environmental consequences. He teaches Disaster Management at the National and Kapodistrian University of Athens, while he has been a guest lecturer in several universities around the world. He has published the following books: "Geology and Environment", "Natural and Technological Disasters", "Principles and applications of Emergency planning for the management of Natural and Technological Hazards", "Earthquake Geodynamics", "Natural Disasters, Learn and Protect". He has participated in field trips in countries affected by large-scale natural and technological disasters such as Japan, Taiwan, Central America, Egypt, India, Italy, Turkey, Algeria, Indonesia, Iraq, Thailand, Sri Lanka, Pakistan and China, and he has published a number of journal articles in peer-reviewed journals. Finally, he is leading an e-learning course of the University of Athens entitled "Natural and Technological Disaster Management".*

Preface

During the last decades the number of losses caused by natural and man-made disasters has increased. Evidently, humans are not always capable of avoiding extreme natural phenomena, technological accidents, or terrorist attacks. There is a need to prepare and plan in advance actions in response to these events in order to support sustainable livelihood by protecting lives, property and the environment. In turn, various disaster management bodies (FEMA, EMA, European Civil Protection, etc) involving authorities at a local, national and international level have been formed to mitigate, prepare for, respond to and recover from such disasters. There are also collaborative research institutes, scientific laboratories and other non-profit organizations studying natural phenomena yet most importantly, the response processes as to advance the disaster management discipline, its practice and application as a whole.

Disaster management is a dynamic and fluid area, which requires the involvement of expertise from different authorities and organizations. It mainly consists of expert individuals and teams from the civil protection, police, fire and rescue services, health and ambulance services, engineering sector, utility companies, local authorities, central government, relief bodies armed forces, monitoring, research and observatory centers. Bringing in expertise from different parties is essential and critical, as these will assist in managing emergency situations in a more informed and holistic approach. Apparently, this type of collaboration bring together the intellectual and physical resources so as to enable the conceptualization, production, utilization and application of disaster management strategies including critical infrastructures, relevant ICT resources, response plans, policies, risk management techniques, recovery and contingency plans.

ICT developments over the last four decades have facilitated organizations with numerous collaborative tools to support various levels of enquiries within the field of application. In particular, the use of advanced distributed technologies has evolved over the years such as to accommodate and advance collaborative endeavors between interested parties (including disaster management stakeholders) scattered across the world. Such utilization of distributed data and resources related to ICT developments – including but not limited to early warning systems and alerting technologies, data mining and advanced decision support systems, data visualization techniques, data and system integration frameworks, next generation collaborative technologies and Web 2.0, service oriented approaches, and grid technologies – should be further aligned for the purpose of augmenting the effectiveness and efficiency of disaster management and risk reduction approaches towards sustainable developments and livelihood.

THE PURPOSE OF THE BOOK

The primary goal of this book is to demonstrate how strategies and state-of-the-art ICT have and/or could be applied so as to serve as a vehicle to advance disaster management approaches, decisions and practices. The achievement of such a goal implies the contribution from various practitioners, scholars in the area and researchers from other disciplines who are willing to offer their expertise and skills in advancing disaster management discipline both as theory and practice.

It aims to provide both conceptual and practical guidance to disaster management stakeholders including ICT and senior managers from relevant organizations. It will help assist in identifying and developing effective and efficient approaches, mechanisms, and systems using emerging technologies to support their effective operation. Specifically, the book aims to build a network of excellence in effectively and efficiently managing advanced strategies and next generation distributed and collaborative ICT for disaster management stakeholders to advance their current practices and approaches. This is achieved by introducing both technical and non-technical details of strategies and ICT demonstrating their application and their potential utilization to the disaster management sector. It also prompts revisiting current approaches and further develops the area for best practice so as to cope with emerging and unforeseen threats.

The book has collected together the vast experience of many leaders demonstrating past and current methods, tools and practices employed for disaster management purposes. As such, the book claims to be a definitive state-of-the-art collection and to prompt the future direction for disaster managers to identify applicable theories and practices in order to mitigate, prepare for, respond to and recover from various foreseen and/or unforeseen disasters.

WHO SHOULD READ THE BOOK?

The content of the book reflects the interests of a broad audience as it offers state-of-the-art information and references for work undertaken in the challenging area of utilizing cutting edge distributed and collaborative ICT to advance disaster management as a discipline to cope with current and future unforeseen threats.

The projected audience ranges from those currently engaged to those interested in joining collaborative work in the field of disaster management utilizing applicable ICT. In particular, audiences currently working in or are interested in joining interdisciplinary, multidisciplinary and transdisciplinary collaborative disaster management related advancements are the primary focus in this book. Specifically, audiences who are: (1) researchers in the areas of disaster management, emerging technologies and collaborative ICT; (2) managers and practitioners in the local authorities, research institutes and scientific centers and the industry; (3) academics, instructors, researchers and students in colleges and universities.

The book can be used as a source for leading edge literature review in the area of emerging and applicable ICT and disaster management, documenting the latest developments in the academia, government and business sectors. It serves as a guide between relevant bodies from different countries providing lessons learnt and paradigms of good practices worldwide. It can also be used as a library reference. Most importantly, specialist training providers, colleges and universities (having relevant courses) could use it as a course supplement. Finally, the book serves as a source of reference material and as a source of ideas for further research and development activities for academics and researchers in the field. Similarly, it serves as a valuable source for researchers willing to join in relevant collaborative works.

The potential impact of this book is to educate, sustain or even enable the formation of communities and teams (like research teams, charities, voluntary bodies, etc) as to support interdisciplinary and collaborative multidisciplinary research and practices towards an effective and efficient protection of human lives, property and the environment.

ORGANIZATION OF THE BOOK

Seventeen self-contained chapters, each authored by experts in the field, are included in this book. The book is organized into three sections according to the thematic topic of each chapter. Thus, it is quite possible that a paper in one section may also address issues covered in other sections. However, the following three sections reflect most of the topics sought in the initial call for chapters.

The first section, Section 1: Current Approaches in Disaster Management includes six chapters. This section introduces concepts and principles of disaster management and ICT, such as systemic approaches to managing disasters, rapid onset natural disasters and disaster informatics. These cover past and recent methods and techniques for ICT based decision-making in disaster management. In addition, some chapters present scenarios and approaches related to the role of education, training and media in disaster reduction. As such, they underpin future development and implementation of relevant approaches.

The second section, Section 2: Advanced Collaborative Technologies for Disaster Management includes six chapters. This section is concerned with the use of various collaborative Information and Communication Technologies (ICT), such as Early Warning Systems, Alerting Technologies, Web 2.0 innovations, Second Life® and simulation, which aim to advance disaster management processes and risk reduction.

The last section, Section 3: Next Generation Approaches and Distributed Frameworks for Disaster Management includes five chapters. This section goes beyond and builds upon current theory and practice, providing visionary and applicable directions on how next generation technologies – based on current state-of-the-art service oriented and grid computing frameworks – could be used in the future to the benefit threat detection, disaster management and risk reduction.

A brief introduction to each of the chapters follows.

In Chapter 1, *A Systemic Approach to Managing Natural Disasters*, J. Santos-Reyes, and A. N. Beard present a Systemic Disaster Management System, that is able to be used proactively, as well as reactively, as a means to manage disasters in a more effective manner.

In Chapter 2, *ICT Approaches in Disaster Management: Public Awareness, Education and Training, Community Resilience in India*, S. Kuppuswamy discusses the important role that ICT plays in the public awareness, education, training and preparedness for disasters. This takes place via the presentation of various ICT approaches and initiatives taken in the coastal districts of Tamilnadu.

In Chapter 3, *Multimedia Educational Application for Risk Reduction*, A. Iztúriz, Y. Barrientos, M. A. González, L. Rivas, M. V. de Bezada and S. Ruíz focus on the application of modern ICT in the learning process, in order to enhance a preventive culture in disaster risk reduction. In particular, they use cognitive learning theories and existing software to create a multimedia version of a boardgame promoting and sensitizing school community members to different urban and rural areas.

In Chapter 4, *Natural Hazards: Changing Media Environments and the Efficient Use of ICT for Disaster Communication*, H. Zemp considers media coverage of floods in Switzerland from 1910 to 2005, in order to examine how disaster reporting is relating to the public's behavior about risk perception and prevention. Further to this, the chapter highlights the opportunities and limitations arising from the emergence of digital, internet-based information and communication technologies used in disaster communication.

In Chapter 5, *United Arab Emirates: Disaster Management with Regard to Rapid Onset Natural Disasters*, H. Al Ghasyah Dhanhani, A. Duncan and D. Chester, examine the structure and procedures for management of natural disasters in the United Arab Emirates. The chapter has a particular focus to issues of governance, accountability and communication within states that are part of a federal system, via the use of case studies. These in turn, highlight the need for clear delineation between regional and federal roles and for effective channels of information to relevant agencies.

In Chapter 6, *Disaster Informatics: Information Management as a Tool for Effective Disaster Risk Reduction*, J. Subedi focuses on information management during decision making for disaster management. In particular, the chapter presents an overview of disaster informatics, and a conceptual framework for information management for disaster risk reduction. Finally, a review on existing approaches of information dissemination through Internet and on the use of combined potential of Internet with tools, which are widely available at grassroots is offered.

In Chapter 7, *Efficient Deployment of ICT Tools in Disaster Management Process*, A. Sagun focuses on the efficient integration of various Information and Communication Technologies in disaster management processes that can help mitigate the impacts of disasters on people and the environment, by minimizing failures and maximizing collaboration.

In Chapter 8, *Current State and Solutions for Future Challenges in Early Warning Systems and Alerting Technologies*, U. Meissen and A. Voisard discuss disaster prevention and mitigation via the use of Early Warning Systems and Alerting Technologies, by presenting general approaches, representative systems, interoperability aspects and the most critical issues regarding efficient alerting.

In Chapter 9, *MedISys: Medical Information System*, J. P. Linge, R. Steinberger, F. Fuart, S. Bucci, J. Belyaeva, M. Gemo, D. Al-Khudhairy, R. Yangarber and E. van der Goot present both the technology and the user perspective of the Medical Information System (MedISys), that is a fully automatic 24/7 public health surveillance system. The MedISys monitors human and animal infectious diseases and chemical, biological, radiological and nuclear (CBRN) threats using open-source media.

In Chapter 10, *Social media (Web 2.0) and Crisis Information: Case Study Gaza 2008-09*, M. Dandoulaki and M. Halkia discuss how social media technologies interact with formal and informal crises communication and information management via the use of a case study and they point out directions for the future development of collaborative intelligence systems for crisis management.

In Chapter 11, *Utilizing Web 2.0 for Decision Support in Disaster Mitigation*, K. Rajan, R. Chen, H. R. Rao and J. Lee describe Web 2.0 innovations through the use of Activity Theory. They examine both the benefits and drawbacks of each technology, highlighting that the principles of Web 2.0, and in particular social collaboration, are well suited to help effectively manage the effects of disasters.

In Chapter 12, *Incident and Disaster Management Training: Collaborative Learning Opportunities Using Virtual World Scenarios*, A. M. Hewitt, S. S. Spencer, D. Mirliss and A. R. Twal present a review of ICT appropriate learning theories and a synopsis of the educational benefits and practices via the use of a case study. This is offered as part of a Master of Healthcare Administration (MHA) course for health care managers, demonstrating the application of a virtual world training scenario hosted in Second Life® and using a Play2Train simulation.

In Chapter 13, *Mathematical Models Generators of Decision Support Systems for Help in Case of Catastrophes: An Experience from Venezuela*, J. G. Hernández R., and M. J. García G. present a series of mathematical models created to support the development of advanced Decision Support Systems, that are able to minimize the impacts of pre-catastrophe, impact and post-catastrophe stages of disasters.

In Chapter 14, *Integrating Scenario-Based Reasoning Into a Multi-Criteria Decision Support System for Emergency Management*, T. Comes, M. Hiette, N. Wijngaards, and F. Schultmann discuss the multi-criteria decision analysis and scenario-based reasoning as decision making techniques for disaster management. The chapter proposes a theoretical framework as a mechanism of scenario integration, able to facilitate the assessment of consequences within a scenario, and to propose new scenarios during emergency management.

In Chapter 15, *POP-C++ and Alpine3D: Petition for a New HPC Approach*, P. Kuonen, M. Bavay, and M. Lehning appreciate the method of numerical modeling, which offers the possibility of simulating complex scenarios for understanding the causes of a disastrous situation. The dramatic increase in the complexity of the processes challenges the traditional HPC approach and thus, the chapter presents the POP-C++ as a method for providing a better understanding of the processes, which lead to natural hazards.

In Chapter 16, *Sensor and Computing Infrastructure for Environmental Risks: The SCIER System*, O. Sekkas, D. V. Manatakis, E. S. Manolakos, and S. Hadjiefthymiades present the SCIER platform, with particular reference to its key software components architecture, as an integrated system of networked sensors and distributed computing facilities. The SCIER platform aims to detect and monitor a hazard and also, to predict its evolution and assist the authorities in crisis management for hazards occurring at Wildlife Urban Interface areas.

In Chapter 17, *A Personalized Forest Fire Evacuation Data Grid Push Service: The FFED-GPS Approach*, E. Asimakopoulou, N. Bessis, R. Varaganti, and P. Norrington present the concepts, architecture and implementation of a Personalized Forest Fire Evacuation Data Grid Push Service (FFED-GPS), as a method able to effectively and efficiently produce personalized forest fire evacuation plans. The innovative feature of the FFED-GPS is to automatically inform potential victims about the most relevant evacuation routes in the most-timely fashion so they can escape a forest fire safely. The service is based on integrated data from heterogeneous and distributed sources that change over time very quickly.

We wish you to find this book an inspirational read.

Eleana Asimakopoulou
Loughborough University, UK

Nik Bessis
University of Bedfordshire, UK

Acknowledgment

It is our great pleasure to comment on the hard work and support of many people who have been involved in the development of this book. It is always a major undertaking but most importantly, a great encouragement and somehow a reward and an honor when seeing the enthusiasm and eagerness of people willing to advance their discipline by taking the commitment to share their experiences, ideas and visions towards the evolvement of a collaboration like the achievement of this book. Without their support the book could not have been satisfactory completed.

First and foremost, we wish to thank all the authors who, as distinguished scientists despite busy schedules, devoted so much of their time preparing and writing their chapters, and responding to numerous comments and suggestions made from the Editorial Advisory Board, the reviewers and ourselves. We trust this collection of chapters will offer a solid overview of current thinking on these areas and it is expected that the book will be a valuable source of stimulation and inspiration to all those who have or will have an interest in these fields.

Our sincere thanks go to Professor Efthymios Lekkas, National and Kapodistrian University of Athens, who wrote the foreword and to all the members of the Editorial Advisory Board who provided excellent comments, expert help and continuous support. These are some of the very same people who are creating these scientific changes. We have been fortunate that the following have honored us with their assistance in this project: Dr Steve Bloomer, Faculty of Engineering & Technology, University of Teesside; Dr Andre Clark, University of Glamorgan, Business School; Professor Angus Duncan, University of Bedfordshire, Institute for Applied Natural Sciences; Dr Stathes Hadjiefthymiades, Department of Informatics and Telecommunications, University of Athens; Professor Fuad Mallick, BRAC University, Bangladesh and Kathmandu University; Dr Ralf Steinberger, Joint Research Centre of the European Commission Institute for the Protection and Security of the Citizen Global Security and Crisis Management Unit; Dr Tim Thompson, Teesside University, School of Science & Technology. Special gratitude also goes to all the reviewers for their most critical comments.

Last but not least, we wish to gratefully acknowledge that we were fortunate to work closely with an outstanding team at IGI Global. Elizabeth Arder, Kristin M. Klinger, Jamie Snavely and Jan Travers were everything someone should expect from a publisher: professional, efficient and a delight to work with. Thanks are also extended to all those at IGI Global who have taken care with managing the design and the timely production of this book.

Finally, we are deeply indebted to our family for their love, patience and support throughout this rewarding experience.

Eleana Asimakopoulou
Loughborough University, UK

Nik Bessis
University of Bedfordshire, UK

Section 1
Current Approaches in Disaster Management

Chapter 1
A Systemic Approach to Managing Natural Disasters

Jaime Santos-Reyes
SEPI-ESIME, IPN, Mexico

Alan N. Beard
Heriot-Watt University, Scotland

ABSTRACT

The objective of this chapter is to present a Systemic Disaster Management System (SDMS) model. The SDMS model is intended to provide a sufficient structure for effective disaster management. It may be argued that it has a fundamentally preventive potentiality in that if all the subsystems (i.e., systems 1-5) and channels of communication are present and working effectively, the probability of failure should be less than otherwise. Moreover, the model is capable of being applied proactively in the case of the design of a new 'disaster management system' as well as reactively. In the latter case, a past disaster may be examined using the model as a 'template' for comparison. In this way, lessons may be learned from past disasters. It may also be employed as a 'template' to examine an existing 'disaster management system'. It is hoped that this approach will lead to more effective management of natural disasters.

INTRODUCTION

Natural disasters may be defined as events that are triggered by natural phenomena or natural hazards (e.g., earthquakes, hurricanes, floods, windstorms, landslides, volcanic eruptions and wildfires). Throughout history, natural disasters have exerted a heavy toll of death and suffering and are increasing alarmingly worldwide. During the past two decades they have killed millions of

people, and adversely affected the life of at least one billion people. For example, recent disasters, such as the quake that triggered a tsunami in the Indian Ocean (United Nations Development Programme [UNDP], 2005); earthquake in Pakistan (Kamp et al., 2008); the Wenchuan earthquake in China (Zhao et al., 2009) and more recently the L'Aquila earthquake in Italy (Owen & Bannerman, 2009). On the other hand, hurricanes have shown how vulnerable coastal communities could be to such events. For instance, Hurricane Katrina caused an estimated $35 to $60 billion in damage

DOI: 10.4018/978-1-61520-987-3.ch001

and resulted in at least 1000 deaths in the United States alone. More recently, on November 2007, the State of Tabasco, Mexico, has been flooded and it has been regarded as one of the worst in more than 50 years. It is believed that the disaster left more than one million people homeless. Finally, it is thought that 2008 has been one of the most devastating years on record; i.e., more than 220,000 people have been killed in 2008 alone.

The above stresses the importance of prevention, mitigation and preparedness including evacuation planning in order to mitigate the impact of natural disasters. Disaster prevention includes all those activities intended to avoid the adverse impact of natural hazards (e.g., a decision not to build houses in a disaster-prone area). Mitigation, on the other hand, refers to measures that should be taken in advance of a disaster order to decrease its impact on society (e.g., developing building codes). Finally, disaster preparedness includes pre- and post- emergency measures that are intended to minimize the loss of life, and to organize and facilitate timely effective rescue, relief, and rehabilitation in case of disaster (e.g., organizing simulation activities to prepare for an eventual disaster relief operation).

Given the above, natural disasters present a great challenge to society today concerning how they are to be mitigated so as to produce an acceptable risk is a question which has come to the fore in dramatic ways in recent years. As a society we have tended to shift from one crisis to another and from one bout of crisis management to another. There is a need to see things in their entirety, as far as we are able. In relation to disaster management, it becomes vital to see disaster risk as a product of a system; to have a 'systemic' approach. Despite this, very little emphasis has been given by academe, international organizations, NGO (Non Governmental Organizations), and practitioners as to what constitutes and defines an effective disaster management system, both in terms of structure and process, from a systemic point of view. This chapter presents a Systemic

Disaster Management System (SDMS) model. The model is intended to help to maintain disaster risk within an acceptable range whatever that might mean. The model is intended to provide a structure for an effective disaster management system. It may be argued that it has a fundamentally preventive potentiality in that if all the sub-systems and channels of communication are present and working effectively, the probability of a failure should be less than otherwise. It is hoped that this approach will lead to more effective management of natural disasters

BACKGROUND

A great deal of effort has been made, by academe, international organizations, and governments, practitioners, to investigate and develop approaches to address disaster risk. For instance, during the 1990s the United Nations (UN) sponsored the International Decade for Natural Disaster Reduction (IDNDR) with the aim of reducing losses caused by natural hazards (Annan, 1988). The IDNDR Scientific and Technical Committee identified five challenges to guide future programs: (1) Integrate natural disaster management with overall planning; (2) anticipate mega disasters due to population concentrations; (3) reduce environmental and resource vulnerability; (4) improve disaster prevention capabilities of developing countries; and (5) assure effective coordination and implementation. The UN has also established the International Strategy for Disaster Reduction (ISDR) which serves as an international information clearinghouse on disaster reduction, developing awareness campaigns and producing articles, journals, and other publications and promotional materials related to disaster reduction; the publication of "Living with risk: A global review of disaster reduction initiatives" document (ISDR, 2004) is an example of these.

Other world organizations and countries have published a vast amount of reports and publications on the management of disasters; inter alia, (Colombo & Vetere Arellano, 2002; ECLAC, 1991; Freeman et al., 2002; Jayawardane, 2006; Kazusa, 2006; Kreimer & Arnold, 2000). Other authors, such as Vakis (2006) discusses natural disasters within the general framework of 'social risk management' and highlights the complementary role that "social protection" can play in the formation and response of an effective strategy for natural disasters management system. The author proposes a number of "social protection" issues that can be used in practice to address natural disasters. On the other hand, it is now recognised that 'development' and disasters have a close and complex relationship. For instance, Mileti et al. (1995) argue that "losses from natural disasters occur because of development that is unsustainable". Similarly, Stenchion (1997) emphasises that "development and disaster management are both aimed at vulnerability reduction". Some authors, such as Cuny (1994) argues that development is often set back by disasters and others assert that post-disaster operations should take into account a development perspective (see also Berke et al., 1993; McAllister, 1993). The United Nations Development Programme published the document "Reducing disaster risk: A challenge for development" (UNDP, 2004). The report in a way summarizes the above points; i.e., natural disaster risk is connected to the process of human development and that disasters put development at risk. Furthermore, it emphasizes that human development can also contribute to reduction in disaster risk. Finally, the report argues that disaster risk is not inevitable and offers examples of good practice in disaster risk reduction that can be built into ongoing development planning policy.

Other researches have concentrated on several issues regarding disaster management; i.e., organizational, technological, early warning systems, economic, emergency, etc. For instance, Granot (1997) reviews the diverse cultures of different organizations and a number of findings regarding emergency services and suggests directions that

may improve inter-organizational relationships. Kouzmin et al. (1995), on the other hand, discusses the efficiency of disaster management policies and programmes in Australia. The authors argue that there are longstanding deficiencies in strategic and operational planning and forecasting approaches; they argue the need for more co-operation and co-ordination between the various emergency services, and finally, the authors discuss the development of terrestrial and space technologies which could be used in disaster management. Other authors have concentrated their research on emergency response preparedness issues. For example, Wilson (2000) examines small group training for those in charged with responding in an emergency situation. Wilson argues that to ensure both effective and efficient training it is important to understand that people learn in different ways. Cosgrave (1996) proposes that decision making is part of all management tasks and that it is particularly important for emergency managers as they often need to take decisions quickly. The author reviews some of the particular problems of emergency decision and looks at the usefulness of Vroom and Yetton's decision process model for emergencies (Vroom & Yetton, 1973), before proposing a simplified problem classification based on three problem characteristics. Cosgrave concludes by reviewing a collection of "emergency" decisions and analysing some of the common factors to suggest a number of simple action rules to be used in conjunction with the proposed simplified decision process model.

Fisher (1998) has investigated the role of the new information technologies in emergency mitigation, planning, response and recovery. The author illustrates the utility of multimedia, CD-ROM, e-mail and Internet applications to enhance emergency preparedness. Technologies such as 'remote sensing', GIS (Global Positioning System) and GPS (Geographical Information System), also known as '3S' technology, have been used in the process of monitoring disasters. Murai (2006) has developed a system for monitoring

Table 1. Fundamental characteristics of the SDMS model

1	A recursive structure (i.e., 'layered') and relative autonomy (RA)
2	A structural organization which consists of a 'basic unit' in which it is necessary to achieve five functions associated with systems 1 to 5. (See Figure 1). (a) system 1: disaster-policy implementation (b) system 2: disaster- national early warning coordination centre (NEWCC) (c) system 2*: disaster-local early warning coordination centre (LEWCC) (d) system 3: disaster-functional (e) system 3*: disaster-audit (f) system 4: disaster-development (g) system 4*: disaster-confidential reporting system (h) system 5: disaster-policy Note: whenever a line appears in Figure 1 representing the SDMS model, it represents a channel of communication.
3	The SDMS & its 'environment'
4	The concept of MRA (Maximum Risk Acceptable), Viability and acceptable range of risk.
5	Four principles of organization
6	'Paradigms' which are intended to act as 'templates' giving essential features for effective communication and control.

disasters using 'remote sensing', GIS and GPS. The author argues that the developed monitoring system records the real status of damages due to natural disasters and analyzes the "cause" of a disaster and predicts its occurrence. Following the tsunami disaster in 2004, the General Secretary of the United Nations (ONU) Kofi Annan called for a global early warning system for all hazards and for all communities. He also requested the ISDR and its UN partners to conduct a global survey of capacities, gaps and opportunities in relation to early warning systems (Annan, 2005). The produced report, "Global Survey of Early Warning Systems", concluded that there are many gaps and shortcomings and that much progress has been made on early warning systems and great capabilities are available around the world (Egeland, 2006). However, it is argued here that it may be not enough to have such systems without concentrating on 'wider' issues, such a system where an EWS may be just part of it.

More recently, there has been considerable interest on the concepts of vulnerability and resilience. However, there are multiple definitions of these two concepts in the literature and there is not an accepted definition (Klein et al., 2003; Manyena, 2006). For instance, Cutter et al. (2008)

defines vulnerability as the "pre-event, inherent characteristics or qualities of social systems that create the potential for harm". On the other hand, numerous frameworks, conceptual models, and vulnerability assessment techniques have been developed in order to address the theoretical underpinnings and practical applications of vulnerability and resilience (Adger, 2006; Burton et al., 2002; Eakin & Luers, 2006; Fussel, 2007; Gallopin, 2006; Green & Penning-Rowsell, 2007; Klein et al., 2003; McLaughlin & Dietz, 2008; Polsky et al., 2007).

A SDMS MODEL

The Systemic Disaster Management System (SDMS) model is intended to maintain disaster risk within an acceptable range in an organization's operations in relation to disaster management. It may be argued that if all the sub-systems and channels of communication and control are present and working effectively, the probability of a failure should be less than otherwise; in this sense the model has a fundamentally preventive potentiality. Table 1 summarizes the main char-

Figure 1. A SDMS model

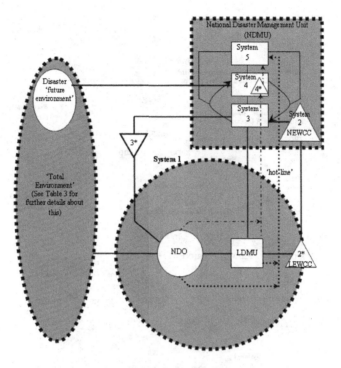

NDMU = National Disaster Management Unit
NDO = National Disaster Operations
LDMU = Local Disaster Management Unit
NEWCC = National Early Warning Coordination Centre
LEWCC = Local Early Warning Coordination Centre

acteristics of the model and Figure 1 shows the structural organization of the SDMS model.

Recursive Structure of the SDMS Model

A Recursion may be regarded as a 'level', which has other levels below or above it. The concept of recursion is intended to help to identify the level of the organization being modelled or being considered for analysis. Figure 2 is intended to show three levels of recursion for an organization.

System 1 at level 1 contains the sub-system of interest; i.e., the 'National Disaster Operations' (NDO) which may be taken to be the highest level of the system of interest (e.g., level of a country). The sub-system is represented as an elliptical symbol that contains two essential elements:

1. The 'National Disaster Management Unit' (NDMU) represented by a parallelogram symbol which is concerned with the 'disaster risk management' in the 'National Disaster Operations' (NDO) of the organization, and

2. The NDO, which is where the disaster risks are created, within system 1, due to the interaction of all the processes that take place within a country, region or community. There may be other risks due to interaction with the 'environment' (see section 'the SDMS & its environment' for further details about these). Note that the double arrow line connecting (1) & (2) represent the managerial interdependence.

Increasing the level of resolution of the system of interest, i.e., NDO at one level below recursion

Figure 2. Recursive structure of the SDMS

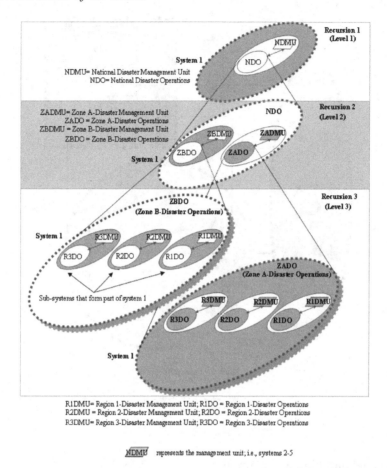

RlDMU= Region 1-Disaster Management Unit; RlDO = Region 1-Disaster Operations
R2DMU = Region 2-Disaster Management Unit; R2DO = Region 2-Disaster Operations
R3DMU= Region 3-Disaster Management Unit; R3DO = Region 3-Disaster Operations

NDMU represents the management unit; i.e., systems 2-5

1 will result in the 'Zone A-Disaster Operations' (ZADO) & 'Zone B-Disaster Operations' (ZBDO) and this is shown at level 2 in Figure 2. It must be pointed out that each of these sub-systems can be de-composed into further sub-systems depending on our level of interest. For example, 'Region-1 Disaster Operations' (R1DO), 'Region-2 Disaster Operations' (R2DO) and 'Region-3 Disaster Operations' (R3DO) are shown as sub-systems of the 'Zone A Disaster Operations' (ZADO) at level 3. In principle, each sub-system that forms part of system 1 at level 3 can be de-composed further depending on the level of interest of the 'disaster management system' modeller or analyst.

Relative Autonomy (RA)

The SDMS is intended to be able to maintain disaster risk within an acceptable range at each level of recursion, but this safety achievement, at each level, is conditional on the cohesiveness of the whole organization. The SDMS contains a structure that favours relative autonomy and local safety problem-solving capacity. Relative autonomy means that each operation of system 1 of the SDMS is responsible for its own activity with minimal intervention of systems 2-5. The organizational structure of the SDMS allows decisions to be made at the local level. Decision making is distributed throughout the whole or-

Figure 3. Disaster management system-in-focus at recursions 1&2

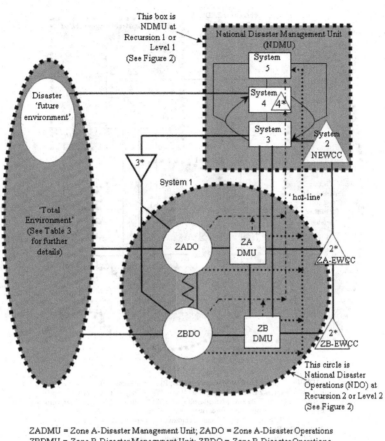

ZADMU = Zone A-Disaster Management Unit; ZADO = Zone A-Disaster Operations
ZBDMU = Zone B-Disaster Management Unit; ZBDO = Zone B-Disaster Operations
NEWCC = National Early Warning Coordination Centre
ZA-EWCC = Zone A - Early Warning Coordination Centre
ZB-EWCC = Zone B - Early Warning Coordination Centre

ganization. This means that distributed decision making involves a set of decision makers in each operation of system 1 and at each level of recursion. These decision makers should be relatively autonomous in their own right and act relatively independently based on their own understanding of safety and their specific tasks. However, it should be recognised that they have interdependence with other decision makers of other operations of system 1 (see Figures 3 & 4). Therefore, each operation of system 1 should be endowed with relative autonomy so that the organizational safety policy can be achieved more effectively. Relative autonomy must not be confused with isolation;

it must be within an adequate system of control and communication.

STRUCTURAL ORGANIZATION OF THE SDMS MODEL

The structural organization of the SDMS model consists of a 'basic unit' in which it is necessary to achieve five functions associated with systems 1 to 5. Systems 2 to 5 facilitate the function of system 1, as well as ensuring the continuous adaptation of the disaster management system as a whole. The operations identified at recursion 2 (see Figure 2) have been represented in the format of

Figure 4. Disaster management system-in-focus at recursions 2&3

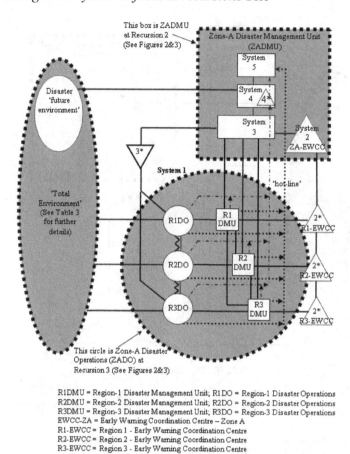

R1DMU = Region-1 Disaster Management Unit; R1DO = Region-1 Disaster Operations
R2DMU = Region-2 Disaster Management Unit; R2DO = Region-2 Disaster Operations
R3DMU = Region-3 Disaster Management Unit; R3DO = Region-3 Disaster Operations
EWCC-ZA = Early Warning Coordination Centre – Zone A
R1-EWCC = Region 1 - Early Warning Coordination Centre
R2-EWCC = Region 2 - Early Warning Coordination Centre
R3-EWCC = Region 3 - Early Warning Coordination Centre

the structural organization of the model. Figure 3 shows what is called here 'disaster management system-in-focus' at recursions 1&2; similarly, Figure 4 illustrates the 'disaster management system-in-focus' at recursions 2&3. It should be emphasized that both Figures should be seen in the context of Figure 2. Referring to Figures 1&3:

System 1: Disaster- Policy Implementation

System 1 may be regarded as the core of the SDMS model. That is, it is where all the daily activities within an organization (i.e., country, region, community, etc.) take place and therefore, it is where disaster risks are created. How system 1 might be broken down is a key question; for example,

it might be de-composed on a basis of geography or functions. For the purpose of the present case system 1 has been de-composed on a basis of geography as shown in Figures 2, 3&4.

As illustrated in Figure 1, system 1 is inter-related with systems 2, 3&3*; i.e., system 1 consists of several subsystems or operations, such as ZADO, ZBDO, etc. Table 2 presents some examples of the information that flows through these channels of communication.

System 2: Disaster- National Early Warning Coordination Centre (NEWCC)

The function of system 2 is to co-ordinate the activities of the operations of system 1. System

Table 2. Examples of the sort of information that flows through the channels of communication

Communication channel (see Figure 1)		Description/Examples
System 1	System 1 to System 2 channel	Information about the maintenance programmes of physical infrastructure, such as early warning systems; training programme of evacuation of the population, etc.
	System 1 to System 3 channel	Information about: the lack of maintenance of the physical infrastructure; compliance and enforcement of the legal and regulatory requirements; lack of forecasting systems; the need of new methodologies for disaster risk identification, analysis and evaluation; the need to improve technologies, for example, to control flood, etc.
	System 1 to System 3* channel	Compliance of public and private buildings with codes and standards as well as with land use plans; whether the planned performance associated with the population's response to an emergency (e.g., the effective response of the people, fire-fighters and police in an exercise based on the scenario of an earthquake occurring) is being achieved or not.
System 2	System 2 to System 1 channel	A wide range of stakeholders need to be coordinated in the operations of system 1; for instance at government level this means ensuring cross-departmental co-ordination; across society as a whole it requires better links between the NGOs, the private sector and academia, etc. Coordination amongst the main actors involved in the early warning chain to provide optimum conditions for informed decision-making and response actions.
	System 2 to System 3 channel	Malfunctioning or failure of a local early warning system (EWS); deficiencies on the channels of communication between forecast and the intended recipient; i.e., the people from the communities, etc.
System 2*	System 2* to System 1 channel	Monitoring of data related to any particular sensor system; e.g., ocean bottom pressure sensors buoys, tide gauging, etc. The communications may be achieved via wire line, wireless, satellite, etc.
	System 2* to System 2 channel	If a deviation from an accepted criterion occurs then this is reported quickly to system 2.
System 3	System 3 to System 1 channel	Resource allocation for disaster reduction; i.e., financial, human, technical, material; legal and regulatory requirements; i.e., laws, acts, regulations, codes, standards. For example, national disaster risk reduction policies; standards (e.g., public and private building codes and standards); education and training programmes: e.g., inclusion of disaster reduction at all levels of education (curricula, education material), national and local training programmes; public awareness programmes, etc.
	System 3 to System 2 channel	The performance of early warning systems,; the population's awareness on how to react in case of an earthquake, hurricane, etc.
	System 3 to System 3* channel	The population's safety culture, etc.; the adequacy of the design and construction of public and private houses; the adequacy of the training of evacuation programmes, etc.
	System 3 to System 4 channel	System 3 communicates its needs to system 4; i.e., information about new developments on risk assessment analysis techniques, new technologies, reassessment of process changes, new development of means of escape, etc.
System 3*	System 3* to System 1 channel	Inadequacy of the design and construction of physical infrastructure; inadequacy of the critical infrastructure; lack of maintenance of the physical infrastructure; deficiencies in the land use planning, etc.
	System 3* to System 3 channel	Deficiencies in the design and construction of public and private houses; deficiencies of the population on how to react in case of a natural hazard; i.e., an earthquake, etc.; lack of training of evacuation programmes, etc.
System 4	System 4 to System 3 channel	Research programmes aiming to risk reduction; new methods in disaster risk identification and assessment; new technologies aiming to improve the physical and technical measures, for example, flood control techniques, soil conservation practices, retrofitting of building, etc.; modern methods of monitoring, e.g., crop production, etc.
	System 4 to System 5 channel	System 4 could, for example, communicate to system 5 about: the new technologies and regulations related to the design of buildings identified in the 'environment'; the development of new technologies related to the prediction of earthquakes; new techniques in order to improve the flood control, etc.; the development of new tools for risk assessments that reflect the dynamic nature of danger, such as, climate change, urban growth, disease, etc.

2, along with system 1 management units, implements the safety plans received from system 3. It informs system 3 about routine information on the performance of the operations of system 1. To achieve the plans of system 3 and the needs of system 1, system 2 gathers and manages the safety information of system 1's operations. Moreover, it also coordinates other local early warning co-ordination centres (LEWCCs).

As illustrated in Figure 1, system 2 is inter-related with systems 1&3. Table 2 presents some examples of the information that flows through these channels of communication.

System 2*: Disaster- Local Early Warning Coordination Centre (LEWCC)

System 2* is part of system 2 and it is responsible for communicating advance warnings to other early warning coordination centres and to key decision makers. This action is intended to help to take appropriate actions prior to the occurrence of a major natural hazard event. Santos-Reyes (2007) gives some details about how this might be achieved. Table 2 presents some examples of the information that flows through these channels of communication.

System 3: Disaster- Functional (Monitoring, Assessment)

System 3 is directly responsible for maintaining risk within an acceptable range in system 1, and ensures that system 1 implements the organization's safety policy. It achieves its function on a day-to-day basis according to its own safety plans and the strategic and normative safety plans received from system 4. The purpose of these plans is to anticipate and act proactively to maintain the disaster risk, arising from the operations of the sub-systems that form part of system 1.

As illustrated in Figure 1, system 3 is inter-related with systems 1, 2, 3*& 4. Table 2 presents some examples of the information that flows through these channels of communication.

System 3*: Disaster- Audit

System 3* is part of system 3 and its function is to conduct audits sporadically into the operations of system 1. System 3* intervenes in the operations of system 1 according to the safety plans received from system 3. System 3 needs to ensure that the accountability reports received from system 1 reflect not only the current status of the operations of system 1, but are also aligned with the overall objectives of the organization. The audit activities should be sporadic (i.e., unannounced) and they should be implemented under common agreement between system 3* and system 1.

As illustrated in Figure 1, system 3* is inter-related with systems 1&3. Table 2 presents some examples of the information that flows through these channels of communication.

System 4: Disaster- Development

System 4 is concerned with safety research and development (R&D) for the continual adaptation of the disaster management system as a whole. By considering strengths, weaknesses, threats and opportunities, system 4 can suggest changes to the organization's safety policies. This function may be regarded as a part of effective safety planning. System 4 achieves its function according to the safety policy of system 5; i.e., to maintain disaster risk within an acceptable range in the organiza-tions operations. System 4 should sense, scan and attempt to respond appropriately to the various threats and opportunities identified in the system's 'total environment' (see Section 'the SDMS & Its environment' for details of the environmental factors). There are two main safety issues which system 4 has to deal with regarding the 'total environment'. First, the large broken line elliptic symbol represents the 'total environment' of the system (see Figures 1, 3&4). Second, system 4 should deal with the 'disaster future environment'. The 'disaster future environment' is concerned with threats and opportunities relating to future development of safety that may be relevant for the organization. Therefore, the SDMS deals not only with current safety problems, but also anticipates or prevents future disasters.

As illustrated in Figure 1, system 4 interacts with the 'total environment', systems 5 &3. Table 2 presents some examples of the information that flows through these channels of communication.

System 4*: Disaster- Confidential Reporting System

System 4* is part of system 4 and is concerned with confidential reports or causes of concern from any employee, about any aspects, some of which may require the direct and immediate intervention of system 5. This means that system 4* analyses all information coming through this channel and develops and plans actions to act upon what has been reported so that these or similar incidents or causes of concern do not occur in the future.

System 5: Disaster- Policy

System 5 is responsible for deliberating safety policies and for making normative decisions. According to alternative safety plans received from system 4, system 5 considers and chooses feasible alternatives, which aim to maintain disaster risk within an acceptable range throughout the life cycle of the total system. Furthermore, these safety policies should: reflect the safety values and beliefs of the whole organization; address the anticipation of disasters due to natural hazard; promote safety culture throughout the organization. System 5 also

monitors the interaction of system 3 and system 4, as represented by the lines that show the loop between systems 3 and 4 as shown in Figures 1&3. The safety policies that are deliberated and decided by system 5 for implementation should address, for example, the following issues:

- It should also promote safety culture throughout the organization.
- Establishment of policy in development planning: e.g., poverty reduction or eradication, social protection, sustainable development, climate change 'adaptation', natural resource management, health, education, etc.
- Promotion of disaster risk awareness through education at all levels of the organization.

Hot-Line

Figures 1, 3 & 4 show a dashed line directly from system 1 to system 5; it represents a direct channel of communication or 'hot-line' for use in exceptional circumstances; e.g., during an emergency. It represents 'initially' one-way communication channel but they may become two way communication channels between systems 1 and 5.

The SDMS & Its Environment

'Environment' may be understood as those circumstances to which the SDMS response is necessary. 'Environment' lies outside the SDMS but interacts with it (see Figures 1, 3&4). It influences and is influenced by the system. Thus, it is important to consider it. For instance, natural hazards such as earthquakes, hurricanes, etc, threaten the system; so that these hazards and the associated risks should be eliminated or controlled. In addition, table 3 lists some 'environmental' factors that should be considered by the SDMS.

Table 3. Some 'environmental' factors that should be considered by the SDMS

External factors that may influence the performance of the SDMS	
Climate change	Poverty
National & local cultures	Cities in a continuous change
Learning from past disasters	Lack of regulations
Unplanned urbanization	Isolation & remoteness
Improper construction of buildings	Armed conflicts
Technology	Epidemics
Weather conditions after a disaster	Politics
Geographical location and settlements	Corruption
	Other

A brief description of each of the above is presented in the subsequent paragraphs.

Climate Change

There is evidence that suggests that emissions of greenhouse gases are already changing our climate (Aalst, 2006; Black, 2006; Helmer & Hilhorst, 2006; Intergovernmental Panel on Climate Change (IPCC), 2001; Trenberth, 2005); e.g., it is believed that the global warning is the main cause of the worsening of floods around manila Bay (Kelvin et al., 2006).

National and Local Cultures

National and communities' cultures on crisis and response management should be considered by the disaster management system, although caution is needed to avoid simplistic and stereotypic judgements. It may be argued that such behaviour is likely to slow down response management and consequently it may create time lags. The disaster management system should take into account such cultural behaviour when assessing risks associated with, for example, an emergency response (Casse, 1982; Heath, 1995; Hofstede, 1980).

Learning from Past Disasters

Past disasters should be analyzed in order to learn from them; i.e., to find out what went wrong and what went right so that lessons can be incorporated into the disaster management system. However, there is evidence that shows that this issue has not been addressed by local communities, governments, etc.

Unplanned Urbanization

The complexity and sheer scale of humanity concentrated into large cities creates a new intensity of disaster risk. For instance, the fast and uncontrolled growth of Mexico City with a population of more than 20 million inhabitants is reflected in dangerous construction of homes. In some areas of Mexico City it is common place to see houses built on steep hillsides (British Broadcasting Corporation [BBC], 2006a, March 16).

Improper Construction of Buildings

Another contributing factor in disasters is related to the materials and methods used to build homes and other buildings. Very often in developing countries public and private buildings are built without taking into account potential hazards. The above highlights the need on inherently safer design houses against natural hazards and these issues should be considered by the disaster management system.

Technology

Technology is bound to affect organization's disaster management systems since there are usually safety implications. The technology related to tsunami early warning systems has already existed such as the website http://www.prh.noaa. gov/ptwc. However, the countries from the Indian Ocean lacked of such systems and were unable to prevent the tsunami disaster in 2004 (UNDP, 2005). This should be considered by the disaster management system.

Weather Conditions After a Disaster

The weather conditions may affect the relief efforts after an natural hazard and this may escalate into disaster. For instance, heavy rain and snowfall hampered relief efforts in Kashmir, where three million people were left homeless by the South Asian earthquake in 2005; roads were closed and helicopters grounded by bad weather and landslides. In addition, survivors' tents were flooded and these made the communities vulnerable to disaster (BBC, 2006b, April 8).

Geographical Locations and Settlements

The geographical location of cities may contribute to disasters; i.e., those that have been founded in highly hazardous locations. For instance, the city of Lima, Peru, was founded in an area of very high seismicity; the city has been severely damaged by earthquakes, such as those that occurred in 1966 and 1970 (McEntire & Fuller, 2002). More recently, the flooding of the city of New Orleans, US, due to Hurricane Katrina in 2005 illustrates the inappropriate location of settlements (Jackson, 2005). On the other hand, when the population expands faster than the capacity of city authorities or the private sector can supply housing or basic infrastructure, informal settlements can explode. For example, some 50% to 60% of residents live in informal settlements in Bogota (Colombia), Bombay and Delhi (India), Buenos Aires (Argentine), Lagos (Nigeria), and Lusaka (Zambia). Similarly, 60% to 70% in Dar Es Salaam (Tanzania) and Kinshasa (DR Congo); and more than 70% in Ad-

dis Ababa (Ethiopia), Cairo (Egypt), Casablanca (Morocco) and Luanda (Angola) (United Nations Human Settlements Programme [UN-HABITAT], 2006). The above highlights the vulnerability of these cities to disasters.

Poverty

Poverty may be another factor that contributes to disaster risk. Moser (1998) argues that disaster risk in cities is shaped by greater levels of social exclusion and the market economy. Social exclusion is associated to the high number of migrants to a city where they are exposed at high risk from disaster.

Cities in a Continuous Change

Cities may be regarded as complex systems which are in a continuous change. They transform their surroundings and hinterlands and these processes may generate and create new hazards. For instance, the destruction of mangroves in coastal areas may increase hazard associated with 'storm surge'; the urbanisation of watershed through settlement, land use change and infrastructure development may contribute to the increase of flood and landslide hazard; see for example, Zevallos (1996).

Lack of Regulations

Very often in developing countries, governments have been ineffective in regulating the process of urban expansion through both land-use planning and building codes. Unregulated low income settlements are the most hazard prone areas; low building standards may be reflect a lack of control, supervision, resources in order to build resistant structures in such areas. It may be argued that hazard prone areas are often preferred by the poor because they may gain greater accessibility to urban services and employment, even though natural hazard risk may be increased. For example, in central Delhi (India), a squatter settlement in the flood plain of the Yemura River has been inhabited for more than 25 years (Sharma & Gupta, 1998; UNDP, 2004).

Isolation and Remoteness

Deficient rural infrastructure and its vulnerability to natural hazards can increase livelihood risks and food insecurity in isolated communities. For instance, the Neelum valley with an estimated 160 000 inhabitants was cut off from the rest of Pakistani-administered Kashmir and became one of the most inaccessible areas hit by the South Asian earthquake in 2005. The mountain people of the valley are dependent on roads; however, the massive landslides at the valley entrance made it completely dependent on helicopters for supplies (BBC, 2006b, April 8).

Armed Conflicts

According to the UNDP (2002) Human Development Report, during the 1990s a total of 53 major armed conflicts resulted in 3.0 million deaths which nearly 90% are believed to be civilians. In 2002, there were approximately 22 million international refugees in the world and another 20 to 25 million internally displaced people. The fact of being a refugee or an internally displaced person raises vulnerability. When the displaced settle in squatter settlements in cities, very often they are exposed to new hazards because dangerous locations where they can find shelter. For example, Afghanistan suffered three years of drought and a major earthquake on top of decades of armed conflict, creating a particularly acute humanitarian crisis (UNDP, 2002).

Epidemics

Epidemic diseases may be seen as disasters in their own right but they also interact with human vulnerability and natural disasters. Following a disaster, for example, the population is influenced by the type of hazard and the environmental conditions in which it takes place, the particular characteristics of those people exposed to the disaster and their access to health services. Natural hazard events, such as, flooding or temperature increase in highland areas can extend the range of 'vector-born' diseases such as malaria. In El Salvador, for example, local health centres were destroyed by an earthquake in the year 2002; as a result, people had to travel for hours to reach medical care. Despite the arsenal of vaccines and drugs that exist today, infectious diseases are on the increase, particularly in the developing countries (UNDP, 2002)

Politics

Politics also contributes to disasters. McEntire & Fuller (2002) argue that the concentration of political power may have limited the capacity of local leaders and emergency managers to undertake the steps they felt were necessary to prevent calamity in Peru. For instance, officials in the city and department of Ica asked the central government as early as November 1997 to take preventive measures or release funds, so potential hazards could be addressed locally but this plea was denied or ignored by the government (La Fernandez, 1998a, February 3). However, when the full strength of El Niño arrived a few months later, Ica was largely unprepared to deal with such event. The centralization of decision making was regarded as one of the main reasons why the city of Ica was devastated by the severe floods on 30 January 1998. Similar problems were evident in other parts of the country as well (Fernandez, 1998b, February 5; McEntire & Fuller, 2002).

Corruption

Humanitarian relief is often needed in countries which are usually corrupt. The risk of aid diversion is high and very often occurs at any point in the response by any or all of the actors involved in: donor contracting, public fundraising, by national officials, UN staff, international NGO (Non Governmental Organizations) and local NGOs, and by recipients themselves (Willitts-King & Harvey, 2005). The term "corruption" is used as a shorthand reference for a large range of illicit or illegal activities. Although, there is no universal or comprehensive definition as to what constitutes corrupt behaviour, the most prominent definitions share a common emphasis upon the abuse of public power or position for personal advantage. Corruption can thrive in times of disaster and when it is already entrenched, the possibilities for abusing emergency aid are even greater. For instance, the province of Aceh is among Indonesia's wealthiest in terms of natural resources; it is also widely considered one of the most corrupt provinces in Indonesia. It is believed that extortion is being reported to be rampant across the province, especially on main highways and carried out almost entirely by the military (TNI) and the police (Clark et al., 2005). It has been reported that TNI was selling freely donated food to homeless people immediately after the 2004 tsunami disaster (James, 2006). Indonesian Corruption Watch said that bureaucrats were reselling donated rice in Aceh and aid supplies were been pilfered before arriving in Banda Aceh (James, 2006).

It should be pointed out that most of the factors mentioned above overlap and the order given is not meant to imply any kind of order of importance but it is simply a list of some of the factors which might be considered by the SDMS. Other factors may also be relevant.

FUTURE RESEARCH DIRECTIONS

A Systemic Disaster Management System (SDMS) has been presented. The SDMS aims to maintain disaster risk within an acceptable range whatever that might be in the operations of any organization (country, community, etc.) in a coherent way. The future research includes:

1. The numerical assessment of the effectiveness of the SDMS model by employing the concept of viability. Viability has been defined as the probability that the SDMS will be able to maintain disaster risk within an acceptable range for a given period of time (see Table 1).

2. To apply the model to the analysis of past natural disasters such as the following:

   a. The Mexico City earthquake. On September 19, 1985, at 7:19 local time, an earthquake with a magnitude of 8.1 on the Richter scale struck Mexico's Capital City. It is believed that more than 10,000 people were killed, 30,000 were injured, and large parts of the city were destroyed. It is thought that about 6,000 buildings were flattened and a quarter of a million people lost their homes. The Mexico City earthquake is being regarded as the most catastrophic in the country's history (Pan American Health Organization [PAHO], 1985).

   b. The Tabasco's flood disaster. On November 2007, torrential rains caused the worst flooding in the southern Mexican state of Tabasco in more than 50 years. It is believed that more than one million people were affected. Some preliminary results have been presented in Santos-Reyes and Beard (2009).

   c. The Tsunami disaster. On 26 December 2004 the biggest earthquake in 40 years occurred between the Australian and Eurasian plates in the Indian Ocean. The quake triggered a *tsunami;* i.e., a series of large waves that spread thousands of kilometres over several hours. It is believed that the disaster left at least 165,000 people dead, more than half a million more were injured and up to 5 million others in need of basic services and at risk of deadly epidemics in a dozen Indian Ocean countries (UNDP, 2005).

These cases may help to illustrate some of the features of the model such as:

a. The possible advantages or disadvantages of the concept of relative autonomy (RA). That is, RA may have the advantage in terms of helping to make local organizations more effective; e.g., in helping to try to get the message to the people 'on the ground'. On the other hand, it may be problematic if the local organization is corrupt, or 'incompetent'. In that case, it would be better to have a strong control from outside (i.e systems 2-5), to try to ensure the effective implementation of safety policies.

b. The need for a direct channel of communication from the NDO to System 4* (i.e., the confidential reporting system, see Figure 1); that is, avoiding the need for people 'on the ground' to always go through the Management Units (e.g., LDMU; see Figure 1), especially as a person 'on the ground' may be complaining about the LDMU (e.g., because of 'corruption' or 'incompetence' or nepotism or partiality).

c. The decomposition of System 2. In the present application, System 2 has been broken into NEWCC (National Early Warning Coordination Centres) and LEWCC (Local Early Warning Coordination Centres). However, it is not clear how the decomposition of System 2 might be at the next higher level of recursion; i.e., at international level. The analysis of the tsunami disaster may help to illustrate this.

d. The channels of communication's effectiveness or lack of it. It has long been known that an organization's communication system has a significant impact on the organization's performance. Moreover, multiple distributed decision-making may be impossible without

communication. The 'Four principles of organization' and the 'Paradigms' (see Table 1) which are intended to give essential features for effective communication and control may help to illustrate the above.

CONCLUSION

The natural disasters described briefly in the introduction section have highlighted that the existing approaches to the management of disaster risk may be inadequate in dealing with such catastrophic events. In addition, they have elucidated the need to improve radically the performance of the existing 'disaster management systems'. A great deal of effort has been made, by academe, international organizations, and governments, practitioners, to investigate and develop approaches to address disaster risk. However, the approaches reviewed in the background section may represent a step forward to managing disaster risk but may not be enough to address the management of natural disasters effectively. Furthermore, it may be argued that they still tend to address disaster risk from an 'isolation' point of view and this will ultimately fail to fundamentally understand the nature of risk (Beard, 1999; Santos-Reyes & Beard, 2001). That is, the cause of a natural disaster may be found in the complexity of the relationships implicit in the physical location of the settlements, the design of the houses, communication systems, Early Warning Systems (EWSs), national infrastructure, climate change, etc. These have been recognised by some researchers, such as McFadden (Kettlewell, 2005a, January 6), who argues that: "there's no point in spending all the money on a fancy monitoring and a fancy analysis system unless we can make sure the infrastructure for the broadcast system is there….that's going to require a lot of work. If it's a tsunami, you've got to get it down to the last Joe on the beach. This is the stuff that is really very hard". Similarly, McGuire (Kettlewell, 2005b, March 25) argues that: "I have no doubt that the technical element of the warning system will work very well but there has to be an effective and efficient communications cascade from the warning centre to the fisherman on the beach and his family and the bar owners". In order to gain a full understanding and comprehensive awareness of disaster risk in a given situation it is necessary to consider in a coherent way all the aspects that may contribute to natural disasters. In short, there is a need for a systemic approach to natural disasters management. Systemic means looking upon things as a system; systemic means seeing pattern and inter-relationship within a complex whole; i.e., to see events as products of the working of a system. System may be defined as a whole which is made of parts and relationships. Given this, 'failure' may be seen as the product of a system and, within that, see death/injury/property losses and losses to the economy as results of the working of systems.

REFERENCES

Aalst, M. K. (2006). The impacts of climate change on the risk of natural disasters. *Disasters*, *30*(1), 5–18. doi:10.1111/j.1467-9523.2006.00303.x

Adger, W. N. (2006). Vulnerability. *Global Environmental Change*, *16*, 268–281. doi:10.1016/j.gloenvcha.2006.02.006

Annan, K. (1988). *International decade for natural disaster reduction* (Report of the Secretary-General). Report A/43/723.

Annan, K. (2005). *In larger freedom: towards development, security and human rights for all* (Report of the Secretary-General). Report A/59/2005, paragraph 66.

Beard, A. N. (1999). Some ideas on a systemic approach. *Civil Engineering and Environmental Systems*, *16*, 197–209. doi:10.1080/02630259908970262

Berke, P. R., Kartez, J., & Wenger, D. (1993). Recovery after disaster: achieving sustainable development. *Disasters*, *17*(2), 93–108. doi:10.1111/j.1467-7717.1993.tb01137.x

Black, R. (2006). 'Clear' human impact on climate. *British Broadcasting Corporation NEWS*. Retrieved May 3, 2006, from http://news.bbc.co.uk/go/pr/fr/-/2/hi/science/nature/4969772.stm

British Broadcasting Corporation (BBC). (2006a, March 16). Quenching Mexico City's Thirst. *BBC NEWS*. Retrieved March 16, 2009, from http://news.bbc.co.uk/go/pr/fr/-/2/hi/americas/4812352.stm

British Broadcasting Corporation (BBC). (2006b, April 8). Quake survivors 'still need aid'. *BBC NEWS*. Retrieved April 8, 2009, from http://news.bbc.co.uk/go/pr/fr/-/2/hi/south_asia/4890252.stm

Burton, I., Saleemul, H. L. B., Pilifosova, O., & Schipper, E. L. (2002). From impacts assessment to adaptation priorities: the shaping of adaptation policy. *Climate Policy*, *2*(2-3), 145–159. doi:10.1016/S1469-3062(02)00038-4

Casse, P. (1982). *Training for the multicultural manager: a practical and cross-cultural approach to the management of people*. Washington, DC: SIETAR International.

Clark, J., Stephens, M., & Fengler, W. (2005). *Indonesia: Rebuilding a better Aceh and Nias* (Six Month Report). Washington, Jakarta / Indonesia. Retrieved June 25, 2005, from http://www.reliefweb.int/rwarchive/rwb.nsf/db900sid/KHII-6DU466?OpenDocument

Colombo, A. G., & Vetere Arellano, A. L. (2002). *Dissemination of lessons learnt from disasters. NEDIES workshop*. Italy: Ispra.

Cosgrave, J. (1996). Decision making in emergencies. *Disaster Prevention and Management*, *5*(4), 28–35. doi:10.1108/09653569610127424

Cuny, F. C. (1994). *Disasters and development. Oxfam*. Dallas, TX: Oxford University Press.

Cutter, S. L. (1996). Vulnerability to environmental hazards. *Progress in Human Geography*, *20*, 529–539. doi:10.1177/030913259602000407

Eakin, H., & Luers, A. L. (2006). Assessing the vulnerability of social - environmental systems. *Annual Review of Environment and Resources*, *31*, 365–394. doi:10.1146/annurev.energy.30.050504.144352

Economic Commission for Latin America and the Caribbean (ECLAC). (1991). *Manual for estimating the socio-economic effects of natural disasters*. United Nations. Retrieved from http://www.reliefweb.int/rw/lib.nsf/db900SID/LGEL-5E2CLJ?OpenDocument

Egeland, J. (2006). Opening Address. In *Proceedings of Third International Conference on Early Warning*, March 27-29, 2006, Bonn, Germany.

Fernández, A. M. (1998a, February 3). Gobierno pudo evitar tragedia de Ica si la declaraba en estado de emergencia. *La República* (p. 5).

Fernández, A. M. (1998b, February 5). Ica, ciudad devastada. *La República* (p. 24).

Fisher, H. W. (1998). The role of the new information technologies in emergency mitigation, planning, response and recovery. *Disaster Prevention and Management*, *7*(1), 28–37. doi:10.1108/09653569810206262

Freeman, P. K., Martin, L. A., Mechler, R., Warner, K., & Hausmann, P. (2002). *Catastrophes and development: Integrating natural catastrophes into development planning* (Working papers series No. 4). Washington, DC: World Bank.

Fussel, H. M. (2007). Vulnerability: a generally applicable conceptual framework for climate change research. *Global Environmental Change*, *17*(2), 155–167. doi:10.1016/j.gloenvcha.2006.05.002

Gallopin, G. C. (2006). Linkages between vulnerability, resilience, and adaptive capacity. *Global Environmental Change*, *16*, 293–303. doi:10.1016/j.gloenvcha.2006.02.004

Granot, H. (1997). Emergency inter-organizational relationship. *Disaster Prevention and Management*, *6*(5), 305–310. doi:10.1108/09653569710193736Green, C., & Penning-Rowsell, E. (2007). More or less than words? Vulnerability as discourse. In McFadden, L., Nicholls, R. J., & Penning-Rowsell, E. (Eds.), *Managing Coastal Vulnerability*. Amsterdam: Elsevier.

Heath, R. (1995). The Kobe earthquake: some realities of strategic management of crises and disasters. *Disaster Prevention and Management*, *4*(5), 11–24. doi:10.1108/09653569510100965

Helmer, M., & Hilhorst, D. (2006). Natural disasters and climate change. *Disasters*, *30*(1), 1–4. doi:10.1111/j.1467-9523.2006.00302.x

Hofstede, G. (1980). Motivation, leadership, and organization: do American theories apply abroad? *Organizational Dynamics*, ▪▪▪, 42–63. doi:10.1016/0090-2616(80)90013-3

Intergovernmental Panel on Climate Change (IPCC). (2001). *Climate Change 2001: The Scientific basis (Contribution of working group I to the Third Assessment Report of the IPCC)*. Cambridge: Cambridge University press.

International Strategy for Disaster Reduction (ISDR). (2004). *Living with risk: A global review of disaster reduction initiatives*. Geneva, Switzerland: United Nations Publications.

Jackson, P. (2005). Hard task of draining New Orleans. *British Broadcasting Corporation NEWS*. Retrieved September 8, 2009, from http://news.bbc.co.uk/go/pr/fr/-/2/hi/americas/4209394.stm

James, E. (2006). *Clean or corrupt: tsunami aid in Aceh*. Canberra, Australia: The Australian National University, Asia Pacific School of Economics and Government. Retrieved from http://apseg.anu.edu.au

Jayawardane, A. K. W. (2006). Disaster mitigation initiatives in Sri Lanka. In *Proceedings of the International Symposium on Management System for Disaster Prevention*, March 9-11, 2006, Kochi, Japan.

Kamp, U., Growley, B. J., Khattak, G. A., & Owen, L. A. (2008). GIS-based landslide susceptibility mapping for the 2005 Kashmir earthquake region. *Geomorphology, 101*, 631–642. doi:10.1016/j.geomorph.2008.03.003

Kazusa, S. (2006). Disaster management of Japan. In *Proceedings of the International Symposium on Management System for Disaster Prevention*, March 9-11, 2006, Kochi, Japan.

Kelvin, S., Rodolfo, S., & Fernando, S. (2006). Global sea-level rise is recognized, but flooding from anthropogenic land subsidence is ignored around northern Manila Bay, Philippines. *Disasters, 30*(1), 118–139. doi:10.1111/j.1467-9523.2006.00310.x

Kettlewell, J. (2005a, January 6). Early warning technology – is it enough? *British Broadcasting Corporation NEWS*. Retrieved January 6, 2005, from http://news.bbc.co.uk/go/pr/fr/-/2/hi/science/nature/4149201.stm

Kettlewell, J. (2005b, March 25). Tsunami alert technology – the iron link. *British Broadcasting Corporation NEWS*. Retrieved March 25, 2005, from http://news.bbc.co.uk/go/pr/fr/-/2/hi/science/nature/4373333.stm

Klein, R. J. T., Nicholls, R. J., & Thomalla, F. (2003). Resilience to natural hazards: how useful is this concept? *Environmental Hazards, 5*(1–2), 35–45.

Kouzmin, A., Jarman, A. M. G., & Rosenthal, U. (1995). Inter-organizational policy processes in disaster management. *Disaster Prevention and Management, 4*(2), 20–37. doi:10.1108/09653569510082669

Kreimer, A., & Arnold, M. (Eds.). (2000). *Managing disaster risk in emerging economies*. Washington, DC: World Bank. doi:10.1596/0-8213-4726-8

Manyena, S. B. (2006). The concept of resilience revisited. *Disasters, 30*(4), 433–450.

McAllister, I. (1993). *Sustaining relief with development: Strategic issues for the Red Cross and Red Crescent*. Boston, MA: Marinus Nijhoff Publishers.

McEntire, D. A., & Fuller, C. (2002). The need for a holistic theoretical approach: an examination from El Niño disasters in Peru. *Disaster Prevention and Management, 11*(2), 128–140. doi:10.1108/09653560210426812

McLaughlin, P., & Dietz, T. (2008). Structure, agency and environment: toward an integrated perspective on vulnerability. *Global Environmental Change, 18*(1), 99–111. doi:10.1016/j.gloenvcha.2007.05.003

Mileti, D. S., Darlington, J. D., Passarini, E., Forest, B. C., & Myers, M. F. (1995). Toward an integration of natural hazards and sustainability. *Environment and Progress, 17*(2), 117–126.

Moser, C. (1998). The asset vulnerability framework: Re-assessing ultra-poverty reduction strategies. *World Development, 26*(1), 1–19. doi:10.1016/S0305-750X(97)10015-8

Murai, S. (2006). Monitoring of disasters using remote sensing GIS and GPS. In *Proceedings of the International Symposium on Management System for Disaster Prevention*, March 9-11, 2006, Kochi, Japan.

Owen, R., & Bannerman, L. (2009). Italy in desperate race to save the buried after the earthquake. *The Times*. Retrieved April 7, 2009, from http://www.timesonline.co.uk/tol/news/world/europe/article6047691.ece

Pan American Health Organization (PAHO). (1985). *Disaster Chronicles No. 3-Earthquake in Mexico September 19 and 20, 1985*. Program of Emergency Preparedness and Disaster Relief Coordination. Retrieved June 28, 2009, from http://www.helid.desastres.net/

Polsky, C., Neff, R., & Yarnal, B. (2007). Building comparable global change vulnerability assessments: the vulnerability scoping diagram. *Global Environmental Change, 17*(3-4), 472–485. doi:10.1016/j.gloenvcha.2007.01.005

Santos-Reyes, J. (2007). Early warning coordination centres: a systemic view. In S. Hernandez, & C.A. Brebbia (Eds.), *Engineering Nature-2007: First International Conference on the Art of Resisting Extreme Natural Forces* (pp.111-120). Sussex, England, UK: WIT PRESS.

Santos-Reyes, J., & Beard, A. N. (2001). A systemic approach to fire safety management. *Fire Safety Journal, 36*, 359–390. doi:10.1016/S0379-7112(00)00059-X

Santos-Reyes, J., & Beard, A. N. (2009). Analysis of Tabasco's flooding by applying the SDMS model. In *Proceedings of the 2nd International Conference on Risk Analysis and Crisis Response*, 19-21 October, 2009, Beijing, China.

Sharma, A., & Gupta, M. (1998). Reducing urban risk through community participation (TDR Project Progress Report). Delhi, India: Sustainable Environment and Ecological Development Society (SEEDS).

Stenchion, P. (1997). Development and disaster management. *Australian Journal of Emergency Management, 12*(3), 40–44.

Trendberth, K. (2005). Uncertainty in hurricanes and global warming. *Science, 308*, 1753–1754. doi:10.1126/science.1112551

United Nations Development Programme (UNDP). (2002). *Deepening democracy in a fragmented world* (Human Development Report 2002). New York, USA. Retrieved June 28, 2009, from http://hdr.undp.org/en/reports/global/hdr2002/

United Nations Development Programme (UNDP). (2004). *Reducing disaster risk: A challenge for development*. New York, USA. Retrieved June 28, 2009, from http://www.undp.org/cpr/disred/rdr.htm

United Nations Development Programme (UNDP). (2005). *Survivors of the tsunami: One year later*. Retrieved June 28, 2009, from http://www.iotws.org/ev_en.php?ID=1685_201&ID2=DO_TOPIC

United Nations Human Settlements Programme (UN-HABITAT). (2006). Retrieved August 25, 2006, from http://www.unhabitat.org/

Vakis, R. (2006). *Complementing natural disasters management: the role of social protection*. Social Protection Advisory Service, Washington, DC: World Bank. Retrieved June 26, 2009, from http://www.preventionweb.net/english/professional/publications/v.php?id=2491

Vroom, V. H., & Yetton, P. W. (1973). *Leadership and decision making*. Pittsburgh, PA: University of Pittsburgh Press.

Willitts-King, B., & Harvey, P. (2005). *Managing the risks of corruption in humanitarian relief operations*. Overseas Development Institute (ODI). Retrieved June 28, 2009, from http://www.odi.org.uk/publications/index.html

Wilson, H. C. (2000). Emergency response preparedness: small group training. Part I: Training and learning styles. *Disaster Prevention and Management, 9*(2), 105–116. doi:10.1108/0965 3560010326987Zevallos, O. (1996). Ocupacion de laderas: Incremento del riesgo por degradacion ambiental urbana en Quito, Ecuador. In Fernandez, M. A. (Ed.), *Ciudades en riesgo: Degradacion ambiental, riesgos urbanos y desastres en America Latina* (pp. 165–178). Lima: ITDG Publishing.

Chapter 2
ICT Approaches in Disaster Management:
Public Awareness, Education and Training, Community Resilience in India

Sunitha Kuppuswamy
Anna University Chennai, India

ABSTRACT

We live in a world where there are potential sources of disasters with a potential to cause a loss all around us. We could be living close to a coastline that is prone to cyclones or mountainous region vulnerable to earthquakes. On the other hand, we might be living in a place where there may be frequent communal tension. Whatever is the area we live in; we need to be aware of our vulnerability to hazardous events. Better awareness about various hazards, proper education, training and preparedness will enhance the community resilience. Information and communication technologies in the form of audio and video (through community radio, village information centres, video awareness programmes through DVDs, etc) play a vital role in creating public awareness, giving education and training to vulnerable communities. This chapter aims to discuss the various ICT initiatives taken in the coastal districts of Tamilnadu, one of the seven coastal states of India.

INTRODUCTION

We live in a world where there are potential sources of disasters with a potential to cause a loss all around us. We could be living close to a coastline that is prone to cyclones or mountainous region vulnerable to earthquakes. On the other hand we may be living close to an industry which could be dangerous or there may be communal tension prevailing in the area we live in (Arnold J.P, 2006). Whatever is the area we live in; we need to be aware of our vulnerability to hazardous events. Better awareness about various hazards, proper education, training and preparedness will enhance the community resilience. Information and communication technologies in the form of audio and video (community radio, village information centres, video awareness programmes through DVDs) play a vital role in creating public awareness, giving education and training to vul-

DOI: 10.4018/978-1-61520-987-3.ch002

nerable communities. Before discussing the ICT initiatives taken in the coastal districts of India, it is important to have a clear understanding on the natural disaster profile of India.

An event or hazard is called a disaster when it threatens property and lives and is unforeseen and often sudden. The World Health Organization (WHO) defines a disaster as 'A severe disruption, ecological and psychological, which greatly exceeds the coping capacity of the affected community' (World Health Organization, 1992). It causes great damage, destruction and human suffering. A disaster is a very complex multi dimensional phenomenon and along many dimensions like social, economic, material, psychological or social, but unlikely to be one along all of these in a specific direction. Often the number of human lives lost is an important criterion for defining a disaster (Arnold J.P, 2006). Disaster is a sudden, calamitous event bringing great damage, loss, and destruction and devastation to life and property. The damage caused by disasters is immeasurable and varies with the geographical location, climate and the type of the earth surface/degree of vulnerability. This influences the mental, socio-economic, political and cultural state of the affected area. It may also be termed as "a serious disruption of the functioning of society, causing widespread human, material or environmental losses which exceed the ability of the affected society to cope using its own resources." (UNDHA, 1992).

India's increase in the vulnerability to disasters, in recent years has been serious threat to the overall development of the country. Around 57% of the land vulnerable is to earthquakes, 28% is vulnerable to droughts, 12% is vulnerable to floods and 8% of the land is vulnerable to cyclones. Subsequently, the development process itself has been a contributing factor to this susceptibility. Coupled with lack of information and communication channels, this had been a serious impediment in the path of progress. Figuratively speaking,

around one million houses are damaged annually, compounded by human, economic, social and other losses (Anil K. Sinha, 2003).

Natural Disaster Profile: India Basic Facts

India is with an area of 3,287,590 km2 having 7,000 km coast line with a population of 1,065,070,607. The GDP (PPP) is $ 3.033 trillion, GDP Per Capita is $ 2,900 and 25% of the population is below the poverty line. Around 1,642,855 of the total population live within 1km of coast, 3,398,071 people live within 2km of coast and the Infant mortality rate is 57.92. The geographical setting of India makes the country vulnerable to natural disasters.

Climatic condition: Covering an area of more than 3,000,000 sq. km, India shares its borders with Pakistan, Nepal, China, Bangladesh, Burma and Bhutan. It has a long coast line with the Bay of Bengal in the east, the Arabian Sea in the west and the Indian Ocean in the south. Geographically and climatically too the country is very diverse, including snow-capped Himalayas in the north, tropical maritime climate in the south, desert in the west, alluvial plains in the east and a plateau in the central region. The River Ganges rising is the life source of the people in the north. The country on the whole has four seasons: winter, summer, spring and monsoon rains.

The land of unique climatic regime, India, has two monsoon seasons (southwest & northeast monsoons), two cyclone seasons (pre & post monsoon cyclone seasons), hot weather season characterized by violent convective precipitation and cold weather season characterized by violent snow storms in the mountainous regions. It is one of the most vulnerable nations in the world, susceptible to multiple natural disasters owing to its unique topographic and climatic conditions. Its coastal states, particularly the eastern coast and Gujarat are exposed to cyclones, 40 million hectares (eight per cent) of land mass is flood prone, 68 per cent faces drought threat, 55 per cent of the area is in seismic zones III-IV and falls under earthquakes (induced tsunami) - prone belt and sub-Himalayan region and Western Ghats are threatened by landslides. Thus different parts of India are affected by different calamities from time to time. Among these, the major coastal hazards are Cyclones, Floods and Earthquake induced tsunami.

Cyclones: The coastline of India extends over up to about 7,000 km, and is affected by 5 to 6 cyclones every year, out of which 2 to 3 are more severe. About 8% of the land is vulnerable to cyclones. Cyclonic activities on the east coast are more severe than on the west coast. Cyclones occur mainly in the months between April and May and October and November. $ 2,900 and 25% of the population is below the poverty line. Around 1,642,855 of the total population live within 1km of coast, 3,398,071 people live within 2km of coast and the Infant mortality rate is 57.92. The geographical setting of India makes the country vulnerable to natural disasters.

Floods: The country is divided into four flood regions according to the river system. They are the Brahmaputra region, Ganga region, Indus region and the Central and Deccan region (comprising the river Narmada, Tapti and all rivers flowing south eastwards). The types of floods in India are Snow-melt floods, Flash floods / cloudburst floods; Monsoon floods of Single & multiple events, Cyclone floods and Floods due to dam bursts / failure. It is estimated that an average of 40 million hectares are subjected to floods annually. On an average, about 30 million people are affected, a few hundred lives are lost, millions are rendered homeless and several hectares of crops are damaged every year. Recent flood in 2004, 2005 affected the states Assam, Bihar, West Bengal, Gujarat, Orissa, Uttaranchal, Tamil Nadu, and Maharashtra in India.

Earthquake induced Tsunami: 56% of the total area constitutes an active seismic zone. The northern regions are the most susceptible to earthquakes. Of the earthquake-prone areas, 12% is prone to very severe earthquakes, 18% to severe earthquakes and 25% to damageable earthquakes. The biggest quakes occur in the Andaman and Nicobar Islands, Kutch, Himachal and the North-East. The Himalayan regions are particularly prone to earthquakes. India has also become much more vulnerable to earthquake induced tsunamis since the 2004 Indian Ocean tsunami.

Impact of Natural Disasters on India

Cyclones: Cyclone is one of the most disastrous natural hazards which is caused by the storm surge and high tidal waves. This is aggravated by torrential rains, leading to coastal flooding. India is worst affected by cyclones as it has a long coastline of 5700 kms, which is exposed to tropical cyclones from Bay of Bengal (more) and Arabian Sea (less) (Arnold J.P, 2006).

In India, annually millions of people are affected by cyclones. Cyclones occur in coastal areas particularly before and after summer monsoon i.e., from April to June or from October to December. Cyclone warnings are provided in two stages. First stage: 'Cyclone Alert' is issued 48 hours before the expected commencement of bad weather along the coast. Second stage: Cyclone warning is issued at 24 hours before the cyclone strikes.

Some Recent Cyclones in India

Floods: In India annually 40 million hectares of land is prone to floods and millions of people are affected. Heavy rainfall combined with improper drainage systems lead to overflowing of the river water resulting in flood situations (Arnold J.P, 2006).

Table 1. Recent cyclones in India

Year	Place	Impact
1999	Orissa	Orissa Super Cyclone - close to 10,000 people were killed
1998	Gujarat	1261 casualties
1996	Andhra Pradesh	1,057 casualties
1997	Andhra Pradesh	8,547 people killed
1874	Bengal	80,000 people killed

Details of Floods in the Last Few Years

South East Asian Tsunami 2004: Impact on the Tamil Nadu Coast

The earthquake that struck northwest of Sumatra, Indonesia, at dawn on December 26, 2004 was a perfect wave-making machine, and the lack of a tsunami warning system in the Indian Ocean essentially guaranteed the devastation that swept coastal communities around southern Asia (Andrew C. Revkin, 2004).

Damage to the Chennai Coast and the Tamil Nadu Coast

The Tsunami of 26.12.2004 which struck Tamil Nadu caused enormous damage to life and property in the coastal areas of Tamil Nadu. The state suffered a loss of Rs.47 billion, accounting for two thirds of the total losses suffered in south India, followed by Kerala (Rs.13 billion), Pondicherry (Rs.5 billion) and Andhra Pradesh (Rs.3.4 billion) due to Tsunami (Indian Government Release, 2005). The death toll crossed 8030, more than 17000 animals perished, thousands of people lost their houses and lakhs of people lost their livelihood. The coastal economy has been devastated due to loss of fishing gear and damage to infrastructure.

Thus natural disasters can neither be predicted nor prevented. The problem before us is how to cope with them, minimizing their impact. Tamil Nadu has witnessed havoc caused by cyclones and storm surge in the coastal regions, earthquakes, monsoon floods, landslides, and recently the Tsunami. Increase in urban population coupled with the construction of man-made structures often poorly built and maintained subject cities to greater levels of risk to life and property in the

Table 2. Recent floods in India

Year	Place	Impact
2005	Tamil Nadu	Millions of people in 22 districts were affected. Many lost their house and properties
2003	Assam	Less than a million hectare land submerged, 30 lives lost, millions of people were affected
2003	Uttar Pradesh	More than 2 million hectare land submerged. 980 lives lost. Millions of people were affected
2003	Gujarat	106 lives lost. More than 2 million affected
2002	Bihar	Nearly 2 million hectare land submerged. 434 lives lost. Millions of people were affected
2001	Orissa	Nearly 1 million hectare crop area destroyed. 99 lives lost. Millions of people were affected

event of earthquakes and other natural hazards. One of the main objectives is to reduce the risk of loss of human life and property and to reduce costs to the society. We have to recognize that in such cases of natural disasters, we deal with phenomena of enormous magnitude that cannot be controlled by any direct means of human intervention. But what we try to do is to reduce the impact on human beings and property (CMDA, Chennai).

NEED FOR COMMUNITY BASED DISASTER PREPAREDNESS AND EDUCATION

Preparedness is usually regarded as comprising measures which enable government organizations, communities and individuals to respond rapidly and effectively to disaster situations (Ishak .R, 2004).

One aspect of preparedness which is not always given adequate priority is individual and/ or family preparedness. In many circumstances where government resources and emergency services are limited, such individual and family preparedness may be vital for survival. We can see three building blocks of disaster preparedness for the people, make disaster management part of development, make people 'disaster aware', and strengthen communities to withstand disaster.

Examples of preparedness measures are: the formulation and maintenance of valid, up-to-date counter-disaster plans which can be brought into effect whenever required, special provisions for emergency action, such as the evacuation of populations or their temporary movement to safe havens, the provision of warning systems, emergency communications, public education and awareness.

Awareness-raising is a process which opens opportunities for information exchange in order to improve mutual understanding and to

develop competencies and skills necessary to enable changes in social attitude, activities and behavior. As UNESCO's Memory of the World Program reminds us in relation to our documentary heritage, "education plays a crucial role in raising awareness" (UNESCO, 2005). The same may be said of environmental, public health and disaster management awareness where education clearly "accelerates the progress of societies toward disaster resilience". Also, "A fully aware, well informed and properly trained population is the best guarantee of safety and of successful response to any disaster." (APELL and UNEP, 2005).

Some common methods, techniques and approaches to education in rising public awareness include:

- Indigenous traditional knowledge could be disseminated through performances of especially composed folk arts, songs like villuppattu, dances, plays, stories and poems.
- Individual communication with community members through public meetings, seminars, presentations, workshops and informal social events.
- Training programmes in the form of mock drills targeted at the community level disseminated through Village Information Centres (VIC) in coastal villages.
- Structured education and training programs in schools, colleges, universities, community centres and village information centres.
- Stagnant and mobile exhibitions, vehicles carrying messages and displays.
- Printed materials - for example, brochures, billboards, cartoons, comics, pamphlets, posters, and resource books.
- Audiovisual resources - for example, prerecorded cassettes, videos, CDs and DVDs which include audio, video, text, graphics and animations.

- Websites, email discussion lists and Web Logs (blogs).
- Mass media interviews and articles in newspapers, magazines and electronic publications accessible via the Internet and on radio and television.
- Celebrity spokespeople - for example, celebrities like A.R. Rahman could be made to support the Disaster Awareness Campaigns in India.

Importance of Communication in Disaster Management

Communication raises awareness of the hazards about the disaster. It provides a means of alert and early warning to the people in the disaster prone zones. It will be helpful in taking preventive measures to avert disasters. It allows vulnerable population and disaster management persons to be aware of the details of their vulnerability. It helps for mitigation decisions to be made. Builds support for programmes and activities which support mitigation. It facilitates planners to have an in depth understanding of vulnerable population, vulnerable areas, hazards and sectors at risk. It allows for appropriate planning measures to be put in place. Communication is needed for giving training and public awareness to the people of the disaster prone zones regarding how to give first aid to the people affected in disasters and where to find the safe place during the disaster period. It also provides data for integration and analysis of spatial and temporal disaster data, modeling and simulation disasters more precisely. Communications helps in real-time decision making and enhance emergency response capabilities.

ICT stands for Information and Communications Technology (ICT) used almost synonymously with IT or Information Technology. ICT is used in almost all phases of the disaster management process. In the disaster mitigation and preparedness process, ICT is widely used

to create early warning systems. An early warning system may use more than one ICT media in parallel and these can be either traditional (radio, television, telephone) or modern (SMS, cell broadcasting, Internet). As demonstrated by AlertNet, the on-line media play an important role. In the immediate aftermath of a disaster, various ICT initiatives for the purpose are introduced by the government and NGOs for activities such as registering missing persons, administrating online requests and keeping track of relief activities or shelters of displaced persons. ICT, like any other tool, can deliver its best when the other necessary ingredients are in place. Some other examples of ICT include: Use of editable Web sites (wikis), blogs, and data-mining tools to capture, analyze, and share lessons learned from operational field experiences. Use of database, Web, and call center technologies to establish a service to provide information about available equipment, materiel, personnel, volunteer organizations, etc.

BROADBAND WIRELESS COMMUNICATION IN DISASTER MANAGEMENT: A CASE STUDY

Communication systems connected by wire and fiber can be partially or completely wiped out in seconds by an attack or a natural disaster. Cellular telephone systems may enable communication to continue, but these are often limited to voice and are quickly saturated. There comes the requirement for Broadband wireless communication systems that are used to restore the problem of communications during emergencies and catastrophic events as it can carry large amounts of visual information. It is also used for communication services deployment in rural suburban areas typically affected by 'digital divide' which refers to the gap between people with effective access to digital and information technology and those with very limited or no access at all.

Broadband Wireless Access and the Digital Divide

Broadband wireless access provides high rate wireless communications between a fixed access point and multiple terminals. These systems were initially proposed to support interactive video service to home, but the application emphasis then shifted to providing both high speed data access (tens of Mbps) to the Internet and the World Wide Web as well as high speed data networks for homes and businesses.

WiMax is an emerging broadband wireless technology based on the IEEE 802.16 standard. The core 802.16 specification is a standard for broadband wireless access systems operating at radio frequencies between 2 Ghz and 11Ghz for non line of sight operation, and between 10 Ghz and 66 Ghz for line of sight operation. Data rates of around 40 Mbps will be available for fixed users and 15 Mbps for mobile users, with a range of several Kilometers. Many manufacturers of laptops and PDAs (Personal Digital Assistants) are planning to incorporate WiMax once it becomes available to satisfy demand for constant internet access and email exchange from any location. WiMax will compete with wireless LANs, 3G cellular services, and possibly wireline services like cable and DSL (Digital Subscriber Line). The ability of WiMax to challenge or supplant these systems will depend on its relative performance and cost, which remain to be seen. The Tsunami in the south of Thailand in December 2004 raised the need for a large capacity ad-hoc communication system for emergency response teams to report losses and coordinate rescue missions because the Communication systems failed completely due to broken connections, collapsed radio towers and power outage. Search and Rescue team usually comes with limited communication capability, not sufficient to serve the desperate public communication systems as they are overloaded with calls.

ICT Initiatives in the Tsunami Affected districts of Tamil Nadu

Ease of access to ICTs is the basic requirement for full participation in the modern digital economy. It is also important for taking part in the global social network formed by the world-wide universal reach of the Internet. Improving connectivity for individuals and communities thereby providing access to critical transformational information is therefore the major means of correcting the present imbalance in the use of ICTs, which has been called the digital divide, and to equalizing opportunities across gender, socio-economic status, and regional lines. A number of NGOs have played pioneering roles in the use of innovative means of strengthening connectivity, through community tele-centres, school-nets, and local radio. Further support for these efforts and for their alignment with broader connectivity initiatives underway statewide, would greatly strengthen the development of communities in the Tsunami affected areas.

In Tamil Nadu we have few pertinent players forming the community level multi-utility tele-centres, which are called Village Knowledge Centers (VKC) or Village Information Centers (VIC) to expand the horizon of ICT applications by fostering more diverse and stronger stakeholder partnerships involving national agencies (like ISRO, IMD), private ICT actors, NGOs, academic and research institutions etc. These centers are focusing on the potential for enhanced capacities to utilize ICT for disaster preparedness, management and sustained recovery.

Village Knowledge Centres / Village Information Centres

The Village Knowledge Centre (VKC) is a place to render distant services from a single window point to rural masses especially in remote areas of the country through modern Information and

Communication Technology. The knowledge centre will be connected to a central studio using technologies Web portal/leased line. There will be live interactive sessions in real time by the central speaker with audience at remote villages or content already prepared on any subject that the rural communities might need or desire, will be disseminated (CAPART, 2006). The main intention behind developing such VKC is to bring access to a range of services, content and information to people who are living in remote villages or areas which do not provide such access otherwise. The VIC/VKC offers a wide range of Information and Communication Services, which include computer training, tailoring, e-mail and browsing, job work, sending applications for obtaining birth and death certificates, disaster preparedness programmes, disseminating disaster warning, sending queries to hospitals and specialist doctors about health problems, entrepreneurship activities, queries about crops and animals, creating awareness on various disasters etc. They are also involved in video conferencing in Health-Care, Agriculture, Legal Consultation, Education, etc.

Puducherry is the first state where in the Government and various Non-Governmental Organizations initiated steps to bring Media–Information Technology to the tsunami have affected villages. With the help of the Village Information Centres [VICs], people can be warned in advance of any imminent disaster. Through field research and secondary search, it was found that VICs have been established in all 20 target villages of Pondicherry, Karaikal and Cuddalore regions which serves mainly for disaster preparedness.

It was also found from the field visit that the VICs are equipped with a computer with broadband internet connection, UPS, printer, scanner and a Public Address System (PAS). A display board is kept outside the VIC. The VIC makes announcements through PAS thrice a day on the weather, wave heights, wind speed, potential fish findings and the daily market price. Thus VICs are considered as one another form of radio

through which varied information could be disseminated. This information is displayed on the display boards in each village, which act as a front line delivery system to the villagers. Apart from these, the community is given training on how to use internet to get various information on natural disasters, and other developmental activities that enhances their standard of living. It is to be noted that women are the lead role players in these VICs, ie., it is only women who run these information centres. Interviewing a local, "Through these Village Information Centres, we get information about the changes in sea: wave height, weather information and potential fish finding zones that help us to yield better income" says Mr. Babu, a fisherman in Cuddalore district of Tamilnadu.

Challenges

Disaster recovery and response require a timely coordination of the emergency services. ICT provides a tremendous potential to increase efficiency and effectiveness in this area by propagating information efficiently to all the right locations. While ICTs have a crucial role to play in disaster management, there are tough challenges in making use of ICTs for the betterment of communities. Some of them are: Generation of local and relevant content; social dimension of ICTs; dominant use of English on the Internet; ownership and trust. These issues need to be sensitively tackled in order to enable the most marginalized groups to reap the fullest benefit of the information society.

Usually information systems for disaster response are divided into three phases: the pre-phase addressing the preparations before, the post-phase analyzing what happened during the disaster (lessons learnt e.g. for training) and the phase in between, i.e. the situation during the emergency, our main focus.

User requirements Analysis: Apart from communication and information management, the following areas were addressed: optimization and simulation, decision support, visualization, geographical information systems, and simulation and training. One of the findings was that maintaining communications is the "primary challenge" during a disaster and that the following major requirements were not yet met in a satisfactory way: • Integration and linking of information • Availability of communication • Redundancy of links • Fast data access • Timeliness and updating of information • Standardization of information (Meissner. A, 2002).

System Architecture: Data Distribution System (DDS) manages data information distribution by controlling who gets what, at in different time, at different places and under different situations. DDS have three main capabilities; Data verification, Quality control and Data storage services to all incoming and outgoing data (UNPAN, 2008). There have to be a sub system as portal to manage all incoming and outgoing transaction. Data Management System (DMS) is a collection of state-of-the art hardware and software that can be used for the management of disaster at every stage of the crisis before, during and after. Decision Support System (DSS) is an information system for disaster management and relief. It is a central database, where data and information can be made available on-line basis. It is an intelligent system, to help planning activities. It is also an electronic based correspondence system report generator that can be modified according to the user. Whole responsibility of this part are; damage assessment, thematic hazards maps, proposed solutions, early warning, Decision support, risk prediction, and situational analysis. With respect to Village Information Centre, the challenges include ensuring reliable Internet bandwidth to provide services, and maintaining uptime of the systems.

The contemporary era of Information and Communication Technology (ICT) is too complicated for the traditional State to handle single handedly. The State needs "external expertise" to meet effectively and efficiently the challenges of ICT (Dalal. P, 2007). Indian Government must appreciate that we need to capitalize "collective expertise" and

an "ideal public-private partnership" base in India for meeting the contemporary challenges of ICT. The Government alone cannot provide a viable solution to the problems associated with ICT. We have to take care of both technological as well as legal issues associated with the use of ICT. Thus, a techno-legal base is the need of the hour.

COMMUNITY RADIO: A CASE STUDY

While personal communication tends to be the most successful means of raising awareness of issues in smaller communities, it is not always the most effectual strategy for communicating a message widely (Ornager.S, Bangkok). To achieve this, we must rely on mass communication through the 'mass media' like radio, television, newspapers, internet etc. Radio is one of the information and communication technology which is known best for its speed, summaries, breaking news, discussion - talk-back radio, satellite radio and community radio, Television having the great impact among all kinds of people, Newspapers with a more considered approach and the web as a complementary media to radio and TV.

There are many remarkable examples where public education and the rapid, widespread dissemination of early warnings saved thousands of lives. In 1977, a devastating cyclone struck Andhra Pradesh, India, resulting in severe impact on the whole state with more 10,000 losses of lives. A similar storm struck in the same area 13 years later during which media played a vital role in emergency preparedness and early warning measures that saved lives and money.

Undaunted by the 2004 tsunami devastation virtually wiping it out, the coastal Vizhundamavadi hamlet in Nagapattinam, a coastal district of Tamil Nadu has now emerged as a model village with all-round community development. Vizhundamavadi, the most-backward village, located 20 km from the district and in close proximity to the sea, was one of the worst affected by the killer waves of December 26 tsunami which claimed 19 lives and left thousands homeless, washing away their belongings. The village has a population of 5,810, including 2,845 women and 415 Dalits. The panchayat comprised four hamlets of Vadapathy, Thenpathy, Thambiran Kudi Iruppu, and Manalmedu. And life became a question mark for the 5,000 odd residents, most of them being fishermen. Kalanjiam Samuga Vanoli (Kalanjiam Community Radio), the first community radio in South India for Disaster Management played a vital role in supporting the tsunami affected fishermen to get back their strength and vigor and to resume their fishing activities. Women in the tsunami-hit villages are now keen in improving their economic condition by starting several new income generating business with the assistance and support extended by the community radio players of Kalanjiam Samuga Vanoli, set up by DHAN Foundation, a Madurai based NGO.

It is a mixed media model that combines radio with video and web-based technologies. With its production centre at Vilunthamavadi village in Nagapattinam, Kalanjiyam has made 900 minutes of different programmes everyday (Kuppuswamy. S, 2009), through a mix of AIR broadcasts, cablecasts and loudspeaker narrowcasts through the VICs, besides webcasting through the internet. The interactive programmes covers self-help groups, woman and child health, education, farming etc.

Funded by the UNDP, the community radio is a joint initiative of the local public and two NGOs – Bangalore-based VOICES and Madurai-based DHAN Foundation. The foundation has already trained radio volunteers from among local youngsters, many who bore the brunt of the tsunami. The foundation says the radio station would work as a warning system and would seek to inform the local community on coping with disasters with a focus on sustainable development.

On 13th September 2004, a major earthquake of the magnitude 8.4 Richter scale, off the coast of Indonesia triggered a series of tsunami warnings. Tsunami warnings had been issued for the south west coast of Sumatra, as well as for Bangladesh, India, Sri Lanka, the Maldives and Mozambique. In Tamil Nadu, communities in close proximity to the sea in Vizhundamavadi village were evacuated once tsunami warning was given through Kalanjiam Samuga Vanoli. Also, the community radio played a major role in controlling the panicking of villagers by constant announcements on Disaster Preparedness. According to interviews with locals, "We are not worried about our lives as Kalanjiam Samuga Vanoli is there to inform us all in all about natural calamities", says Mr. Muthu, a 42 year old fisherman of Vizhundamavadi hamlet. Mrs. Amulamal, an entrepreneur in Vizhundamavadi says that "They give us hands on training about solid waste management, mat-bag-weaving, sanitation measures, importance of hygiene, disaster preparedness etc". "The sustainable livelihood programme is found to be very useful and I get revenue by weaving bags which was learnt from the same programme", she adds.

Thus, information and communication technologies like community radio are an effective tool through which we can fulfill the community disaster information needs. Education and training on various topics like what are the different types of disasters? What are its causes? Which region is vulnerable to which type of disaster? When will it happen? How to prepare? When to stay and when to go? Where to go? How to travel? How long will it last? How to mitigate? How to prepare? Who can advise and help? What help will be given and when need to be given to the community members through such ICTs.

MULTIMEDIA CAMPAIGNS: A CASE STUDY

Awareness is the basic tool for disaster preparedness and protection of the environment. It is vital to take necessary steps to raise awareness towards various natural disasters, their effects, environmental problems, etc particularly among children as they are the capable actors in the post-disaster scenario. Moreover, raising environmental awareness among the children also assists in strengthening child participation at the local level.

Media plays a vital role in raising public awareness and influencing public perceptions. Therefore, a well-organised media campaign should be considered as a prerequisite in any public awareness programme. In certain instances the media may have to be motivated to play a pro-active role in environmental issues.

Many mass media campaigns on Disaster Preparedness have been implemented in the developing countries in the past several decades. The goals of these campaigns are generally to persuade individuals to either take personal steps to protect themselves, their belongings, and their environment or to avoid Environmental degradation, loss of lives and properties, etc. The campaign has targeted wide range of public including children, adults, men and women.

This part of the chapter aims to discuss a Media Campaign on disaster preparedness for children conducted by the Science & Technology Communication Division, Department of Media Sciences, Anna University during March 2009 in the coastal town of Puducherry, a former tiny French colony surrounded by Tamil Nadu, India. The campaign was as a part of Science & Technology Communication project funded by National Council for Science and Technology Communication (NCSTC) and Department of Science and Technology (DST), Government of India. The campaign targeted IInd and IIIrd Standard Corporation Middle School Children, Puducherry.

The campaign uses information and communication technologies in the form of both traditional and electronic media to train and create awareness among children regarding Disaster Preparedness. Extensive creative pre-testing was undertaken prior to the campaign. The pre-study was undertaken with the corporation middle school students of Puducherry to find out the prevalent

attitudes and behaviors prior to the start of the campaign. This was followed by post-study few weeks after the campaign which aimed to identify any changes in their activities as a result of the campaign. Comparisons were then made between the findings from the pre-study and those from the post-study based on observation and interviews.

From the pre-study, it is understood that the students had minimal knowledge about disasters. The actual cause for various types of disasters was not known. They were unaware of the reason for their vulnerability. Taking all these points into consideration, the content of the campaign was designed to have all the basic science concepts with respect to disasters. Information communication technologies in the form of audio programmes in CDROM and video programmes in DVD were produced and showed to them. Digital banners, posters with colorful pictures were also used to communicate messages. Apart from all these, traditional media in the form of Villupaatu and puppetry were also used.

All the students in the school were interesting in participating in the campaign but we had to restrict the number to 100. The Multimedia Campaign grabbed the attention of children with the use of three main media tools, namely Audio, Video and Traditional Media. It was visual media which had a maximum reach among school children, followed by the traditional media in the form of Villupaattu and puppetry.

Majority of students had a clear understanding on the message that was communicated and majority of the respondents made an attempt to follow the disaster preparedness tips communicated through the campaign and almost all the children discussed about such tips to their family members, neighbors and friends. Video programmes, specially the graphics and animations used in such programmes made the children sit, watch and understand the message. From the observation, the post-study revealed a clear increase in the understanding of the science concepts behind natural disasters. Child participation in disaster preparedness enables children as future actors in the post-disaster scenario.

From the campaign, raising awareness was found to be a long-term approach. It requires continuous enforcement to ensure sustainability. The teachers had to repeatedly remind about the campaign and the message that was communicated for the children regarding disasters. Hence sustainability is still a doubtful factor which could be answered by developing and executing such child-centric awareness programmes repeatedly which can be a solid ground for disaster management and sustainable development of the country.

The Multimedia campaigns should be flexible and properly designed to specific target groups. Design of the campaign and its activities can be diversified to suit the background and interests of each target group. This could be achieved by performing a pre-test before the execution of campaigns studying their needs and wants with respect to the campaign. Moreover, awareness and capacity building schemes also provide opportunities for children to learn and adopt skills and knowledge to take part in improving environmental situations.

FUTURE DIRECTIONS AND CONCLUSION

With the help of the information disseminated using ICT (in the form of Village Information Centre, Community Radio) fisher folks got back their vigor and strength to resume their fishing activities. Fishermen who never had any links with banks or financial institutions have started transacting with them for financial support. The Weather forecast and Ocean state forecast disseminated by the public address system is useful for the fishermen practically. The training programmes on various disasters really made them aware of various types of disasters and paved way to know how to prepare, react during disasters. Agriculturists for whom the desalination process

was initiated by way of sustainable agriculture through radio programmes began to realize the importance of natural farming. Unlike the fishermen, farmers could not be rehabilitated in a short span of time since desalination is a continuous and time consuming process. Having realized the havoc played by chemical manure and pesticides in the past they have now switched over to vermin-compost and other natural pesticides. Women who had been confined to their kitchens most of the time have now come out to discuss common issues and problems that confront them through radio programmes. Also they have now become supporters of the family in their own little way by taking part in the SHG group activities. Women empowerment has been achieved, but to a very limited degree. There is still a long way to go.

Young people who were initially scared of venturing into the sea after Tsunami, being given some alternate skills training and awareness through radio programmes, are now more confident of their future. Children are now turning to their original form and composure. They are slowly getting out of the traumatic experiences they had been undergoing. In short everyone knows how to react before and after the disaster and very well prepared.

1. One gap identified through the review was the need for rapid, reliable, flexible, multimedia communications-there remains much to do to ensure that ICT is contributing to rapid, effective humanitarian action

2. Internet Communication Technology (ICT) problems had in some cases significantly hampered agency response

3. Certain partnerships with the public sector, academic institutions or other NGOs demonstrated a very open approach to solving problems, and are a very positive sign. The report concludes that ICT alone will not sustain an emergency response, but it should be considered the starting point for nearly all activities and stakeholders including the

government and communities need to begin treating it as the critical resource that it is.

With respect to multimedia campaigns, learning from both the successes and failures of past mass communication campaigns really helps in using the mass media to promote social issues. Providing access to information alone will not automatically lead to a significant increase in the behavioral change among children. A bridge needs to be built, linking the information pillar with the participation pillar, to stimulate the involvement of children in this cause. If the child along with the public surrounding them is well-informed about the disaster preparedness, and environmental issues, properly consulted and encouraged to participate in management process, then changes will occur. Also, studies have demonstrated that when long-term mass communication campaigns are designed and executed, they can play a meaningful role in changing behavior, either directly or indirectly. Thus, Multimedia Campaigns with formative research with a neat design and execution had always been successful.

REFERENCES

Anil, K. S. (2003). *Development of an Integrated Disaster Management System in India: Importance of Reliable Information*. Paper Presented at the International Conference on Total Disaster Risk Management. 2-4 December 2003.

Arnold, J. P. S. (2006). *Disaster Management – A Hand Book for NGOs*. Chennai: TNVHA.

CAPART. (2006). Retrieved from Available at http://capart.nic.in/scheme.vrc.pdf

Dalal, P. (2007). *Cyber law in India needs rejuvenation*. Indian Attorney.

http://www.cmdachennai.gov.in/Volume1_English_PDF/Vol1_Chapter10_Disaster%20Management.pdf

http://www.pmsss.org

http://www.unep.org/tsunami/apell_tsunamis.pdf

http://www.unesco.org/webworld/mdm/administ/MOW_fin9.html

Indian Government Release.

Ishak. R. (2005). Special Report: Disaster Planning and Management. *NCD Malaysia 2004*, 3(2).

Kuppuswamy, S., et. al. (2009). Women, information technology and disaster management:

Meissner, A. et.al. (2002). *Design Challenges for an Integrated Disaster Management Communication and Information System*. Paper presented at the First IEEE Workshop on Disaster Recovery Networks (DIREN 2002), June 24, 2002, New York City, co-located with IEEE INFOCOM 2002.

Ornager, S. (n.d.). *Media & Communication: How can partnership with the media support ESD? UNESCO Communication & Information, Bangkok*.

Revkin, A. C. (2004). With No Alert System, Indian Ocean Nations Were Vulnerable. *The New York Times*, Retrieved October 30, 2009, from http://www.nytimes.com/2004/12/27/science /27science.html

tsunami affected districts of Tamil Nadu. *International Journal of Innovation and Sustainable Development, 4*.

UN DHA. (1992). *Internationally Agreed Glossary of Basic Terms Related to Disaster Management. UN DHA (United Nations Department of Humanitarian Affairs) (1992, December)*. Retrieved from.

unpan1.un.org/intradoc/groups/public/.../UNPAN025913.pdf

Websites

World Health Organization, Division of Mental Health. (1992). *Psychological Consequences of Disaster: Prevention and Management*. Geneva: World Health Organization.

Chapter 3
Multimedia Educational Application for Risk Reduction

Ana Iztúriz
Universidad Pedagógica Experimental Libertador, Venezuela

Yolanda Barrientos
Universidad Pedagógica Experimental Libertador, Venezuela

María A. González
Universidad Pedagógica Experimental Libertador, Venezuela

Larry Rivas
Universidad Pedagógica Experimental Libertador, Venezuela

Matilde V. de Bezada
Universidad Pedagógica Experimental Libertador, Venezuela

Simón Ruíz
Universidad Pedagógica Experimental Libertador, Venezuela

ABSTRACT

The 21ˢᵗ century has brought changes and innovations in educational technology that should be incorporated in the curriculum at all educational levels. Therefore, it is necessary to apply modern technologies in learning processes. This work was based on basic cognitive learning theories and principles, and used software, such as Freehand, Adobe Photoshop, Jigsaw Puzzle Creator, Brainsbraker and Jigsaw Puzzle Lite, to create a multimedia version of the SALTARIESGOS board game. The game was applied and validated with students from elementary school of 2nd, 4th, 5th and 6th grades. The results we obtained can be considered satisfactory based on the student's opinion, and the pedagogical strategy was as an effective tool for achieving our goals. The educational contribution of this interactive version of the game will promote and sensitize school community members at different urban and rural areas, in order to enhance the preventive culture in disaster risk reduction.

DOI: 10.4018/978-1-61520-987-3.ch003

INTRODUCTION

Education in the 21st century outlines poses some inherited and new challenges as response to the population growth, the marginalization, the migrations, the economic-political blocks in which the world has been divided. As a result we have led to the present environmental crisis due to the impact of threats and societal vulnerabilities. UNESCO declared, in the Millennium Objectives, the need that half of the planet inhabitants should be alphabetized and therefore obligatory primary education be reached as one goals to achieve for 2015 (UN, 2005).

Education for disaster risk reduction (DRR) is a trans disciplinary exercise aimed at developing knowledge, skills and values which will empower people of all ages, at all levels (Dey, 2006).

During any catastrophe children are the most affected and vulnerable to such adversities. Safety culture promotion and disaster preparedness among school members will aware them on environmental dynamics and safety (Abhyankar, 2006). According to the International Federation of Red Cross and Red Crescent Societies the education is a much broader concept in its theoretical and practical implications and a more systematic vision of the concepts, skills and values offered on DRR.

Based on these commitments, the incorporation of formal and informal education for risk management in Latin America has had a non harmonious and stable advance in spite of the regional efforts carried out by the EDUPLANhemisférico, PREANDINO, CAPRADE, PREDECAM among others (UN-ISDR, 2007). Part of the obstacles have been the limitations in the incorporation of the transverse core on disaster risk reduction into school curricula, as well as the inappropriate school infrastructure, and the background of teachers and students to reach the safety school standards existing in other continents.

BACKGROUND

Instruccional games are used to develop cognitive processes, values, emotion control and group exclusion handling. Jensen (1999, 2000) indicated that educational activity improved attention, planning, memory, language, creative and divergent thought. Also, it allows societal integration and worldwide understanding (Szczureck, 1989, Amaya, 2005).

Recently, new educational developments involving ITC's has been reported in Argentina, such as: ¡Alerta Sismo! Prevención Sísmica en las Escuelas, ¡Alerta Sismo II! Plan de Emergencia Familiar, designed to increase the preventive culture towards the main dominant geologic threats (Malmold and Balmaceda, 2005 a; 2005 b; 2006); and the ABC Desastres (UN-ISRD, 2006). This last author points out that Colombia, Venezuela, Peru, Chile, Cuba, Nicaragua, El Salvador, and Costa Rica have had successful experiences in previous preparation for disasters. It is important to note that, in Mexico, the teaching on DRR is obligatory from the legal viewpoint, they have several study cases and instructional materials related to this study matter. It is also important to point out that not all these efforts develop the application of ITC's as teaching and learning strategies.

The International Strategy for Reduction of Disasters has produced a very significant educational resource production in multimedia format on DRR (UN-EIRD-UNICEF, 2005) such is the case of the game "Stop Disaster" (UB-ISRD, 2007). The UK High School Teacher Association made an interactive website in 2007 for multi users that allows the exploration and discovery of drills related to adverse events, as well as learning about the nature of hazards and risks (University of Nothhumbria, UK-Disaster and Developing Center) (ISRD-IDRC-Prevention Web, 2008).

The Venezuelan educational system has limited instructional resources related to risk and disaster reduction subjects, especially those associated with natural hazards and socio-natural events. This situation should attract more attention to teachers in order to design and apply several pedagogical strategies about these topics in the educational processes to mitigate the hazards and risks in vulnerable schools and create a preventive culture in the educational community (Barrientos., et al 2006).

In that regard, some efforts have been made on pedagogical strategies to create threat and socio natural risks awareness by FUNDAPRIS-CEAPRIS (Cárdenas, E. et. al, 1990), CENAMEC (2005); Barrientos et al., (2006); Iztúriz et al., (2007 a, b); Iztúriz et al., (2008 a, b). Recently a board game, Ruta por la paz, has been applied in several Venezuelan schools in order to promote the culture of peace (Dos Reis, 2009). The instructional resources designed so far are printed, only to be used in classroom and home, limiting their potential for the kind of massive distribution needed to meet the community demands for information about the geological, hydrometeorological, technological, biological, and social as well as other complex human threats (Urbina-Medina, 2006).

Why do we teach this subject at school? The country has a long record on catastrophic events, due to natural hazards such as earthquakes, landslides, flash floods, tropical storms, fires as well as social, technological and biological risks. Venezuela has a high seismic activity due to the main fault system along 70% of the territory. This situation should draw more attention to teachers in order to design and apply pedagogical strategies during educational processes to mitigate seismic hazards particularly at vulnerable schools.

During December 1999 when torrential rains hit Venezuela northern regions and affected the educational sector by destroying school infrastructure and hampering synergy. Therefore there was a need to produce and apply massively educational experiences in order to increase capacity building in DRR and communities resilience.

The need to increase the application of the instructional resources in favor of Venezuelan and Latin American education on DRR, has motivated our group a research experience, in order to create ITC's version of several printed pedagogical games to improve and facilitate safety learning drills and knowledge at the different educational levels and for local community members. In that sense, the instructional games in both formats (printed and multimedia) will be valuable tools to teach on DRR, and will allow teachers to overcome difficulties in the classroom and some sectors of society.

However, the use of educational ITC's in learning processes requires the appropriate hardware and software tools, also it is important to remark that teachers and other users must have the necessary training on this matter (Szczurek, 1989; Bolívar and Ríos, 1990; Mazzarella and Ríos, 2006).

SALTARIESGO: Printed Version Description

The risks are very complex events to teach and this didactic proposal is a useful methodological tool in order to develop and promote topics related to natural hazards and socio-natural risks contained in the National Basic Curriculum on education, risk management and preventive culture. For this reason, it is necessary to apply pedagogical strategies that will facilitate the learning and teaching of these matters.

The present game is an interactive educational strategy which allows students from primary education to learn about local hazards, socio natural and technological risks through playing; stimulating environmental and human values in order to promote preventive and safety culture. It is a board game that contains definitions, processes and safety drills (before, during and after for each adverse event), including social risks under a complex and environmental perspective.

SALTARIESGOS resembles the school community vulnerability features to be applied in mountain or coastal landscapes; therefore it allows individual or group participation. The kit contains a board, one dice and several tokens. The game has 68 colored squares: 48 whites of free access; 9 blue to reward; 6 reds for penalty and 5 yellows to prevent. All board squares, except the whites, have a message to be read by the participant, who will continue, stop or go backwards (Iztúriz et al., 2007b, 2008). The following game rules are found at the board front:

In order to start the game, the players throw the dice and choose who will play first and select the colored tokens.

Each participant moves forwards as squares as the dice indicated.

The player should read the message present in the colored squares to advance, remain or return along the game board.

In order to reach the final square (LLEGADA), the participant must obtain the precise number in the dice. Otherwise the player should wait the next turn.

In this game all the participants are winners and safety culture promoters.

The objective of this manuscript was to turn into a multimedia version a printed puzzle called SALTARIESGOS to be applied and validated in schools provided with the hardware/ requirements. The educational games are included in the priority action N° 3 of the Plan of Action of Hyogo (2005-2015) which points out:

[The use of knowledge, the innovation and the education to build a culture of safety and resilience at all the levels through: The standard use of DRR terminology, the inclusion of the DRR in formal and informal education into school curricula, training and learning about disaster reduction at communitarian level, local authorities, special stakeholders, with the same access opportunities].

Methodology

The present work was framed in the theories and principles of basic cognitive learning, based on an eclectic approach according to Piaget, Gagné and Ausbel (Ogalde y González, 2008). The research had a quasi-quantitative and descriptive character. It was a case study and a field work. The methodology implied techniques and instruments such as: Observation, field notes, photographic and video records, diagnostic forms and pool opinion (Martinez, 1994).

The didactic planning of the game (printed and digital) involved several pedagogical strategies carried out with the students according to school curricular levels. This game was designed for primary school students from 2nd, 3rd, 4th, 5th and 6th levels. The curricular objectives involved the understanding of conceptual, behavioral and valuable contents on hazards, threats and risk prevention and safety culture promotion, as well as, reading comprehension, values acquisition, actions and responses associated with environmental education, risk proactivity, school community cooperation and capacity building in risk reduction.

SALTARIESGO: Digital Version Description

The game design on its digital version has required the programs Freehand and Photoshop. A multidisciplinary team of specialists conformed by a programmer, a graphical designer, experts on DRR and pedagogy was responsible for its creation.

The software's used are included in the category of declarative or heuristic programming as the adequate for pedagogical applications, since they allow the design and the implementation of learning aids and the development of educational contents (Ruíz and Ríos, 1990; Mazarella and Ríos, 2006). The computerized system designed required the selection of the commercial programs Jigsaw Puzzle Creator 2.01, Jigsaw Puzzle Lite

Figure 1. The board game SALTARIESGOS as a puzzle for children at Elementary School First Level.

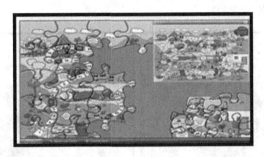

version 1.7.3 and Brainsbreaker with animation and sound effect. These programs turn the puzzle into 9 and 35 pieces puzzle (Figure 1). The game contains definitions, processes and links on risk reduction strategies to an integral and complex vision of hazards and threats, helping the students to understand and to handle the environment around them (Amaya, 2005).

The game's interactive presentation was made with three Power Point slides, with texts in Microsoft Word, with the invitation to play, the instructions and the complete image of SALTARIESGOS. Next, the screen showed the connection icon to execute the program, and finally the recognition of the activity end and run time used.

The didactic planning (application and validation) of the digital game involved three phases: (1) Theory (motivation towards the theme). (2)

Praxis (game playing) and (3) Evaluation; the activity was carried out at the Computer Science Laboratory at San Antonio de Padua II school with 45 students, the classroom teacher and the game facilitators. **Phase 1:** The classroom teacher set up all the hardware- software and explained to the group the activity purpose. A short presentation on natural hazards, socio natural and technological risks was provided by one of the game facilitators. **Phase 2:** The students were organized in groups of two for each PC. The invitation appeared in the screen: LET'S PLAY! (Figure 2a) followed by game instructions (Figure 2b). In the next slide appeared a complete picture of SALTARIESGOS (Figure 2c). The third slide showed the puzzle pieces in a number according to the curricular level or grade: 9 for 2nd and 35 for 4th, 5th and 6th grades. The teacher or game facilitator plays

Figure 2. Slides sequence for screen presentation (a) Invitation to play (b) Puzzle game instructions (c) Puzzle set and end of the game.

Figure 3. (a) Brain storming activity on hazards and socio natural risks during phase 1; (b) Students interactions during puzzle set on phase 2.

a   b

an important role focusing and, discussing the contents according to the pupil cognitive level.

The game facilitators elaborated a check list to register the student's performing times (also included in the software), concentration, interest, cooperation, communication, instruction understanding and help requested. **Phase 3:** It was organized in two sections. First, once the puzzle was assembled by the students, a short evaluation questionnaire was applied of 10 dichotomous items (yes or no) about game design, participant valuation, thematic and implemented logistics; two (2) open questions: one related to the identification of natural threats, socio naturals and technological risks in the game. The other item indicated possible changes in puzzle elements and features. The second part of the game validation consisted of making a classroom dynamics, with the students, facilitators and classroom teacher, on questions and answers around the natural threats identification, student perception of school, home and community socio natural and technological risks.

Game Application and Validation: Phases 1 and 2

The application and validation of this educational development allowed us to state that the partici-

pants had previous knowledge on natural threats, socio natural and technological risks, acquired resulting from the formal and non formal education.

How did the pupils played the game? The puzzle was set by the participants using three routes: The first one was used by 41 students, placing the 35 game pieces in the image edge and then they selected each one by trial and test. When the right piece was placed, the program emitted a sound to reinforce the student's attention and a happy face appeared, as a rewarding signal. The second route was chosen by four participants setting up first the puzzle edges. This solving procedure probably obeyed to the player's ability to identify the straight puzzle pieces. In the third route, four boxes were placed at each screen corner, in order to place in them the puzzle pieces according to each attempt. This last option was used by 12 students of 2nd, 4th and 5th levels.

During the game activity there was a suitable communication between the players, exchanging opinions, sharing time with the PC mouse and finger pointing on the monitor (Figure 3a and 3b). Also, the students asked few times for help to their teacher or facilitators. This situation could indicate a good understanding of the initial instructions, grasping the subject by the pupils. The learnt to recognize main local risks and signs

Table 1. Run times of SALTARIESGO puzzle with 9 and 35 pieces for 2nd, 4th, 5th and 6th level student.

School level	N° students	N° of pieces	Maximum run time (min)	Minimum run time (min)	Average (min)	Standard Deviation
2°	12	09	16' 00"	10' 01"	12' 36"	2.2
4°	10	35	23' 48"	08' 58"	17' 47"	5.18
5°	14	35	17' 02"	10' 59"	13' 40"	2.6
6°	09	35	16' 02"	08' 17"	12' 55"	3.8

of them, and also previously developed capacities related to the use of ITC's as part of education curriculum and home activities.

In relation to the quantitative outcome, the required time to set the puzzle was variable according to the student's academic level, interest, visual and motor capabilities. In the 35 puzzle version pieces, the minimum run time corresponded to two 6th level students with 08' 17" and the maximum of 23' 48" was done by two 4th level students. For 2nd level the maximum run time was of 16' 00' and the minimum of 10' 01" (Table 1).

In the puzzle run time averages, we found a slight difference among 2nd and 6th levels (12' 36" and 12' 55" respectively). This may be an interesting teaching and learning indicator in terms of game versatile possibilities for further development on the theme of RRD; (considering the computing class duration of 45 minutes weekly).

Evaluation of the Didactic Development: Phase 3

Once the game was over, an evaluation questionnaire with 10 dichotomous items was applied to the students for opinion registration. The items 1 and 2 were about the instructions and the affability of the game. They had unanimous acceptance. This evaluation agreed with the one previously made on the printed version of the game with different participants (Iztúriz et al., 2008). In item 3 related to the game name, the answers were divided, having the 100% of acceptance in the

students of 2nd, 50% in 4th, 64% in 5th and 6th levels respectively.

The items 4 and 5 that referred to the simplicity measure and to the run time obtained 100% of positive answers among the 2nd level students whereas those from 4th to 6th grade, the positive answers ranged between 50 and 70%, perhaps due to increase in the puzzle piece numbers that was almost 4 times greater than for the first group.

Questions 6, 7, and 8 related to the game design had a high acceptance range between 70 and 100% in the students of 2nd, 5th and 6th levels; and question 7, those students from 4th grade divided their opinion between 50% of affirmation and 50% of negation. Questions 9 and 10, on returning to play and the disadvantages that could be presented/displayed in the development of the activity, the answers were between 70% (4th level) and 100% (2nd, 5th and 6th levels) for the first item. In relation to item 10 their answers were between 60% and 100% of no limitations during the activity.

The last two questionnaire items were open: One related to game threats, socio naturals and technologic risks identification, and the other one gave the student the opportunity to suggest adding or taking any element from the board design. The main natural threats identified by all the participants were fires and rain; water pollution and floods were mentioned by the students of 2nd, 4th and 5th levels; earthquakes by 2nd, 5th and 6th levels; landslides were indicated by 4th, 5th and 6th grades; thunders by 2nd and 6th

Table 2. Advantages, limitations and aims comparison of printed and digital SALTARIESGOS formats.

Printed *SALTARIESGO*	Digital *SALTARIESGO*
Collective game	Personalized/ individual game
Increase group interaction and learning	Limited group synergism
Limit information retention in short/long term	Improve information retention in short/long term
No hardware or software requirements	Hardware and software needed
Use in classroom or any educative spaces	Computing laboratory required
No website access	Website availability
Traditional teaching methods	ITC's employment in the classroom
Rigid format	Enhancing format: Links, hypertexts, etc
Game time shorter	Game time longer

levels and environmental pollution by 5th and 6th grades. The threats of low frequency in their recognition were lightning and falling trees by 4th level; electrical towers, storms, eroded mountains by 5th level and flash floods by those of 6th grade.

Perhaps the above mentioned percentages do not represent a logical sequence order. Obviously a fallen tree represents a great treat after a storm, hurricane or earthquake causing a lot of collateral damages. However, we may understand pupils answer due to the low amount of trees close by school and home in this part of the city. A completely different response may be expected in a more rural area or in temperate latitudes.

The students could approximately identify 50% of the total threats, socio natural and technological risks presented in the game, recognizing the most common ones.

In relation to question N° 12 considering which game aspects they would add or eliminate, 2nd level students did not find any change and were satisfied. The participants of 4th and 5th grade answered that font size was small and limited text reading. The 6th level participants indicated to increase the multimedia effects and to play individually. Others, on the contrary were totally satisfied.

ISSUES, CONTROVERSIES, PROBLEMS

The educational experience allowed us to compare both game developing, the printed and digital formats. Therefore, it is necessary to contrast the advantages, limitations and aims.

The use of both game formats depends on the teacher training, equipment facilities, cognitive pupil level and the timetable. The formats are easily reproducible and involve much interaction between students and teachers. Both game versions allow a direct feedback, to make necessary changes according to the curricular needs in each school and the specific impact of any event.

During the game's application and validation, there were found some limitations related to the resolution of PC monitor's. We recommended the use 17" or 19" size screen in order to facilitate students' reading of game texts. An interesting remark made by students in this opportunity was their preference to play individually. The activity was planned as student collaborative pedagogical strategy but it may be reversed in terms of its need or interest.

Another issues that we would like to point out is the aim to increase our team research with more under- postgraduate students from the Computing Science, the Earth and Environmental Science departments for a more multi and trans disciplinary work with the necessary grants, in order to elaborate the game prototypes and future application in urban and rural schools of Venezuela and the South American region.

FUTURE RESEARCH DIRECTIONS

The educational contributions from multimedia educative games on DRR will build teachers and students safety culture and resilience as part of the school and community training at urban and rural schools. The innovative and versatile instructional character and the social and environmental pertinence of these game versions will constitute working tools in the classroom, for the learning process in the preventive culture promotion. This effort is outlined in the Millennium Objectives (UN, 2000), the Hyogo Framework for Action 2005-2015 (UN, 2005) and a new Venezuelan legal document titled Ley de Gestión Integral de Riesgos Socionaturales y Tecnológicos (AN, 2009).

The research group is working in two educational projects: The first one will incorporate hypertext, new animation effects and related links in Internet to SALTARIESGOS multimedia version. The second project deals with an interactive version of a game called SUDOKU-RIESGOS, its structure and way of playing will be based on the Sudoku commercial format (Mepham, 2005), replacing the numbers from 1 to 9 for international icons on hazards and socio natural risks. We believe that schools are the best option to train pupils on disaster management programs.

At the present time, we are developing other ITC's applications such as hypertexts and simulations for the described game considering the multi hazard approach employed.

CONCLUSION

The versatility of the electronic educative game will allow different didactic applications according to curricular level.

The game's interactive version reinforces student abilities, skills and competencies in the use of ITC's.

The obtained results in the use of the interactive Saltariesgos can be considered satisfactory and an effective tool in the promotion of the preventive culture in their familiar environment, school and community.

The employed didactic strategy was developed on natural hazards and socio-natural risks theme in terms of concepts, attitude and procedures related to curriculum contents for all the groups. This topic can be approached by teachers using different techniques as described, to develop creativity and interest during classroom activities involving family members as information sources of local experiences and cultural history from previous adverse events.

The pupils reached the proposed objective, learned preventive drills, DRR terminology, oriented their demands and expressed their ideas.

The participants developed basic cognitive processes such as observation, comparison, memorization, instruction handlings and drills on socio natural risks and safety culture related subject matters: Languages, natural sciences, technology, math, physical education, environmental and value cores present in the Venezuelan Elementary School Curriculum.

This educational game is a motivational strategy that complements the others pedagogical resources during learning and teaching processes.

The described pedagogical experience had as main goal to contribute to a culture of safety and resilience towards natural hazards, socio natural and technological risks through education.

All the participants are considered winners from this experience in order to become promoters of disaster risk reduction. Let our children teach us! (UN-ISRD, 2006).

REFERENCES

Abhyankar, M. (2006). Education, knowledge, innovation-building a culture of safety and resilience. An Indian Experience. In W. Almman et al. (Ed.), IDRC Davos 2006 (Vol. 2, Extended Abstracts, pp. 13-16). Davos, Switzerland: Swiss Federal Research Institute.

Amaya, J. (2005). *Fracasos y falacias de la educación actual*. Ciudad de México, México: Editorial Trillas.

Barrientos, Y. (2006a). Campaña educativa comunidad-escuela para la mitigación de riesgos socionaturales asociados a las cuencas de los ríos Osorio y Piedra Azul, estado Vargas, Venezuela. Informe Técnico N° 3. Caracas, Venezuela. FONACIT-UPEL. Documento Inédito.

Barrientos, Y., Iztúriz, A., García, A., & Ruíz, S. (2006 b). Instructional and Methodological Strategies for Learning about Natural Hazards and Socionatural Risk at Elementary School at Vargas State, Venezuela. In W. Almman et al. (Ed.), IDRC Davos 2006 (Vol. 3, Extended Abstracts, pp. 696-697). Davos, Switzerland: Swiss Federal Research Institute.

Bolívar, C., & Ríos, P. (1990). El uso de la informática en la educación. *Investigación y Postgrado*, 5(2), 59–91.

Centro Nacional para el Mejoramiento de la Enseñanza de la Ciencia (CENAMEC). (1996). *Rompecabezas de Placas Tectónicas*. Caracas, Venezuela: Colsum.

de Cárdenas, E. Sánchez, T. de., & Quintero, N de. (1990). *Revisión de los programas existentes para la Educación Básica con el objeto de incluir en ellos el material actualizado sobre riesgos naturales y prevención sísmica*. Paper presented at meeting III Congreso Venezolano de Geografía, Universidad de Los Andes, Mérida, Venezuela.

de las Naciones Unidas, O. Estrategia Internacional para la Reducción de Desastres., UNICEF., Federación Internacional de las Sociedades de la Cruz Roja y Media Luna y Ayuda Humanitaria de la Comisión Europea. (2005). *Campaña Mundial: La Reducción de Desastres Empieza en la Escuela*. Consulta: 2006. Enero 22. Disponible en: http//www.eird.org

Dey, B. (2006). Building a culture of safety and resilience towards natural hazards through education. In W. Almman et al. (Ed.), IDRC Davos 2006 (Vol. 2, Extended Abstracts, pp. 134-136). Davos, Switzerland: Swiss Federal Research Institute.

Dos Reis, A. (2009, agosto 11). Videojuego dispara para un clic contra la violencia. El Nacional (p. 6). ISRD- IDRC- Prevention Web. (2008). Virtual disaster risk reduction library. CD-ROM.

Iztúriz, A., Barrientos, Y., Ruíz, S., & Vierma de Bezada, M. (2007 b). Structured instructional game SALTARIESGO to promote safety culture at Vargas State, Venezuela. In S. Wang et al., (Eds.), *Strategy and Implementation of Integrated Risk Management. International Disaster Reduction Conference*. (pp. 472-475). Harbin, China: Qunyan Press.

Iztúriz, A., Barrientos, Y., Ruíz, S., & Vierma de Bezada, M. (2008). Juego instruccional estructurado: Ludograma para la prevención de riesgos en el Estado Vargas, Venezuela. *Research in Geographic Education*, 10(1), 33–47.

Iztúriz, A., Barrientos, Y., Ruíz, S., & Vierma de Bezada, M. (2008). *Un ludograma para la prevención de riesgos en el Estado Vargas*. Paper presented at the meeting XIII Congreso Nacional de tecnologías de la Información Geográfica. Universidad de Las Palmas de Gran Canaria y el Grupo de Tecnologías de la Información Geográfica de la Asociación de Geógrafos Españoles. Las Palmas de Gran Canaria, España.

Iztúriz, A., Tineo, A., Barrientos, Y., Pinzón, R., Ruíz, S., & Montilla, J. (2007a). El Juego instruccional como estrategia de aprendizaje sobre riesgos socio-naturales. *Revista Educere, 11*(36), 103–112.

Jensen, E. (1999). *Teaching with the brain and mind*. Alexandria, CA: ASCD.

Jensen, E. (2000). Moving with the brain in mind. *Educational Leadership, 58*(3), 34–37.

Malmold, A., & Balmaceda, M. (2005). *La universidad y la formación para la construcción de planes de contingencia*. Paper presented at the meeting I Encuentro Internacional y 2do Encuentro Nacional Educación Superior y Riesgos. Hábitat y Riesgo el Rol de las Universidades Universidad Central de Venezuela. Caracas, Venezuela.

Malmold, A., & Balmaceda, M. (2005). ¡Alerta SISMO! Prevención sísmica en las escuelas. *EIRD Informa*, (10).

Malmold, A., & Balmaceda, M. (2006). La multimedia ¡Alerta SISMO II! Plan de emergencia familiar. *EIRD Informa*, (13), 71.

Martínez, M. (1994). *La investigación cualitativa etnográfica en educación*. México: Editorial Trillas.

Mazzarella, C., & Ríos, P. (2006). Desarrollo y validación de un sistema computarizado para el aprendizaje de un contenido de genética. *Investigación y Postgrado, 21*(2), 11–42.

Mepham, M. (2005). *Soduko (Traducido por Javier García Sanz)*. Caracas, Venezuela: Editorial Melvin.

Asamblea Nacional. (2009). *Ley de Gestión Integral de Riesgos Socionaturales y Tecnológicos*. Gaceta Oficial de la República Bolivariana de Venezuela Extraordinaria N° 39.095 del 9 enero de 2009.

Ogalde, I., & González, M. (2008). *Nuevas tecnologías y educación. Diseño desarrollo, uso y evaluación de materiales didácticos*. Cuidad de México, México: Editorial Trillas.

República Bolivariana de Venezuela. (2009). *Ley de gestión Integral de Riesgos Socionaturales y Tecnológicos. Gaceta Oficial N° 39.095 del 9 de enero de 2009*. Caracas, Venezuela: Asamblea Nacional.

Szczurek, M. (1989). *Simulaciones y juegos instruccionales. Un hipertexto. Trabajo de Ascenso. Universidad Pedagógica Experimental Libertador. Instituto Pedagógico de Caracas. Trabajo Inédito*. Caracas, Venezuela: Autor.

UN-ISDR. (2007). *Words into action: A guide for implementing the Hyogo Framework*. Geneve: UN- ISRD.

UN-ISRD. (2006). *Let our children teach us!* India: Books for Change.

United Nations. (2005). *UN Millenium Development Goals Report 2005*. New York: United Nations. Retrieved May 14, 2006, from: http://unstats.un.org/unsd/mi/pdf/MDG%20Book.pdf

United Nations. (2005). *Hyogo Framework for Action 2005-2015: Building the Resilience of Nations and Communities to Disasters*, Geneva: International Strategy for Disaster Reduction [ISDR] World Conference on Disaster Reduction (A/CONF.206/6). Consulta: 2007. Marzo 18. Disponible en: www.unisdr.org.wcdr

Urbina-Medina, H. (2006, Enero 14). Los niños y los desastres. Sociedad de Puericultura y Pediatría. *Encartado El Nacional* (pp. 12-13).

Chapter 4
Natural Hazards:
Changing Media Environments and the Efficient Use of ICT for Disaster Communication

Helena Zemp
University Zurich, Switzerland

ABSTRACT

The growing importance of mass media in the 'information society', combined with society's increased dependence on electronic modes of information is important to the perception, regulation and management of risk at a local, national and international level. However, media organisations have their own logic and goals that are not necessarily compatible with the logic and goals of disaster planning and assistance agencies. Using a detailed study of the media coverage of floods in Switzerland from 1910 to 2005, we will illustrate the salient features of disaster reporting and how these relate to issues of risk perception and risk prevention behaviour in the public sphere. The findings are used to discuss the traditional media's shortcomings for the goal of risk reduction, the public's information seeking behaviour, and the opportunities and limitations arising from the emergence of digital, internet-based information and communication technologies (ICT) for disaster communication.

INTRODUCTION

When people are under threat, perceived or actual, information seeking is intensified. In such circumstances the national mass media system has a major responsibility to disseminate news, as well as public perceptions of disasters. Optimal preventative strategies for reducing damage require planned interactions with the media. As

a system and process, risk communication does not take place in a vacuum. Communication is shaped by a variety of contingent and historical factors, including politics, media and culture (Renn, 1992; Dunwoody, 1992). Optimal disaster communication needs to fine-tune all activities related to disaster planning and relief with the logic of traditional and new media. Effective communication needs to be based on profound knowledge of media systems. Not understanding media channels and the public's use of them can

DOI: 10.4018/978-1-61520-987-3.ch004

worsen crisis and disaster situations (Zemp & Bonfadelli, 2008).

We will argue that communication strategies that are not aligned with the potential victim's behaviour have a limited opportunity to raise the level of risk awareness. By producing a comprehensive map of the basic structures of disaster coverage deployed by traditional mass media we can identify which areas of the disaster process are covered and which topics are relatively neglected or ignored.

Findings on the changing logic of news production, its consequences for disaster reporting, as well as the public's information seeking behaviour and its perception of risk, will enable us to identify lessons for the use of new information and communication technologies (ICT)1 for risk communication. Evidently, these developments provide low-threshold access to worldwide information, communication and publication. We conclude with suggestions for the successful adaptation of risk communication in an increasingly commercialised environment and the role of web-based information channels.

BACKGROUND

The Function and Changing Logic of the Media

The core function of the media is not simply to transfer information or to report what has happened and what is being done. Rather, the media is a dynamic interpreter that analyses events and even prescribes what should be done (Peters, 2009). The mass media operates as a critic in democracies, where scrutinizing public officials' performances is a well-accepted practices, along with institutions to judge, punish, compensate and protect the general public. In other words, the publication of information and criticism perceived to be of public interest is understood as one of

the primary roles of mass media in democratic societies (McQuail, 2005).

In this process, the media also select the events and issues to be reported. Journalists can choose among many sources for their reports. Official and expert sources hold powerful positions, sometimes as the sole authority figures, serving to reassure or warn the public, or feeding into an ongoing debate. However, additional views may be sought to counteract or amplify the expert sources (Reese, Grani & Danielan, 1994; Boykoff & Rajan, 2007; Peters, 2009). The interplaying factors of internal norms for editors and journalists, personal judgments in news selection, and organisational and ideological pressures, all lead to the framing of the discourse in important ways. This is inevitable as media practitioners attempt to make sense of our world, which leads to the emphasis on some aspects of the world reality and the relative disregard of others. These "patterns of presentations, of selection, emphasis, and exclusion" are known as framing (Gitlin, 1980, p. 7; Entman, 1993).

Over the last decades, there have been tremendous structural changes in the media system. In most of the developed world there is a history of state-related telecommunications and broadcasting supplemented by party-political and commercial press. The relationship between the political system and media has been weakened and was gradually replaced by an independent media system. In the 1980s and 1990s government monopolies in broadcasting and telecommunications were broken up. With the increasing competition amongst media organisation for attention in the public sphere comes a trend towards commercialisation. Media organisations must package their stories in an increasingly competitive and unprecedented 24/7 real time context (Cottle, 2009). Most importantly, media and journalistic activities that decide what is and what is not newsworthy is increasingly dependent on audience ratings and sales figures. News values for instance, unexpectedness, negativity, dread, personalisation or good visuals, are often regarded as factors that contribute to the newsworthiness of a potential story (Galtung & Ruge, 1965; Schulz, 1990; Ruhrmann & Göbbel, 2007). Due to these changes the coverage shifts from public service orientated information towards more entertainment-oriented content. Even the so-called quality press concentrates more on sensational human interest stories; broadsheet front pages mimic those of the tabloid press using more and bigger pictures, larger print and shorter sentences. This trend is referred to as tabloidization (Holly, 2008). A second but intertwined trend is established through the emergence of new communication channels at the end of the 20th century. Based on digital technology the internet has become increasingly important. This affects journalistic research practices and growing online activities of traditional media organisations. Media are increasingly catering to a mobile consumer and new competitors have arisen (Chaudhary, 2004). Additionally, the new technical opportunities foster the development of globalised news distribution and media ownership. These changes in structure have led to more individualised and commercialised communication. Literature suggests that these developments also weaken of public service oriented goals of media reporting (McManus, 1994; McQuail, 1998; Picard, 2005). In summary, there have been two different developments that have jointly changed the media ecology during the last decades: (1) a long-term trend towards commercialisation, and (2) the emergence of new media based on digital technology (Geser, 1997).

Additionally, the audience and its usage of the media have to be considered. It is important to note that the audience is not a passive and homogeneous receiver of information. Members of the audience have individual characteristics, varying information needs and different information seeking behaviour. These parameters are important meditating factors of media effects (Seeger, 2008). The uses-and-gratifications perspective suggests that the audiences' media choices and usage – whether for instrumental or ritualised reasons – are characterised by the following features: (1) Socio-psychological needs, which generate (2) expectations of (3) the mass media and other sources, therefore leading to (4) differentiated patterns of media exposure, resulting in other consequences, perhaps mostly unintended. This approach – first theorised by Katz (1974) – shifts the emphasis in communication science to the question, "What do people do with media?", and away from the former paradigm of assumed effects, "What does media do to people? (Blumer & Katz, 1974). The audience is weaving together mediated knowledge, institutionally acquired knowledge, along with the information and evaluation resources grounded in personal experiences and local knowledge, in order to make sense of a situation. Acknowledging this active process means there are substantial variations in both the interpretation of and reaction to specific media content (Bonfadelli, 2001). This shifting interplay between information sources constrains and limits understanding. The active audience perspective examines the audience's choices to satisfy information needs alongside different conceptions of new media phenomenon such as internet, cell phones, interactive cable television, etc., in contrast to and competing with traditional media, in particular radio, broadcast television and newspapers.

DISASTER MANAGEMENT AND THE ROLE OF COMMUNICATION

Managing disasters is an integrated and multilayered process. Generally speaking, the functions of disaster management include: (1) Clarifying risk and encouraging preparedness; (2) Issuing evacuation and warning; (3) Enhancing coordination, cooperation, and logistics; (4) Facilitating mitigation on the part of the public and affected communication; (5) Helping make sense of the disaster; (6) Reassuring, comforting, and consoling those affected; (7) Recreating order and meaning, facilitating renewal, and learning and disseminating lessons (Auf der Heide, 2009; Seeger, 2005). Indisputably the quality of communication plays a central role for the fulfilment of the listed functions and thereby influences the vulnerability and resilience of the society.

For the threatened population and individuals, disaster communication – encompassing both direct and mediated forms – is crucial for developing realistic perceptions of risk. In addition, communication can motivate people to prepare for a disaster and enable them to take appropriate actions during the event and for recovery in the aftermath. As a non-routine situation, accurate information is extremely important. Disaster communication can be vital to survival in the face of uncertainties that require interpretation, explanation and consolation. Problems in the communication process between disaster agencies and the public can spread dysfunctional dynamics with destructive consequences (Comfort, Dunn, Johnson, Skertich & Zagorecki, 2004).

For the overall goal of risk reduction, it is useful to divide the communication strategies into three phases:

1. Public awareness (pre-event)
2. Public warning (during the event)
3. Informing and advising the public (immediately following and long-term post-event)

In all three phases the media is extremely important to the communication strategies of disaster agencies as media channels (newspapers, television, radio and – increasingly internet or cell phones) provide easy access to a large public. Before a crisis the media raises the public awareness through reporting on existing risks. During a crisis the media distributes warnings and release specific information on protective measures that need to be taken by the public. In the aftermath media communication disseminates information into the public sphere, which stimulates public debate that may then be used to inform and create

a policy agenda for future planning (Seeger, 2008; Auf der Heide, 2009). Through all phases the level of coverage, exposure, placement, headlines and photographs, contribute to the way in which events and risks are construed by the public in the immediate and the long term (Ashlin & Ladle, 2007).

Policy-makers and disaster agencies acknowledge the increasingly powerful role of the mass media in the process of disaster communication. But the liaison with journalists is not a straightforward exercise and institutions often face difficulties in working with the media. Disaster agencies and media organisations have different and sometimes even conflicting, goals. While agencies must assure public safety through their communication, media organisations want to attract readers, viewers or listeners. From the disaster management point of view, what they expect from the media – especially under extreme urgency – and what they get, may support but can also obstruct the goals of disaster authorities and relief organisations (Peters, 2009). A good understanding of the disparate requirements and the organisations involved in the process of communication is of critical importance for effective disaster communication.

Disaster Communication in a Changing Media Environment

The non-routine nature of disasters or crisis increases the importance of information for the public. The role of disaster agencies is to provide the needed information through different channels. At the same time disasters have everything it takes to gain attention in a commercialised media environment as they are characterised by news values such as unexpectedness, negativity and dread. These circumstances should provide a good basis to transmit important information and achieve the overall goal of risk reduction for the public. However, as described earlier, the changing logic of the media and the evolution of new communication technologies complicate the flow of information, requiring a thorough analysis of the relationship between disaster agencies, media and the public. There are two issues that deserve special attention: (1) journalistic routines and their impact on disaster reporting, and (2) the public information seeking behaviour and interests.

Concerning the first issue, the effects of the changing media ecology on disaster communication and media content, we have already pointed out that operational rules differ among the media and disaster agencies, with the former more concerned with business than public communication of the type desired by the latter (Nudel & Antokol, 1988; Peters, 2009; Auf der Heide, 2009 Chapter 10). In times of increasing commercialisation where the media and journalists face fierce economic and personal competition this gap can become even larger. There is tremendous pressure on journalists, particularly in the immediate aftermath of a disaster, to bring a story that will interest but also inform the consumer at the least cost (Berington & Jemphrey, 2003). As real world events and information disseminated by the agencies interact with journalistic norms and business practices the original messages that agencies wish to convey may differ considerably from media output. The demands created by varying levels of crisis management to inform and successfully communicate will often be hindered by news production conditions. For example, the selective attention of media to personal interest stories and the focus on tabloid-style journalism of traditional media can be a hindrance rather than a help for the plans of disaster management. In this context the informed use of new media technologies by disaster agencies may balance the traditional media's shortcomings. For example, disaster agencies can use their internet sites to bypass the media's gate keeping process and to have a direct communication link with the public.

Concerning the second aspect, one needs to understand that in disaster situations the public's information needs and information seeking behaviour differ considerably from routine media

choices. This is due to the high level of uncertainty and the perceived threat. At the same time, digital communication technologies enable the establishment of an increasing number of communication channels. Everyday media usage becomes more and more fragmented. Likewise in crisis situation; the public has many options to fulfil its information needs about a disaster. From the perspective of disaster agency, the availability of differentiated media channels and the destabilisation of traditional usage habits complicate the communication of urgent and coherent messages to the public (Liebes, 2005). In this context, disaster agencies need to be well-informed about the public's usage of available communication channels in a crisis situation to strategically plan their information policy.

This is especially important if relatively new communication channels are employed. Literature shows that in the realm of disaster communication strategies proposed ICT solutions often fail to consider the public's information behaviour (Carey, 2003; Crowe, 2008). Although thousands of disaster organisations have created WWW homepages, they do not necessarily reach a critical mass; new technologies by themselves will not result in getting information to everybody (Morris & Ogan, 1996; Neumann, 2002). Also, the majority of emergency management websites lack usability when addressing the general public (Crowe, 2008). Finally, it is necessary to be aware that potential receivers have individual characteristics, varying information needs and information seeking behaviour. These factors are important determinants of message effectiveness (Seeger, 2008).

In the following paragraphs we will present a case study on floods in Switzerland to illustrate the changing disaster coverage of traditional media and the public's information seeking behaviour. The data will give us the opportunity to point the media's production logic and shortcomings. Based on our research results, as well as on literature we

Figure 1. Damages (CHF); death toll and comparison of no. of newspaper articles spanning nine flood events, 1910 to 2005 in Switzerland.

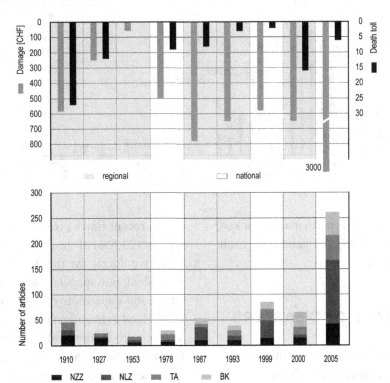

will discuss the advantages and disadvantages of ICT in the context of disaster communication.

CASE STUDY: INFORMATION ACQUISITION, PERCEPTION OF RISK AND PRESS COVERAGE OF FLOODS IN SWITZERLAND

Data Basis

Natural disasters, such as floods, represent one of the most hazardous environmental risks of our time. Like many other highly developed nations, Switzerland has been affected by flooding events fairly frequently since the 1970s and is exposed to a high hydro-geological risk level.

The case study is based on two different data sets. The first data set is a content analysis of the media coverage of floods in Switzerland from 1910 until 2005. The nine analysed floods were selected based on having been determined as highly catastrophic by natural scientists. The major criterion for this assessment is widespread geographic damage with costs exceeding 100 million CHF (see figure 2). Four major newspapers were analysed: Neue Zürcher Zeitung (NZZ), Tages-Anzeiger (TA), Neue Luzerner Zeitung (NLZ). Additionally, the tabloid newspaper Blick (BK) was analysed, but could only be included in the sample after its establishment in 1959. This longitudinal study provides an opportunity to trace and analyse changes in the media system, the conditions under which disaster management works, and the resulting press coverage. The second data set consists of a large telephone survey conducted in 2007. It focuses on a major flood in 2005 and covers issues concerning information sources, perception of the risk and preventative actions taken by the public. The representative

Figure 2. Diffusion of news about the August flood 2005 in Swiss dailies within one month.

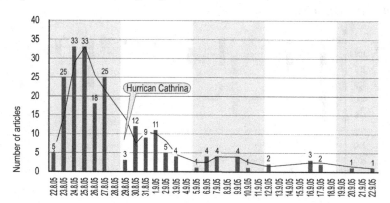

sample consisted of 2063 participants, ranked with respect to age (15-95) and gender for each of the 26 Swiss cantons.

MEDIA COVERAGE OF DISASTERS

Figure 1 gives a quantitative overview of the extent of flood coverage represented in the number of articles, and the financial impact and loss of life.

Considering all newspapers, the data shows a clear increase over the analysed time period. The later trend becomes clear: flood disasters since the 1970s attract more media attention, although with scale-related fluctuations. Media attention started to rise considerably in 1978 and reached its peak in 2005, the year of the most costly flood, which caused damages of 3 billion Swiss Francs and six fatalities. In comparison with similar catastrophes during the last 100 years, the number of articles exceeded previous coverage by a long margin. It is noteworthy that the level of coverage does not correspond with the salience of an event attributed by experts. Climate scientists view of the floods of 1910 (damages: 584 mio. Swiss Francs; death toll: 27) and 1999 (damages: 580 mio. Swiss Francs; death toll: 2) as equivalent events, however nearly twice as many articles appeared in 1999. Further, the disaster in 2000 caused damages of 650 million Swiss Francs and the highest death

toll in recent times (16), yet reached scarcely a quarter (23.5%) of the 2005 coverage.

If we focus on the level of reporting by individual newspapers, data suggests that the enormous expansion of news reporting in the 2005 flood is accompanied by a substantial rise in article production in all four newspapers. These findings indicate that the amount of reporting is not necessarily an objective representation of the real situation. The so-called elitist paper Neue Zürcher Zeitung is an exception; its coverage has been more or less consistent with experts' assessment of the gravity of the floods over the last 100 years. In contrast to this, the other dailies tend to generally increase coverage, especially the Neue Luzerner Zeitung.

Figure 2 shows the quantity of press coverage of the 2005 flood during the month of occurrence

Around 80% of the total press coverage of the flood occurred within this month. Media attention was highest in the first week, and attention reached its peak on the third and fourth day of the flood. More than 60% of all articles about the flooding were produced in the first week. After this initial attention the media coverage continuously decreased. The relatively rapid fall of news coverage may have been due to Hurricane Katrina in New Orleans. So-called 'killer issues' can sweep a news event very quickly from the headlines. There appears to be a disaster coverage rule: "A rare hazard is more newsworthy than a common

Table 1. Framing structure: disaster reporting of three time periods

Framing structures in disaster coverage	1910-1953	1978-1999	2000-2005
General description of event	**19.1%**	16%	12.6%
Safety/ Rescue operations/ Resettling	**13.8%**	8.8%	7.2%
Affected people/ Official impact report	**13.2%**	4%	3.8%
Political reactions/ Consequences/ Laws	**12.6%**	**11.7%**	8.6%
Private aid/ Organised solidarity	**7.8%**	4.5%	2.1%
Economy/ Employment	6.6%	8.5%	7.3%
Human Interest	6.6%	**11.9%**	**15.6%**
Science/ Technology	4.8%	**9.7%**	5.6%
Nature/ Environmental problems	3.0%	4.5%	4.5%
Damage/ Consequences	2.4%	7.2%	**11.6%**
Insurance/ Compensation	2.4%	1.7%	2.5%
Religion/ Church	2.4%	0.5%	0.5%
Retrospect/ History	1.8%	1.8%	2.9%
Entertainment/ VIPs/ Culture	1.8%	2.2%	2.1%
Future expectations	1.2%	4.5%	**8.7%**
Other topics	0.5%	2.7%	1.40%
Total n* (1197)	**100% (167)**	**100% (401)**	**100% (629)**
Frequency of issues (%); *An article may have up to 2 main issues			

one (…); a new hazard is more newsworthy than an old one; and a dramatic hazard – one that kills many people at once, suddenly or mysteriously – is more newsworthy than a long-familiar illness" (Singer & Endreny, 1987, p. 13). As a result of event-focused rather than process-focused orientated disaster reporting, audiences tend not to be able to find complex information about the disaster or the response activities at the regional, national or international level. Nor do they have access to analyses by scientists, insurance companies or affected people beyond the superficial.

In a historical perspective we are able to identify three different periods of disaster coverage as regards the story focus. Table 1 represents the issues and their importance during the different phases.

The first period covers the time from 1910 until 1953. Against the background of a public service oriented media system, press coverage mainly focused on a general description of events (19.1%). Other important issues were public safety, rescue operations and resettlement (13.8%), official reports on people affected (13.2%), as well as consequences and implications for policy (12.6%). Also organised solidarity and private aid is newsworthy (7.8%). In summary, the main focus of disaster coverage during the first half of the century is shaped by social issues. The second period covers the floods from 1978 to 1999. During this time there was growing concern about environmental problems, technological risks, as well as concerns about the future. In the press, coverage of science and technology (9.7%) and issues concerning the scale of loss (7.2%) become important issues. Risk scenarios and risk calculations gained 4.5% of coverage. In addition, human interest stories provided subjective views on the personal lives of victims (11.9%). The third period from 2000 – 2005 is characterised by an

Figure 3. Disaster visualisation from 1910-2005: A comparison of picture area

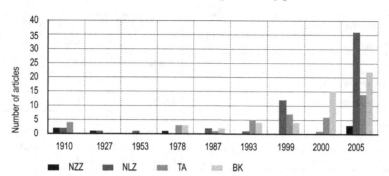

intensified reporting on human interest issues (15.6%). These stories do not illustrate concerns of the general public or have political relevance, but highly individual and private tragedies, focus on sensationalist storytelling and have a 'touchy-feely' tone to the writing. This is in line with the focus on damages (11.6%).

There is a certain trend that can be observed over time. The general description of events and the political reactions are important topics in all phases, but their share of coverage is decreasing over time. The focus of coverage shifts towards human interest stories, which rise to almost 16% of coverage in the 2000-2005 period. Such stories are marginal in the earlier coverage produced during the period of political press and public service oriented journalism. This development indicates the increasing market and reader driven approach in the production of news using a sensational-

ist and voyeuristic point of view. Damage and consequences as well as future expectations also increasingly shape the mediated representation of disaster reporting.

Figure 3 shows the amount of human interest stories in the different newspapers over time.

It becomes clear that all newspapers are increasingly using this frame in their reporting. Only the Neue Zürcher Zeitung (NZZ) seems to be an exception. This paper adheres to the established editorial position among broadsheet papers to respect the private sphere of victims; until 2005 it does not report human interest stories at all. This is in sharp contrast to the development of their other so-called quality papers: Neue Luzerne Zeitung (NLZ) and Tages-Anzeiger (TA). The dissolution of differences between the tabloid and quality press also affects Blick (BK), the only

Figure 4. Human interest stories in disaster coverage 1910-2005

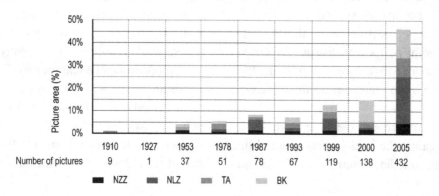

Figure 5. Neutral, alarming vs. reassuring tone in disaster reporting headlines

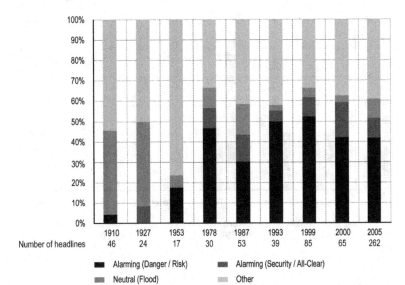

true tabloid paper to also increase its amount of human interest stories.

The number of pictures and the space occupied by visualisations has increased from event to event and reached a peak in 2005: Over 46% of the picture area was devoted to coverage of the 2005 flood. This triples the visual information on the flood in 2000 and is 46 times the space taken in 1910. Newspapers have moved away from the written to the visual. On one hand, this is due to ever improving technical possibilities. On the other, this reflects the changing style of journalism. Visuals have become prevalent in all of today's media presentation: Editors look for expressive images, particularly those that prompt and touch emotions. One points out: "I look at visuals. I look at interests" (McManus, 1994, p. 132). It has been shown that pictures evoking emotions can impact risk estimation of flood disasters (Keller, Siegrist & Gutscher, 2006). While the trend towards highly visual reporting is evident, the change of picture content is less obvious. There is a trend away from photographs of objects to that of people, in particular affected people in personal circumstances. This parallels the trend in human interest stories in all aspects of journalism. Figure 5 illustrates the tone of headlines in the flood coverage over time.

The drama in 2005 is expressed in headlines and alarming tones, compared to the more neutral language in the first half of the 20th century. The flood of 1978 signalled a change in rhetoric, nearly half of the headlines alluded to risk or danger. This trend in dramatic headlines has persisted since then, but with even higher percentages in the 1990s than 2000 or 2005. Headlines using a reassuring tonality are less attractive to the editors, as they may not attract the audience's attention, as well as dramatic headlines, such as "Switzerland submerged!" Of course, we cannot simply assert that changing mass-media norms constitute the only reason for these changes. News media also reflects changes in society as a whole. Information sources may also have taken more alarming tones with regard to hazards.

The results presented on the increasing use of human interest stories, visualisations and dramatised headlines underline the interpretation that there is a trend toward the tabloidization of disaster coverage. This trend has the potential to

Figure 6. Use of news media in the August flood 2005: A comparison of importance

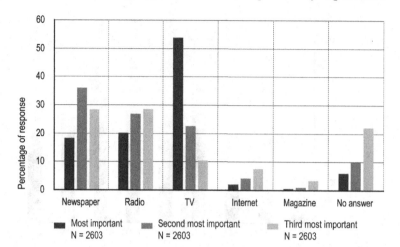

produce unwanted side-effects with regard to how affected people might react prior to, during and after emergencies. Those involved with disaster planning and management need to be aware of these intervening factors and plan communications accordingly. Tabloidisation can cause disproportionate and unbalanced reporting and leads to inaccurate interpretations of events by readers. The reality of disasters becomes distorted by the verbal and visual pictures presented in the media in appeals for money based on emotions and fear. Furthermore, prior coverage of disasters creates pre-conditions affecting the ways in which disasters get perceived and covered. These factors are compounded by the pressures faced by journalists in the context of market competition.

Information Seeking Behaviour and Media use in Disaster Situations

In the case of calamities, the public becomes dependent on the media for important information from public authorities and news. This information may be vital for survival. Also, active information seeking by the audience is increasing. This is supported by our study: Nearly a third of the respondents (30%) reported an increase in media consumption during disaster periods.

The respondents demonstrate pervasive access to different news media during the crisis (see Figure 6).

Television is the primary source of information. More than half of all respondents (53%) first turned to TV, followed by radio (20%) and then to newspapers (18%). In contrast, the internet was not used frequently as a primary source of information. Only about 2 percent considered this new channel of information as the most important during the crisis situation. Although the percentages rise slightly as the second or third most important sources of information, the internet's general influence as an information source during a crisis is marginal. This is coherent with the findings of other analyses (Rogers, 2003; Cohen, Ball-Rockeach, Jung & Kim, 2003; Bucher, 2003).

The population has specific information needs during a crisis. At the top of the list is the desire for expert knowledge and opinion regarding causes and consequences of a flood (48.7%). This is followed by information on how the community is coping with the crisis and political crisis intervention (28.7%). Next comes information about rescue operations and official help (26.9%), followed by advice on what to do (24.3%) and interest in individual stories of victims (24%). The least important information is on donations

Figure 7. Percentage of response to risk awareness and adoption of protective behaviour

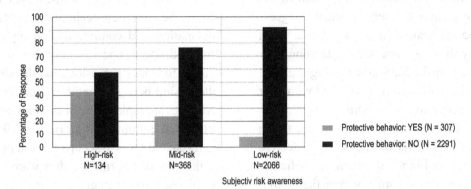

and other assistance (17.3%). Despite this stated public interest, looking at the press coverage of the floods from 2000 to 2005, only one in twenty articles included particularised expert knowledge and information on causes (see Table 1, frame science/technology). Interestingly, the respondents express relatively little interest for information about life support and advice on actions to be taken by potential victims. These are only two examples of the disparity between audience interest, actual coverage, and the goals of disaster agencies.

Risk Perception and Protective Behaviour

Most people know that floods can generally result in serious damage to environment, property and people. However, the perceived personal possibility of being affected by a flood the findings from the survey are sobering. Only a small number of respondents envisaged a high personal risk (5%). Most people (81%) classified their personal situation as low risk concerning flood damage. As expected, the less the perceived risk is the less preventative or protective action gets taken (see Figure 7).

Safety beliefs did have a cross-linked influence on safety behaviour. Nevertheless, most people perceiving a high level of personal risk stated that they had not implemented any kind of precautionary measure (57%).

Numerous studies point out that people without flood experience can not envisage the negative effects of severe damage due to floods (Lowenstein, 1996; Siegrist & Gutscher, 2006; 2008; Miceli, Sotgiu & Settanni, 2008). Fear and helplessness as consequences are particularly strongly underestimated. Not surprisingly, in this group most people did not take preventive actions in case of a flood situation (93%). People who had previous flood experience did take such measures more often; still the small percentage in this group (22%) is surprising. Consequently, past negative experience is not a sufficient indicator for future preparation. Siegrist & Gutscher (2006) assume that reasons for not taking preventive action in spite of personal affectedness are the high costs of measures or knowledge gaps about possible measures.

IMPLICATIONS OF THE CASE STUDY FOR DISASTER COMMUNICATION AND THE USE OF ICT

The case study shows that traditional mass media still play a major role in people's information seeking behaviour in times of crisis. In recent years, rapid changes in mass media structures coupled with new communication technologies have promoted a shift in disaster communications. The analysis indicates that journalism has moved away from public service orientated goals towards the imperative of market logic. Most coverage by the newspapers are extending their focus on human-interest stories, highlighting the personalities of victims combined with ever larger visuals/photographs. Dramatic elements appear as necessary elements in response to the pressures of the media business. The selective processes (and arbitrary decisions on whether or not to cover the big story within the media) prompted by business logic leave out important elements of disaster processes which can have far-reaching serious consequences for disaster management agencies and citizens. One example from our case study is the low perceived risk of the public: Despite the general increase in media coverage on floods the people seem to trivialise the actual risk. This may due to the media's trend towards short-lived human interest stories using tabloid styles of presentation.

In accordance with the findings, we need to consider ICTs as effective tools for disaster agencies to raise and strengthen risk perception, as well as self-protection measures. ICTs facilitate access to official disaster information in times of emergency and have advantages over traditional media for disaster communication agencies. First, information provided by the disaster agency directly and via ICT is independent of the gate-keeping process by classical media outlets. ICTs offer disaster agencies the opportunity to create their own web pages, to constantly update information beyond space and time limitations, and address audiences directly. Disaster agencies can transmit relevant content rapidly, provide content in different languages, and use different forms of presentation. Information on this basis may be more authentic than that processed through the media system and the content can include detailed information of local, national and international services. According to Nudell & Antokol (1988) in crisis situations "it is always the best if the information comes from you!" (p. 68), i.e.,

from disaster managers or designated spokespeople. Second, new technologies offer unique information and communication opportunities. The traditional one-to-many communication without feedback provisions and the hierarchical relationship between media communicators and audiences are replaced by bi-directional or multi-directional communication (Geser, 1997). New media enable users to set up personal preferences for the kind of information they want to receive. For disaster management, personalised forms of information before, during or after an event offer useful applications. In particular, individualised information about the necessary behaviour in the case of a disaster must be emphasised (Winerman, 2009). Additionally, people will no longer be just passive audiences, as web-based software supports interactive tools. People are able to report incidents, post messages and start discussions (Morris & Ogan, 1996; Geser, 2002). Third, the archive function of ICT presents an advantage for disaster communication. After initial publication on the internet, digitalised information is available for days, weeks or even months. In addition, the electronic mode of communication results in abundant information in all domains of disaster and risk knowledge.

There are many opportunities through ICT. However, if we only focus on technological solutions within the field of computer-mediated communication and neglect the audiences' information needs and media usage habits, we ignore limitations in disaster communication. First, it is important to note that new media are only one element alongside many in peoples' daily lives and media choices (Carey, 2003). From the case study it is apparent that the internet is not considered a very important communication channel in the disaster information seeking processes. In a crisis situation people still remain highly dependent upon traditional mass media. Accordingly, there is a need to critically assess statements overestimating the importance of web-based communication in situations of crisis (Arellano, 2008). Nevertheless, many governmental disaster agencies have created web-based services and their remains a gap between availability and actual use. For the effective use of ICT communication the question 'How to lead people to specific web portals?' is of primary importance. Second, ICT and related information sources often lack credibility. Media research suggests the perceived source-credibility, rather than the actual information conveyed, is important for information processing of the audience (Kaufman, Stasson & Hart, 1999). ICT and the surplus of available information place the burden to determine the trustworthiness of sources and news on the user(Morris & Ogan, 1996). In situations of extreme urgency and danger, it is not surprising people still choose traditional media channels or the online version of established media brands where the perceived credibility is high (Bucher, 2003; Winerman, 2009). Third, there may be sources that spread inaccurate or even false information, prior to official assessments. Winerman (2009) points out that the increasing use of social network sites for information seeking can be problematic. Social Network communication may bypass official information. This can lead to rumours and 'wrong' headlines being circulated literally around the world. False information can have side-effects, such as distorted perceptions of the crisis, disproportional fear or reputation losses of crisis management in the affected country. In this context it is of extreme importance for disaster communication to differentiate between official information provided by disaster agencies and unofficial 'rumours going around the internet'. This may establish trust in their online communication and overcome the public's resentment against this form of communication. Forth, ICT raises questions concerning issues of access, exclusion and participation. Although the spread of ICT is ongoing, social and economic differences, as well as unequal distribution of infrastructure, determine unequal access to new communication technology. These differences need to be accounted for and dealt with by disaster management agencies.

For disaster communication an approach which recognises that both, traditional media and ICT, and takes into account varied user styles is important. This includes recognising both the capacities and limits of ICT. However, recent research results point out that professional emergency managers are not knowledgeable regarding the way in which the public uses media during disasters or how to utilise web-based services as an effective tool for communication (Carey, 2003; Arellano, 2008). Good communication in a disaster is much more than posting information on the internet and working with news organisations. An analysis of several factors is necessary in order to produce effective communication systems and to deal with the unwanted side effects of media logic. These include the promotion of emergency management websites as accessible and credible tools providing safety-related information, the enhancement of their usability, and training and advice on how to use these services. These services need to be put into the context of the affected communities and their information seeking abilities, styles and preferences. Disaster communication via ICT should include not only information leading up to and during emergencies, but also in the aftermath of events.

THE NEED FOR FURTHER RESEARCH

Effective disaster risk management depends on risk awareness, good governance, proper technical and communication infrastructure adapted to information needs, and the empowerment of all those who are at risk. Learning how to prepare and take appropriate action without firsthand experience of catastrophes is a big challenge. In this sense, no single medium meets all the communication needs of both disaster managers and the public. The overall goal of risk reduction is influenced by how experts and practitioners handle information in the course of managing disasters.

For this purpose they need special knowledge which must be delivered by scientific research.

Only systematic, country specific and comparative research, accounting for traditional and new media, media audiences as active participants and knowledge of the unintended effects of media reporting will help us to identify and understand the potential offered by ICTs in disaster situations. Furthermore, it is important to take into account not only different geographical areas, between and within countries, but also variables such as age, gender, class, ethnicity, literacy etc. This requires studies at the micro-level in order to produce data that is relevant to a range of media user contexts. Moreover, disaster and communication research must continue to evolve in an integrated fashion. Thus, any discussion of state-of-the-art technologies and methods must ultimately be cast in terms of how they relate to the conditions of the media system as well as the needs of the general public and affected people.

Disaster coverage, the public's information habits and the rapidly changing media environment (with all its technological possibilities) need to be continuously monitored in order to strategically plan and effectively adapt disaster management and communication. ICT may give disaster management a lot of opportunities and enormous resources for crisis communication. However, this success will not be attained simply because of the availability of these channels or the multifaceted functions of ICT. Success will depend more on the form of use and basic principles like audience access, usability, trust and reliability are key elements in order to achieve the potential of these channels in assisting affected people or the general public.

CONCLUSION

To reach its goal disaster management needs to take into account the divergence between media coverage, audience interests, and information necessary for public safety. The mass media have their own logic; they foremost address the general public, not only people directly affected by the disaster, and want to attract audiences for advertisers. This may lead to coverage of disasters that is not necessary aligned with the interests of disaster agencies and the general goal to reduce risk. Perhaps the use of ICT technologies, and especially the internet, can compensate for the mass media's shortcomings. On the one hand, direct ICT communication by disaster agencies can bypass the traditional media and journalistic production logics that may be a hindrance to effective communication. New electronic networks allow alternative structures, which work quite differently to the one-to-many nature of traditional mass media system and allow disaster agencies to spread specific information to the affected people. On the other, the reach of new media technologies and the use of these channels by the general and affected public in disaster situations seem to be the greatest obstacles of such communication strategies. Traditional media still are the most used and most trusted information sources in a crisis.

Disaster prevention agencies must take these factors into account when planning their communication strategies. Concerning the different stages of communication, ICT may be of specific use in the pre-event communication to raise public awareness about risks and motivate mitigating behaviour. As suggest by the case study the low perceived risk by people who are under threat by future floods seems to be related to existing knowledge-gaps about actual risks and this, in turn, might be affected not only but also by the tabloidization of media coverage. Technical advancements and ICT can provide the disaster agencies with the power to take necessary steps to improve risk awareness by the public and motivate individuals to engage in preventative behaviour though improved information. This can happen by way of providing valuable information, establishing special forums or online communities where experts share their own experiences with floods, but also through playful interactions, free online learning programmes for children as well as for adults etc. The possibility of interactive elements offers many opportunities for disaster communication, such as the integration of information provided by lay people or the direct reaction to public concerns. However, risk communication must not focus solely on pre-event information and early warning technology. Trustworthy and reliable information is crucial in disaster situations. Here the internet may have limitations as it offers a multitude of different information, perception and judgments and is as yet not regarded as a generally credible information source by the public. In order to make informed choices between media sources, media and information literacy skills are required and the internet services of official actors need to be promoted more widely to the public.

Generally, agencies need to attend to the specific user habits, differences of tradition, culture and access, as well as media regulations so that communication can be effective. Developing capabilities to reach the majority of the population in a timely way and with the right information before, during, and in the aftermath of disaster, is in itself a challenge. The constantly changing media environment, the new possibilities and challenges of ICT, and the changing journalistic routines add increasing demands on disaster agencies. Handling these challenges is the key to future successes in risk reduction. Web-based technologies must therefore be understood within the broader social context of how they function in everyday life. It is hoped that this chapter will be helpful in prompting further research in this field, highlighting the practical implications, and most of all that this will be useful in generating further interest in this fascinating and important field.

REFERENCES

Arellano, N. E. (2008). Internet Connect with Facebook generation, expert urges Canadian companies. *Itbusiness.ca, Business Advantage through Technology*. Retrieved April 1, 2009, from http://www.itbusiness.ca/it/client/en/home/News.asp?id=48910

Ashlin, A., & Ladle, R. J. (2007). 'Natural disasters' and newspapers: Post-tsunami environmental discourse. *Environmental Hazards, 7*(4), 330–341. doi:10.1016/j.envhaz.2007.09.008

Auf der Heide, E. (2009). The media: Friend and Foe. In *Disaster Response: Principles of Preparation and Coordination*. Retrieved April 13, 2009, from http://orgmail2.coe-dmha.org/dr/DisasterResponse.nsf/section/10?opendocument&home=html

Blumer, J. G., & Katz, E. (1974). *The uses of mass communications: Current perspectives on gratifications research*. Beverly Hill, CA: Sage.

Bonfadelli, H. (2001). *Medienwirkungsforschung. Grundlagen und theoretische Perspektiven*. Konstanz, D: UVK Verlagsgesellschaft mbH.

Boykoff, M.T. & Rajan, R.S. (2007). Signals and noises. Mass-media coverage of climate change in the USA and the UK. *Membo reports, 8*(3), 2007-211.

Bucher, H.-J. (2003). Internet und Krieg: Informationsrisiken und Aufmerksamkeitsökonomie in der vernetzten Kriegskommunikation. In M. Lö (Ed.), Krieg als Medienereignis II: Krisenkommunikation im 21. Jahrhundert (pp. 275–296). Wiesbaden, D: VS Verlag für Sozialwissenschaften.

Carey, J. (2003). The functions and Uses of Media during the September 11 Crisis and Its Aftermath. In M.A. (Ed.), Crisis Communications: Lessons from September 11 (pp. 1-17). Lanhalm, MD: Rowman & Littlefield Publishing Group.

Chaudhary, A. G. (2004): Convergence: Globalisation, Localization and New Communication Technologies In P.J (Ed.), Mass Media in Transition: An International Compendium (pp.11-24). Athens: Athens Institute for Education and Research.

Cohen, E. L., Ball-Rokeach, S. J., Jung, J.-Y., & Kim, Y.-C. (2003). Civic Actions after September 11: A Communication Infrastructure Perspective. In M.A, No (Ed.), Crisis Communications: Lessons from September 11 (pp. 31-43). Lanhalm, MA: Rowman & Littlefield Publishing Group.

Comfort, L. K., Dunn, M., Johnson, D., Skertich, R., & Zagorecki, A. (2004). Integrating Real-Time Information in to Disaster Management: The IISIS Dashboard. In D. Ma & T. Pl (Eds.), Disasters and Society - from Hazard Assessment to Risk Reduction (pp. 227-235). Berlin, D: Logos Verlag.

Cottle, S. (2009). *Global Crisis Reporting: Journalism in the Global Age*. Maidenhead, UK: Open University Press.

Crowe, A. (2008). A closer look at emergency management websites. *Crisis Response Journal*, *4*(3), 44.

Dunwoody, S., & Peters, H. P. (1992). Mass media coverage of technological and environmental risks: a survey of research in the United States and Germany. *Public Understanding of Science (Bristol, England)*, *1*(2), 199–230. doi:10.1088/0963-6625/1/2/004

Entman, R. M. (1993). Framing: Toward Clarification of a Fractured Paradigma. *The Journal of Communication*, *43*(4), 51–58. doi:10.1111/j.1460-2466.1993.tb01304.x

Galtung, J., & Ruge, M. H. (1965). The structure of foreign news. The Presentation of the Congo, Cuba and Cyprus crisis in four Norwegian newspapers. *Journal of Peace Research*, *2*, 64–91. doi:10.1177/002234336500200104

Geser, H. (1997). The System of Public Media in Transition. Some contemporary "Megatrends" and their Implications for Social Theory and Research. Retrieved October, 10, 2009, from http://socio.ch/intcom/t_hgeser05.htm

Geser, H. (2002). Towards a (Meta-)Sociology of the Digital Sphere. Retrieved April 6, 2009, from http://socio.ch/intcom/t_hgeser13.htm

Gitlin, T. (1980). *The whole world is watching. Mass media in the making and unmaking of the new left*. Berkeley, CA: University of California Press.

Holly, W. (2008). Tabloidisation of political communication in the public sphere. In R, Wo & R. Ko (Eds.), Handbook of Communication in the Public Sphere (pp. 317-344). Berlin, New York: Mouton de Gruyter.

Kaufman, D. Q., Stasson, M. F., & Hart, J. W. (1999). Are the tabloids always wrong or it that just what we think? Need for cognition and perceptions of articles in print media. *Journal of Applied Social Psychology*, *29*(9), 1984–1997. doi:10.1111/j.1559-1816.1999.tb00160.x

Keller, C., Siegrist, M., & Gutscher, H. (2006). The Role of the Affect and Availability Heuristics in Risk Communication. *Risk Analysis*, *26*(3), 631–639. doi:10.1111/j.1539-6924.2006.00773.x

Liebes, T. (2005). Viewing and Reviewing the Audience: Fashions in Communication Research. In Cu, J., & Gu, M. (Eds.), *Mass Media and Society* (4th ed., pp. 356–374). London: Edward Arnold.

Lowenstein, G. A. (1996). Out of control: Visceral influences on behaviour. *Organizational Behavior and Human Decision Processes, 65*(3), 272–292. doi:10.1006/obhd.1996.0028

McManus, J. H. (1994). *Market-driven journalism: Let the citizen beware?* London: Sage.

McQuail, D. (1998). Commercialisation and Beyond. In D. McQ & K. Si (Eds.), Media Policy. Convergence, Concentration and Commerce (pp. 107-127) London: Sage.

McQuail, D. (2005). *McQuail's Mass Communication Theory* (5th ed.). Thousands Oaks, CA: Sage.

Miceli, R., Sotgiu, I., & Settanni, M. (2008). Disaster preparedness and perception of flood risk: A study in an alpine valley in Italy. *Journal of Environmental Psychology, 28*(2), 164–173. doi:10.1016/j.jenvp.2007.10.006

Morris, M., & Ogan, Ch. (1996). The Internet as Mass Medium. *The Journal of Communication, 46*(1), 39–50. doi:10.1111/j.1460-2466.1996.tb01460.x

Neuman, R. W. (2002). The futur of the Mass Audience. In D. McQ (Ed.), McQuail's Reader of Mass Communication Theory (pp. 364-374). London: Sage

Nudell, M., & Antokol, N. (1988). *The handbook of effective emergency and crisis management.* Lexington, MA: Lexington Books.

Peters, H. P. (2009). Natural disaster and the media. International Strategy for disaster Reduction ISDR Retrieved April 13, 2009, from http://www.chmi.cz/katastrofy/peters.html

Picard, R. G. (2005). Money, Media and the Public Interests. In O. Ge. & K. H. Ja (Eds.), The Institutions of Democracy: The Press (pp. 337-350). New York: Oxford University Press.

Reese, S. D., Grani, A., & Danielan, L. H. (1994). The structure of news sources on television: A network analysis of 'CBS News, ' 'Nightline', 'MacNeil/Lehrer,' and 'this Week with David Brinkley.'. *The Journal of Communication, 44*(2), 84–107. doi:10.1111/j.1460-2466.1994.tb00678.x

Renn, O. (1992). The social Amplification of Risk. Theoretical Foundations and Empirical Applications. *The Journal of Social Issues, 48*(4), 137–160. doi:10.1111/j.1540-4560.1992.tb01949.x

Rogers, E. M. (2003). Diffusion of News of the September 11 Terrorist Attacks. In No, M. A. (Ed.), *Crisis Communications: Lessons from September 11* (pp. 17–30). Lanhalm, MA: Rowman & Littlefield Publishing Group.

Ruhrmann, G., & Göbbel, R. (2007). *Veränderungen der Nachrichtenfaktoren und Auswirkungen auf die journalistische Praxis in Deutschland.* Retrieved April 2, 2009, from http://www.netzwerkrecherche.de/docs/ruhrmann-goebbel-veraenderung-der-nachrichtenfaktoren.pdf

Seeger, M. W. (2008). Disasters and Communication. In W. Do (Ed.), *The International Encyclopedia of Communication.* Oxford: Blackwell Publishing. Retrieved March 24, 2009, from http://www.communicationencyclopedia.com/subscriber/tocnode?id=g9781405131995_chunk_g97814051319959_ss43-1

Siegrist, M., & Gutscher, H. (2006). Flooding risks: A comparison of lay people's perception and expert's assessments in Switzerland. *Risk Analysis, 26*(4), 971–979. doi:10.1111/j.1539-6924.2006.00792.x

Siegrist, M., & Gutscher, H. (2008). Natural hazards and motivation for mitigation behaviour: People cannot predict the affect evoked by a severe flood. *Risk Analysis, 28*(3), 771–778. doi:10.1111/j.1539-6924.2008.01049.x

Singer, E., & Endreny, P. M. (1987). Reporting hazards: Their benefits and costs. *The Journal of Communication, 37*(3), 10–26. doi:10.1111/j.1460-2466.1987.tb00991.x

Winerman, L. (2009). Crisis Communication. *Nature, 457*(7228), 376–378. doi:10.1038/457376a

Zemp, H., & Bonfadelli, H. (2008). Hochwasserereignisse im Spiegel der Presse. In G.R. Be & C. He (Eds.), Ereignisanalyse Hochwasser 2005, Teil 2 – Analyse von Prozessen, Massnahmen und Gefahrengrundlagen (pp. 347-362). Bern, CH: Bundesamt für Umwelt BAFU, Eidgenössische Forschungsanstalt WSL, Umwelt-Wissen, 2508.

ENDNOTES

[1] Though the term Information and Communication Technology may encompass both traditional electronic media and so-called new media we will concentrate on opportunities arising by the internet.

[2] The study was financed by the Swiss Government; "Flood 2005 in the memory of Swiss" in 2007.

Chapter 5
United Arab Emirates:
Disaster Management with Regard to Rapid Onset Natural Disasters

Hamdan Al Ghasyah Dhanhani
University of Bedfordshire, UK

Angus Duncan
University of Bedfordshire, UK

David Chester
University of Liverpool, UK

ABSTRACT

The United Arab Emirates (UAE) has more exposure to natural hazards than has been previously recognized. In the last 20 years the UAE has been subject to earthquakes, landslides, floods and tropical storms. This chapter examines the structure and procedures for management of natural disasters in the UAE, in particular issues of governance, accountability and communication within states that are part of a federal system. The study involved interviews with officials at both federal and emirate levels and case studies are presented of the impact of recent natural hazard events. Two emirates were selected for more detailed examination, Fujairah the most hazard prone and a rural emirate and Dubai which is a highly urbanized emirate which has undergone rapid development. There is now increasing awareness of natural hazards in the UAR and progress is being made at regional and federal levels. There needs to be a clear delineation between regional and federal roles and an understanding of the need for effective channels of information to relevant agencies.

INTRODUCTION

Natural disasters are defined as sudden events that are triggered by natural factors (such as storms, earthquakes, floods, and landslides) that lead to significant loss of life and/or damage to infra-

structure and economic activity. The magnitude and frequency of natural disasters both show an increasing trend over the last 50 years (Degg, 1992). Over the past two decades, more than 3 million people have been killed and 1 billion affected by natural disasters (UNESCO, 1993a; Chester et al, 2001). In 1991 alone, the economic cost of disasters was estimated at £17 billion and

DOI: 10.4018/978-1-61520-987-3.ch005

the management of disasters is one of the great challenges of the 21st century (UNESCO, 1993b). There is a necessity for a systematic approach to the management of disasters. The principal reasons for the continuing increase in natural disasters are related to the growth of population, the increase in building density and the growing concentration of people in urban areas that are more highly exposed to natural disasters (Chester et al, 2001). Thus, natural disasters exert an enormous toll on society and on development (Sqrensen et al, 2006). The 1990s were designated by the United Nations then, International Decade for Natural Disaster Reduction (IDNDR) and during second half of the IDNDR a new approach to the study of hazards emerged, focusing on the interactions between extreme events and features of human vulnerability to be found in places where disasters are experienced. During the opening years of the twenty-first century, the IDNDR has been superseded by the International Strategy for Disaster Reduction (ISDR) (United Nations, 1999, 2002), which has continued to emphasise a more 'incultured' approach to hazard assessment and post-disaster recovery. More recently this perspective has been enshrined in the Hyogo Framework for Action 2005-2015 (United Nations, 2005), that forms the context under which hazard research is carried out.

Urban areas are growing at a rapid rate all over the world, particularly in developing countries. In most cities, there is influx of population from the surrounding area, mainly in search of employment and better living conditions, so that new arrivals often have no alternative other than to occupy unsafe land, construct unsafe dwellings and this contributes to an increase in vulnerability. The growing urban population in economically less developed countries is concerning since it is taking place in the absence of effective civic services and of proper planning and regulation. This unplanned growth, with a proliferation of poorly engineered buildings, leads to disaster-prone areas becoming more vulnerable (Wisner et al, 1994).

The Middle East has a long record of historical seismicity. Despite advances in understanding the physical threat, however, the Arabic and Islamic societies of the region remain vulnerable to earthquakes and other disasters (Degg and Homan, 2005). Recent earthquakes in Turkey, Iran, Pakistan and the Asian tsunami of 2004 have brought to the fore the impact of natural disasters in Islamic countries.

The United Arab Emirates (UAE) is an Islamic and Arabic country in the Middle East. Located near the edge of the Arabian Plate adjacent to the Iranian plateau and close to the Zagros Fault zone, which is characterized by high seismic activity, the UAE is not as safe from natural hazards as has often been assumed (Wyss and Al-Homoud, 2004). The rapid growth of population and its increased concentration in urban centres, with a lack of a clear hazard planning policy and engineering regulations for seismic resistance contributes to making this area more vulnerable.

For example, in Fujairah the following rapid onset natural hazard events have occurred in the period 1995-2009: Masafi earthquake 2002, Al Qurayah flood 1995, Al Tawaian landslide 2005, Tropical Gonu storm 2007 and Sharm flash flood 2009 (Fujairah Municipality personal communication 2009). The purpose of this chapter is to explore the following issues which are raised by the development plans to manage possible future natural disasters in the Emirates. Specifically these include:

a. Issues of governance, particularly those involved with operating within states that are part of a federal structure.

b. The particular management issues involved in innovating plans within highly urbanised communities with extremely heterogeneous expatriate groupings, which exist alongside more traditional rural populations.

c. Dealing with disasters within a society that has - and continues to develop - at a rapid

Figure 1. Location map of United Arab Emirates

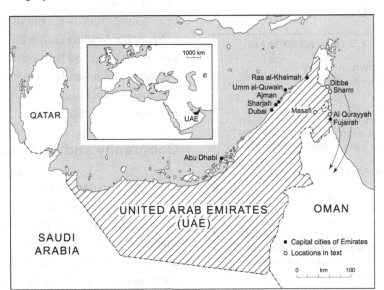

rate and in which there is little 'cultural memory' of past events.

The research involved interviews with selected officials both at the federal level and within individual emirates; the study of official records and academic publications. Case studies will be presented of Fujairah and Dubai emirates. These two emirates have been selected as Fujairah is the most hazard prone and rural emirate, and this will be contrasted to Dubai, which is a highly urbanised emirate which has undergone rapid development in recent years.

THE UNITED ARAB EMIRATES

The UAE in Brief

At the end of 1971, when the British Government announced its intention to withdraw from the Arabian Gulf, Sheikh Zayed governor of Abu Dhabi and Sheikh Rashid governor of Dubai, acted rapidly to initiate moves towards establishing closer ties between the emirates. Agreement was reached between the rulers of six of the emirates

excluding Ras al-Khaimah. The federation to be known as the United Arab Emirates (UAE) was formally established on 2 December 1971 with Sheikh Zayed as its President. The seventh emirate, Ras al-Khaimah, formally acceded to the new federation on 10 February 1972 (Al Hammadi, 2008).

The UAE is a federation of seven emirates (Figure 1), Abu Dhabi, Dubai, Fujairah, Sharjah, Umm al-Qaiwain, Ras al-Khaimah, Ajman. The language is Arabic and the religion Islam. It has a total area of approximately 82,880 km2 along the southeastern tip of the Arabian Peninsula and a total population of 5.06 million (Table 1). The majority of people in the UAE reside in three major cities: Abu Dhabi, Dubai, and Sharjah which have undergone rapid expansion and development (Teadad programme - Ministry of Planning, 2006).

Abu Dhabi city is capital of the UAE and Abu Dhabi is the largest emirate, extending over approximately 65,000 km2. Dubai is the second largest emirate, with an area of around 3,900 km2 and is located on the Arabian Gulf, in the northeast of the United Arab Emirates (Figure 1). It extends from the city of Jebel Ali in the west to the neighbouring emirate of Sharjah to the north-

Table 1. Information on United Arab Emirates (data from Vine, 2009)

Population: 4.488 (2007); 4.76 million (est. 2008); 5.06 million (est. 2009)
Nationals: 864,000 (est. 2007)
Non-nationals: 3.62 million (est. 2007)
Males: 3.08 million (est. 2007)
Females: 1.4 million (est. 2007)
Annual population growth rate: 6.31% (est. 2008–2009)
National population growth rate: 3.4% (est. 2008–2009)
Least populated emirate: Umm al Qaiwain emirate with 52,000 inhabitants.
Most populated emirate: Abu Dhabi emirate with circa 1.5 million people.
Religion: Islam; practice of all religious beliefs is allowed.
Language: Arabic
Percentage of women students at the UAE University: 75%
Percentage of UAE women in labour force: approx. 30%
School enrolment: 648,000 students in 1259 public and private schools (2007/08), of which over half are female
No. of government and private universities: approx. 60
Illiteracy rate: 7%

---

east. Dubai emirate has a multicultural society and is a cosmopolitan city. It is an important tourist destination and its Jebel Ali seaport operates at the centre of the exporting trade in the Middle East. Dubai used a moderate amount of its oil revenues to generate the infrastructure for trade, tourism and manufacturing in order to diversify its economy, and it is a major international commercial centre in the Middle East. Over the decade until 2007 Dubai was one of the world's fastest growing economies but is suffering at present in common with most of the economically developed world, from the effects of global recession. It is the main economic contributor out of the seven emirates, which make up the UAE (Balakrishnan, 2008). Sharjah emirate lies directly to the east of Dubai along the Gulf coast, with further small enclaves in Fujairah. The total land area is around 2,600 km2. Ajman is situated between the emirates of Sharjah and Umm Al-Qiwain. The area of the emirate is 259 km2; it is the smallest of the seven emirates. Umm al Quwain emirate is located on the coast of the Arabian Gulf, stretching over a distance of 24 km, between Ajman and Ras Al-Khaimah. It is second smallest in area and the least populous of

the seven emirates. The total area of the emirate is 777 km2. Ras al-Khaimah is located north of UAE on the Arabian Gulf with a total area of 5,800 km2 and is situated between Um Al Quwain and Fujairah. It is noted for its quarrying, cement and pharmaceutical industries (CAMERAPIX, 2002). Fujairah is discussed below.

Geography and Geology

The UAE is a roughly triangular landmass whose coastlines form the south and south-eastern shores of the Arabian Gulf and part of the western shores of the Gulf of Oman. Except for Fujairah all the emirates have a coastline on the Arabian Gulf, one emirate, Fujairah, lies on the Gulf of Oman. The UAE enjoys a desert climate, warm and sunny in winter, hot and humid during the summer months. The UAE can be divided into three major ecological areas: coastal areas close to the sea; mountainous areas and desert. Desert occupies the greater part of the area of the UAE. Salt marsh is found on the littorals of the UAE, in the east and west. The topography is a low-lying coastal plain which merges into the rolling sand dunes of the Rub al-Khali desert with rugged mountains along the UAE eastern border (Hellyer et al, 2005).

Geologically the UAE is part of the Arabian Platform (Abdul Nayeem, 1994). Tectonically, the UAE is situated in the Arabian tectonic plate near to its southeastern margin. Earthquakes occur along the boundary of the Arabian and Eurasian Plates, the Zagros Mountain belt of Iran, being one of the most seismically active regions in the world (Abdalla and Al Homoud, 2004). In the UAE there are two main faults. One runs along the west coast through the major cities from Abu Dhabi to Ras al-Khaimah and there is no record of felt seismic activity in historic times. The other fault is the Dibba Fault which links with the Zagros belt and runs south west through Fujairah (Wyss and Al Homoud, 2004) and this has been seismically active with a number of earthquakes in recent years.

Governance and Development Strategy

The UAE enjoys a high degree of political stability and is the only state in the Arab world to have a working federal system that has stood the test of time. The UAE constitution was made permanent in 1996. Government operates within a federal framework. The ruler of Abu Dhabi is the President of the United Arab Emirates and the ruler of Dubai is the Vice President and Prime Minister. The relative political and financial influence of each emirate is reflected in the allocation of positions in the federal government. The system of governance in all the emirates is hereditary (Andrea, 2007). The political influences and financial obligations of the emirates are reflected by respective positions in the Federal government. Each emirate retains autonomy over their own territory exercised by a governor. A basic concept in the UAE's development as a federal system is that a significant percentage of each emirate's revenues are devoted to the UAE central budget. The federal system includes the executive branch, which consists of the President, Vice President, and the Federal Supreme Council, composed of the Emirates' seven rulers. After the death of Sheikh Zayed bin Sultan Al Nahyan in 2004, the Federal Supreme Council elected his son, Khalifa bin Zayed Al Nahyan, as governor of Abu Dhabi and president of the UAE.

In foreign aid, the UAE has provided around US$70 billion in grants and loans, as assistance for development projects in some 100 countries. The UAE has also been a major contributor of emergency relief to regions affected by natural disasters, in particular through the UAE Red Crescent Society (Vine, 2009).

The most important recent development in the UAE was the formal launching in early 2007 of a UAE Government Strategy for the future. Covering 21 individual topics, in six main areas. The UAE strategy aims to maintain high standards of living, and achieve sustainable development. The Strategy is subject to review and may be updated as circumstances require. The six major areas are:

1. **Social Development:** An important development goal is to raise the skills base of the emirates' citizens and the principal direction of public policy within this area is to improve student and school performance levels to international standards, enhance the managerial independence of schools, and increase the performance quality of public and private schools, as well as promoting student centred education processes. An additional aim is to upgrade the capacity of the health services and formulate a public policy that sets the priorities for the development of the health service.

2. **Economic Development:** The major direction for public policy within this area is to upgrade regulations and legislation to match current and expected economic growth and stimulate economic growth so as to strengthen the competitiveness of the national economy.

3. **Public Sector Development:** The major direction for public policy within this area is focused on strategic planning. It also includes: upgrading the civil service; designing of an integrated performance tracking system, which concentrates on principles of competency as the main criteria for recruiting and qualifying leadership; and promoting and retaining suitable staff.

4. **Justice and Safety:** With regard to the Judicial System the major direction of public policy in this sector will focus on training, increasing of the proportion of Emirate citizens in professional positions, raising the standards of the judiciary, as well as implementing International Best Practices (IBP) based on local conditions.

The major directions in the development of public safety policy is the formulation of a National Emergency System, so as to identify operations, roles, and responsibilities to ensure prompt response and to boost the readiness of the system. A further aim is to develop the institutional framework and enhance co-ordination between Federal and regional bodies. Some selected initiatives are:

- To establish a National Emergency System
- To identify specialties lacking in the national (i.e. Emirate) workforce with regards to the emergency system, and to coordinate between institutions to train qualified nationals in these areas of expertise
- To introduce the concept of volunteering and provide for a greater role for volunteer staff.
- To devise a comprehensive Federal Emergency Management Regulation System.
- To increase coordination with the private sector in developing and implementing the National Emergency System.

This is the first attempt by the UAE to address the issue of natural disaster management. This reflects the increased level of awareness with respect to natural disasters and the changing needs of a developing nation. An important goal will be putting in place the general policies and reinforcing coordination among the relevant authorities federally and locally. In addition the national comprehensive emergency plan needs to clarify the operations, roles and responsibilities to guarantee effective responses for all expected risks, and to reinforce the readiness of different agencies to support the emergency sector.

5. **Infrastructure:** The major direction of public policy is to enhance cooperation between the federal and regional housing programmes and co-ordinate the transport infrastructure.

6. **Rural Areas Development:** Major direction for public policy within this area includes investment in human resources, providing quality basic services throughout the country, and improving living conditions in rural areas.

Emirate of Dubai

Dubai is the second largest emirate area in the UAE after Abu Dhabi, It has an area of 3885 km2, which is 5% of the UAE total. Dubai's economy enjoys a competitive combination of cost, market and communication that has created an attractive investment climate for local and foreign businesses. This situation has placed Dubai at the forefront of the world's dynamic and emerging market economies. Dubai is considered a regional and global centre in modern tourism and the period from 2000-2008 has shown rapid growth in construction activity (Dubai Statistics Center, 2008). As the global environment becomes more competitive and challenges grow, Dubai Emirate needs to ensure that it continues to build on its success through proper planning and development.

In early 2007 the Government of Dubai prepared the Dubai Strategic Plan 2015 as a 'road map' of the emirate's future development. The plan has five sections (Dubai Government, 2009):

1. Economic and trade.
2. Security, justice and public safety.
3. Social development.
4. Infrastructure and environment.
5. Government excellence.

The aim of this plan is to establish a coherent vision among the various government departments and to ensure a common operational framework. The plan serves as a framework under which all government-related initiatives converge to consistently meet the vision and aims set by Dubai's Government.

Emirate of Fujairah

Fujairah has a long history following the arrival of envoys from the Prophet Mohammed in 630 AD which heralded the conversion of the region to Islam. The Emirate of Fujairah covers 1150km2 (about 2% of the area of the UAE). Its population is around 154,000 inhabitants (Fujairah Emiri Court, personal communication 2008). The governor of Fujairah, His Highness Sheikh Hamad bin Mohammed Al Sharqi, succeeded his father in 1974. It is the only emirate that lies on the eastern side of the UAE, along the Gulf of Oman. Fujairah was known as the land of the sea giants and its people regarded as tough and courageous. The Fujairah economy is based around subsidies and Federal Government grants. Local industry consists of cement, stone crushing and mining. These industries have witnessed a resurgence due to the increased construction activity taking place in other emirates (Fujairah Municipality personal communication 2009).

Fujairah is the only Emirate that is almost totally mountainous. All the other Emirates are largely covered by desert. The temperature averages around 30 0C but during the summer months can reach 460C. This period also coincides with the rainy season, though rainfall is higher than the rest of the UAE the climate is characteristic of a hot, desert zone with mild winters. The principal settlements are Fujairah, Dibba, Siji, Masafi and Tawian (Fujairah, 2008).

Geologically, Fujairah is bound by a number of northeast to southwest-trending high-angle brittle faults, which separate the sediments of deeper water origin from the shelf limestones of the carbonate Musandam platform to the north and the Semail Ophiolite to the south. The rapid growth of the population and its increased concentration often in hazardous environments has escalated the severity of the impact of natural hazards in Fujairah. Unplanned growth and poorly engineered constructions make these areas vulnerable. For example, the increasing number of quarries, there

are around 69 quarry companies in Fujairah, has led to issues of slope stability (Fujairah, 2008). Most of the recent natural hazard events in the UAE have occurred in Fujairah.

NATURAL HAZARDS FACING UAE

Earthquakes, tropical storms, floods and landslides are rapid onset natural hazards which face the UAE. These hazards are well illustrated by recent events in Fujairah.

Al Qurayah Floods 1995

Heavy rainfall in the arid environment of the UAE typically results in high level discharge and flooding. A good example is the flooding that occurred in Al Qurayah in 1995. Al Qurayah town is located in the northern part of Fujairah on the Arabian Sea coast at the foot of the Hajar Mountains and is located at the mouth of the Safad valley. The population is around 5026 and the total number of buildings and houses around 414 (Fujairah, 2008). On Friday December 11 1995 at midnight, following three days of heavy rain and storms, high water levels caused the failure of old dams leading to extensive flood damage (Dubai Police, 2007). The floods of December 1995 were the worst in the history of the Fujairah emirate and almost 90 percent of the Al Qurayah area was affected. The dam collapse resulted in considerable damage to roads, farms, buildings and property. The collapse of the dam on the hillside above Al Qurayah sent a high surge of water rushing down the valley into the village. Many houses were destroyed and the waters damaged several others (Figure 2). Mud and debris were deposited in streets and the whole population was evacuated from their houses. Most of the houses were damaged, households lost their poultry, livestock and other durable assets. In addition a surge from the sea damaged many houses along the waterfront.

Figure 2. Damage to houses caused by Al Qurayah flood in 1995 and still visible in 2009

Figure 3. Stabilisation work on roadside following Al Tawaian landslides

Masafi Earthquake 2002

On March 11 2002 an earthquake of shallow depth and local magnitude 5.1 shook the Fujairah Masafi region in the UAE (Othman et al, 2002). The focal depth was just 10 km. The earthquake occurred on the Dibba fault in Fujairah with the epicentre of the earthquake at 20 km NW of Fujairah (Arthur et al, 2006).

Al Tawaian Landslides 2006

In early 2006, a major rock fall occurred on the newly opened Tawaian to Dibba highway, closing the road for over 6 months (Figure 3). A civil engineering company, Halcrow, was appointed to undertake an independent investigation into the causes of the landslide and the study involved detailed site reconnaissance including geological mapping and a review of the original design, results of the investigation were presented in a report identifying the causes of the landslide (Halcrow Middle East, 2006).

Tropical Gonu Storm 2007

On June 5, 2007, tropical storm Gonu struck Oman and part of Fujairah emirate. This was one of the worst storms recorded in the North Indian Ocean

and the Arabian Sea. The Arabian Sea rarely sees tropical storms, and when they do occur, they typically remain weak and move away from the Arabian Peninsula. As Gonu made landfall in Oman it brought winds in excess of 240 km/h, storm surges, heavy rainfall, thunderstorms and giant waves. The storm resulted in heavy rainfall along its path, peaking at over 100 mm in Oman. Twenty thousand people became homeless and the flooding destroyed many homes. There were power failures, communication links were badly affected, there were transportation difficulties and the forced evacuation of 7,000 people from the coastal areas in Muscat, Oman's capital (Figure 1). Business losses were estimated at c. $4 billion and over 50 people were killed in Oman and significant disruption to IT communications (Ali et al, 2008; El Rafy and Hafez, 2008). Tropical Gonu storm affected Fujairah with fierce winds and torrential rains, causing damage to Fujairah regions Sharm, Al Bedyah and Suhilah. Fujairah experienced torrential rainfall which lasted for two hours. Raging waves breached the Fujairah coast and flooded the streets, causing chaos and panic among tourists and residents. The deluge of rainfall flooded many houses in Fujairah and major roads in the city were under more than 1 m of water, forcing hundreds of motorists to abandon their vehicles for higher ground. In Fujairah the

Figure 4. Impact of the Sharm flood in March 2009

relief operation was conducted by the Ministry of Defence, Fujairah Police and Fujairah Civil Defence. The operation revealed a lack of clarity in the roles of these different agencies and there was ambiguity with regard to responsibilities and this led to some confusion.

Sharm Flood 2009

Heavy rainfall in the arid environment of the UAE typically results in high-level discharge and flooding. In March 2009 strong south-easterly winds brought thunderstorms, lightning and heavy rain which caused flooding and led to the collapse of an old earth dam. Around 30 houses, 10 farms, 2 mosques and more than 10 cars were damaged (Figure 4). Roads to Dibba and Khor Fakkan were cut off (Fujairah Municipality personal communication 2009). Many cars and houses in Sharm and the major road from the Dibba and Fujairah were under more than 1 m of water, forcing hundreds of motorists to abandon their vehicles and seek safety on higher ground. Mud and debris were deposited in streets and all people were evacuated from their houses with considerable damage to farms and buildings. The relief operation conducted by the Fujairah Civil Defence and Fujairah Police was broadly effective. There was a lack of appropriate logistic support, however, and this reflects the need for better preparedness and

proper planning (Fujairah Municipality personal communication 2009).

Tsunamis

Jordan et al (2005) have undertaken an analysis of tsunami hazard along the coasts of the UAE. The geometry of the Gulf with its shallow depth and lack of historical tsunamis is interpreted at low risk from tsunamis (Jordan et al, 2005). However, the regional historical records indicate that the Indian Ocean coast of the UAE has been impacted by tsunamis in the past. The Asian tsunami of 26 December 2004 was noted in neighbouring Oman where wave heights in southern Oman (Salalah) were close to 2 m. In the UAE only small waves of 3 to 30 cm in height were recorded along the Fujairah coast (Kowalik et al, 2005). So it is recommended that any construction in the eastern coast of the UAE including Fujairah city, Dibba, Sharm and Dadnah should involve tsunami mitigation planning, including awareness and public education.

IMPROVING PREPAREDNESS OF RESPONSE TO POTENTIAL NATURAL DISASTERS: REGIONAL LEVEL

The rapid urban development and construction in Dubai has raised significant issues in terms of planning and adjustments with regard to natural hazards. Transformation from a largely dispersed rural population to a modern commercial centre raises significant challenges. Dubai, particularly through its police force, has taken steps to address these issues.

The Dubai police is one of the most forward thinking and progressive of Arab Police Forces today. The personnel has the highest educational standard of any organization in the UAE. In 2006 the Dubai Police Headquarters moved to new building designed to provide an effective organisational centre to handle emergencies of different scales. The new command and control room, which has been equipped with Barco video wall technology, is at the forefront of a modern infrastructure comparable with similar facilities in Finland, Norway, China, Greece and Germany (Dubai Police, 2009). Prevailing and developing situations can be visualised on the Barco video wall in order to assess threats and vulnerabilities. Images generated from aerial and Satellite imagery, allowing the presentation of a 3D model with detailed data about landmarks, buildings and streets with which can be integrated other data sets (e.g. availability of resources, social data). This integrated GIS technology provides a highly sophisticated management tool. The main hall can accommodate up to 87 staff and is designed to enable expansion over the next 15 years. Airborne cameras also offer live broadcast to the operations room. An advanced 3D model of the city of Dubai, allows the Dubai Police control room operators to coordinate operations down to the smallest detail (BARCO, 2008). This is an extremely powerful tool. To exploit its potential to the full it needs to be linked to all the relevant agencies and tested through simulation exercises.

One of the strategic goals of the Dubai Police is to ensure readiness to deal with crises and disasters effectively. In order to achieve the goal of readiness to deal with crises and disasters effectively in Dubai, a specialized strategic police department to deal with crises and disasters was created in 2007, the Crisis and Disasters Management Department. The Crisis and Disasters Management Department is part of the Public Department of operations of the Dubai Police. The primary mission of this department is to reduce the loss of life and property and protect the Emirate from all hazards, including natural disasters and man-made disasters

The terms of reference of the Department are to coordinate, enforce and monitor activities related to,

1. Hazard mapping and risk assessment. information and data collection, research and analysis.
2. Disaster information collection and management.
3. Natural disaster mitigation.
4. Early warning, forecasting and information dissemination.
5. Preparedness to respond to disasters and crisis when and were they occur.
6. Disasters operations management.
7. Preparedness, response, relief, recovery, rehabilitation and reconstruction at national level and all sub levels.
8. Training, education and public awareness.
9. Management of the post-disaster activities.

The new department will coordinate and collaborate with the other departments, agencies, and local authorities, and federal NGO's in managing the process of risk reduction in Dubai (Dubai Police, 2009).

The Dubai Police has gained experience in responding to natural disasters by contributing support teams to relief operations in other countries which have been impacted by natural disasters. A rescue team was sent to Iran to help in relief operations in the Bam earthquake in 2003 where there were 26,000 fatalities. Following the Pakistan earthquake 2005, the Dubai Police rescued a 14-year-old boy in Balakot who had survived 96 hours under the debris of his collapsed house. The Dubai police heat sensing device located the position of the boy. The Dubai police was the first foreign delegation offering relief services to the devastated areas in Indonesia during the Asian Tsunami of 2004. Twenty five personnel from the Dubai Police worked in the area of the worst affected villages. The rescue team of Dubai Police team has been registered on United Nations Office for the Coordination of Humanitarian Affairs (OCHA) website (www.reliefweb.int). In 2008 the Dubai Police Academy established a new

Figure 5. Range of vehicle types of Dubai ambulance service

academic diploma in "Risk, Crisis and Disaster Management".

An ambulance service is a critical service in providing a response to natural disasters and the government of Dubai gives this a high priority. In line with the Dubai Strategic Plan a Consolidated Centre of Ambulance in Dubai was established to assume the function of providing an integrated ambulance service in Dubai. In Dubai the ambulance service has recently undergone a large number of changes following the formation of the Ambulance Service Centre. There has been a significant investment in the training and development of ambulance crews resulting in more effective patient care. The ambulances are equipped with a wide range of emergency care equipment, disaster and rescue equipment. There is a wide range of types of vehicle (Figure 5), for different emergency and disaster conditions including standard ambulance vans, car/SUVs, and motorcycles (which are used for rapid response in an emergency as they can travel through heavy traffic). Helicopters are used for emergency care, either in areas where speed is of the essence, or in places which are inaccessible by road. The Ambulance Service Centre in Dubai has the longest ambulance in the World. The length of this ambulance is 18 m and can hold up to 44 patients at the same time. It features Internet and satellite

facilities providing highly effective information and communication facilities for the doctors on site. With state-of-the-art operating rooms, an intensive care unit, a radiography room and an integrated pharmacy, the ambulance is more like a mobile hospital and will provide on the scene emergency operating facilities in disasters.

Fujairah, a less prosperous emirate, has to rely more heavily on Federal support in responding to disasters. In addition as discussed above, Fujairah is subject to more natural hazards than other emirates. The response to the natural hazard events described above provides a good example of how the regional and federal agencies responded.

The examples that we have selected demonstrate that the wealthy emirates like Dubai can provide the necessary resources and expertise themselves and can provide assistance to other emirates and indeed other countries. Smaller emirates, such as Fujairah, are reliant on federal support when responding to larger events.

DISCUSSION

There is increasing awareness of the potential threat of natural disasters to sustainable development and the need for countries to have an agency coordinating disaster management. These considerations led to the launch in 2007 the National Crisis & Emergency Management Authority (NCEMA) as an organization working on disaster management at the federal level.

The NCEMA came into existence in 2007 by a Government of UAE order upgrading the National Centre for Disaster Management, which was located at Abu Dhabi. NCEMA will build up the national framework in disaster preparation in the UAE and its terms of reference are set out in Table 2. In addition to administration to tackle disasters, it will also coordinate research projects, training programmes and will build a database on natural disasters together with case studies. NCEMA will continue to strengthen current disaster response

Table 2. Terms of reference of NCEMA

1. Building strategic plans for the management of emergencies and crises. Take all necessary measures to be applied in cooperation with relevant organisations.
2. Coordination of role in disaster management of relevant agencies (government and non-government).
3. Development of disasters awareness strategy.
4. Fostering of research and development in disaster management.
5. Coordination and command arrangements in each hazard.
6. Planning and procurement of resources.
7. Development of performance evaluation assessed by conducting emergency management and response exercises.
8. The authority will ensure that appropriate plans, systems, policies and processes are in place to meet national legislation and requirements for emergency planning, response and recovery.
9. The authority will define bodies' responsibilities and ensure that emergency preparedness, training, exercises, job descriptions, policies and procedures.
10. The authority will advise and recommend emergency planning policies and practices in response to national developments.
11. The authority will define; and oversee any related sub committees that are established.
12. Any other task assigned by the government.

capacities by consolidating best practice international systems, procedures and related training. In the coordination of management, greater attention will be paid to situations that are in a transitional stage between emergency situations and subsequent rehabilitation/development work. The aim of NCEMA is to manage the challenge of increasing risk and losses from hazards and other crises. The NCEMA will prioritize vulnerability reduction and

it will provide for the prevention and mitigation of disasters and strengthen capacities to manage risks. This national approach is complemented by research and study to identify best practices and facilitate their sharing. Fostering leadership and cooperation in the region, and to stimulate coordinated actions between different agencies (regional and federal) is critical in effectively

Figure 6. Structure of NCEMA

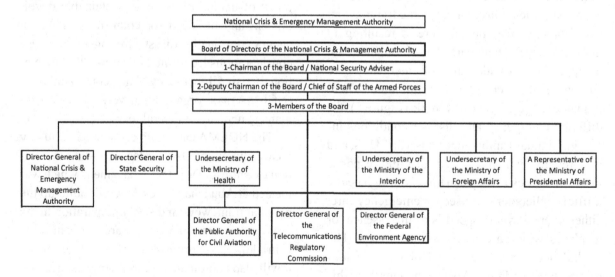

mobilizing national resources. The organisational structure of NCEMA is shown in Figure 6.

CONCLUSION

The United Arab Emirates provides a good example of a region where rapid development and urbanisation leads to a change in the pattern of risk faced by the population. The risk faced by a dispersed predominantly rural population is very different from that faced by a highly urbanised population. The speed of the development presents very special issues in that rate of construction has outstripped the analysis of risk and how this could be mitigated. The development which has taken place over the last thirty years means that there is little "traditional memory" and the population has limited awareness of the natural environment in which they live.

It is clear that there is increasing awareness of natural hazards within the UAE and that steps are being taken at both regional and federal levels to address this issue. There is a need to work on the integration of the roles of regional and federal agencies to ensure that there are clear lines of accountability and management, and effective provision of information sources linked to relevant users.

REFERENCES

Abdalla, J., & Al Homoud, A. (2004). Seismic hazard assessment of the UAE and its surroundings. *Earthquake Engineering*, *8*, 817–837. doi:10.1142/S1363246904001778

Abdul Nayeem, M. (1994). The United Arab Emirates. Prehistory and protohistory of the Arabian Peninsula, Hyderabad, India.

Al Hammadi, M. (2008). *Britain and administrative conditions in the Trucial 1947-1965. ECSSR*. UAE.

Al Khabey, A. (2006). Dibba Tawaian Road Landslides. *Al Bayan Newspaper*. Retrieved July 24, 2009 from http://www.albayan.ae

Ali, A., H., Ashrafi, R., Al-Majeeni, A., O. & Mayhew, P., J. (2009). IT disaster recovery: Oman and Cyclone Gonu lessons learned. *Information Management & Computer Security*, *17*, 114–126. doi:10.1108/09685220910963992

Andrea, B. (2007). *The Political Culture of Leadership in the United Arab Emirates (Annotated edition)*. Palgrave Macmillan.

Arthur, R., Fowler, A., Abdullah, M., & Al Enezi, A. (2006). The March 11, 2002 Masafi, United Arab Emirates earthquake. *Tectonophysics*, *415*, 57–64. doi:10.1016/j.tecto.2005.11.008

Balakrishnan, M. (2008). Dubai – a star in the east. A case study in strategic destination branding. *Place Management and Development*, *1*, 62–91. doi:10.1108/17538330810865345

BARCO. (2008). Dubai Police General Headquarters, A paragon of control room technology. Retrieved July 17, 2009 from http://www.barco.com

CAMERAPIX. (2002). *Spectrum Guide to the United Arab Emirates* (2nd ed.). Interlink Publishing Group.

Chester, D., Degg, M., & Duncan, A. (2001). The increasing exposure of cities to the effects of volcanic eruptions: A global survey. *Environmental Hazards*, *2*, 89–103. doi:10.3763/ehaz.2000.0214

Degg, M. (1992). Natural disasters: Recent Trends and Future prospects. *Geography (Sheffield, England)*, *77*, 198–219.

Degg, M., & Homan, J. (2005). Earthquake vulnerability in the Middle East. *Geography (Sheffield, England)*, *90*, 54–66.

Dubai Government. (2009). The Official Portal of Dubai Government. Retrieved August 1, 2009 from http://www.dubai.ae/en

Dubai Police. (2009). Dubai police official website. Retrieved June 6, 2009 from http://www.dubaipolice.gov.ae

Dubai Statistics Center. (2008). *Annual booklet, 2008-2009. ECSSR.* UAE.

El Rafy, M., & Hafez, Y. (2008). *Anomalies in meteorological fields over northern Asia and its impact on hurricane Gonu.* 28th Conference on Hurricanes and Tropical Meteorology, 28 April-2 May 2008, Orlando, FL. American Meteorological Society.

Fujairah (2008). *Statistical yearbook.* Fujairah.

Halcrow Middle East (2006, July). *Tawaian to Dibba Road Landslide final report.* S:/25 Inter Emirate/PGT 207.

Hellyer, P., Aspinall, S., Al Bowardi, M., & Edmonds, J. (2005). *The Emirates: A Natural History.* Trident Press.

Jordan, B., Baker, H. & Howari, F. (2005). *Tsunami hazards along the coasts of the United Arab Emirates.* Department of Geology, United Arab Emirates University. Arabian coast 2005 Papers, Theme D, Paper 4.

Kowalik, Z., Knight, W., Logan, T., & Whitmore, P. (2005). Numerical modeling of the global tsunami: Indonesian tsunami of 26 December 2004. *Science of Tsunami Hazards, 23,* 40–56.

Lippard, S., Smewing, J., Rothery, D., & Browning, P. (1982). The geology of the Dibba zone, northern Oman mountains; a preliminary study. *The Geological Society of London, 139,* 59–66. doi:10.1144/gsjgs.139.1.0059

Ministry of Public Works. (2007). *Evaluation of Gonu storm (project).* UAE.

Othman, A., Othman, F., Fawler, J., & Bedh, A. (2002). Impact of Fujairah earthquake study. University of Emirates report (pp.1-20).

Sqrensen, J., Vedeld, T., & Haug, H. (2006). *Natural hazards and disasters drawing on the international experiences from disaster reduction in developing countries. Report, Norwegian Institute for Urban and Regional Research.* NIBR.

Teadad Programme. (2006). *Ministry of Planning.* UAE.

UNESCO. (1993a). *Medicine in the International Decade for Natural Disaster Deduction* (pp. 7–15). IDNDR.

UNESCO (1993b). *Environment and development-Disaster Deduction, 5,* 2-13.

United Nations. (1999). *International Decade for Natural Disasters Reduction: Successor Arrangements.* New York: United Nations.

United Nations. (2002). *Living with Risk.* Geneva: United Nations.

United Nations. (2005). *Report on the World Conference on Disaster Reduction.* Kobe, Hyogo, Japan. United Nations, Geneva GE.05-61029

Vine, P. (2009). *UAE at a glance.*

Wisner, B., Blaikie, P., Cannon, T., & Davis, L. (1994). *At Risk* (2nd ed.). Routledge. doi:10.4324/9780203428764

Wyss, M., & Al Homoud, A. (2004). Scenario of Seismic risk in the UAE an approximate estimate. *Natural Hazards, 32,* 375–393. doi:10.1023/B:NHAZ.0000035556.17601.1f

KEY TERMS AND DEFINITIONS

Natural Hazard: Probability of a potentially dangerous natural phenomenon or activity.

Risk: Probability of a hazard event causing human casualties and/or damage to property and economic activity. To mitigate risk it is necessary to reduce the vulnerability of people, property and economic activity to the impact of likely hazards. If risk cannot be reduced then it may be necessary to relocate residences and economic activity to zones of lower hazard.

Vulnerability: Susceptibility to loss from a natural hazard event.

Chapter 6
Disaster Informatics:
Information Management as a Tool for Effective Disaster Risk Reduction

Jishnu Subedi
Tribhuvan University, Nepal

ABSTRACT

From risk identification to emergency response and recovery, information plays a vital role and the effective use of information is instrumental to reduce the impact of disasters. With the advancement of information and communication technology in the last few decades, lack of information is no longer a major issue for disaster risk reduction. The major issue, rather, is managing the information, translating it into a comprehensive knowledge for decision making and disseminating it to the communities at risk for action. The advancement of technology and reach of communication tools at a grassroots level have created an opportunity to increase effectiveness of disaster risk management with the optimum use of disaster informatics. This chapter presents an overview of disaster informatics, a conceptual framework for information management for disaster risk reduction, a review of existing approaches of information dissemination through the Internet and a review of the combined potential of Internet with tools which are widely available at grassroots levels.

INTRODUCTION

The significance of gearing up efforts towards disaster risk reduction and mainstreaming it in the sustainable development agenda is ever increasing in today's world. There are three basic reasons for its increasing significance: First, the hazards affecting the human beings from time immemorial have not decreased even with the advancement in technology. Natural hazards like earthquakes, floods, landslides and tsunamis are still very frequent and remain major challenge for human civilization. The frequency and severity of some of the hazards such as floods, droughts and landslides have increased because of impact of global warming and climate change. The efforts so far have been able to contain human losses from these disasters to some extent in developed

DOI: 10.4018/978-1-61520-987-3.ch006

countries; however, the losses in terms of affected people and property have increased in the past years both in developed and developing countries. A comparison of the numbers of disasters in the three decades from 1973 to 2000 points to the fact that the numbers of disasters from natural hazards have doubled in each decade. Although the numbers of people killed in the last decade from these disasters is less than that of previous decades, the numbers of affected people and amount of economic losses have increased by double in each decade (UNISDR, 2004).

The second reason for increasing significance of disaster risk reduction is that the risk has been accumulating historically because of inappropriate development choices and more so in the developing and the least developed countries. The 2004 UNDP report states that "while only 11 percent of the people exposed to natural hazards live in low human development countries, they account for more than 53 percent of total recorded deaths" (UNDP, 2004, pp. 13). Third reason for increasing significance of disaster is management is because of the significant surge in the new threats such as armed violence, terrorism and other man-made disasters.

On the one hand there is increasing concern for reducing the impact of growing frequency and severity of disasters – natural and man-made - and on the other hand the phenomenal advancement on Information and Communication Technology (ICT) has laid out promising possibilities for effective and optimum use of information resources for building resilient communities to disasters. The challenge, then, lies on how to capitalize the potential of ICT for reducing impact of disasters and building resilient communities to disasters.

This chapter presents an overview of disaster informatics, conceptual framework for information management for disaster risk reduction, review of the existing approaches of information dissemination through internet, integrated system of ICT and future direction of disaster informatics on use of combined potential of internet (large information available at one click) with mobile phone and radio (widely available at grassroots).

BACKGROUND

Disaster risk reduction is "the conceptual framework of elements considered with the possibilities to minimize vulnerabilities and disaster risks throughout a society, to avoid (prevention) or to limit (mitigation and preparedness) the adverse impacts of hazards, within the broad context of sustainable development" (UNISDR, 2004, pp. 17). Disaster risk reduction has become a mainstream agenda in the sustainable development (UNDP, 2004) and efforts are underway to put more money on prevention which not only saves development efforts being washed from disasters but also will result in reduction of the resources spent on relief and recovery. The evolution and progress in disaster management (Alexander, 1997) has shifted the focus from emergency response to building resilient communities to disasters (UNISDR, 2005).

In the evolution of disaster management, the importance of information for effective disaster management has been firmly grounded. Information is a vital form of aid in case of disasters (IFRC, 2005) and people need information as much as water, food, medicine or shelter. Information can save lives and information can save resources. Information management is collection of the information, processing it, translating the information into knowledge and action and disseminating them to the communities in need. However, the emphasis of institutions working in the field of disaster management is much on collection of information and the later stages of information management are not in priority. ICT has advanced by leaps and bounds in the last couple of decades and its advancement has opened the possibility of efficient and effective information management for disaster risk reduction.

The increased possibility for efficient information management will have a direct impact to reduce the disaster risk to communities. However, it has not been so in the past as the work toward interlinking information management through ICT to disaster risk reduction has not been given a priority and the attention has been given in this direction only recently. The approach so far has been more focused on emergency management rather than integrated disaster management. Furthermore, how best to utilize information technology in a disaster situation poses a number of problems for which there is lack of necessary and relevant informatics research. Informatics is the science concerned with gathering, manipulating, storing, retrieving and classifying recorded information. Disaster informatics is "the theoretical and practical operation of processing information and communication in a disaster situation (Glossary of Risk Management, 2009)." The definition has limited scope as it envisages disaster informatics in the narrow concept of disaster situation which may be understood only as the emergency situation. In order to encompass the holistic approach of enabling the communities to reduce impacts from disasters during pre-disaster, disaster and post-disaster phases, disaster informatics can be defined as operating, processing and ensuring access to information in order to achieve effective disaster risk reduction and building communities resilient to disasters. ICT should suffice to the overarching goal of reducing disaster impact and building resilient communities to disaster for sustainable development.

Realizing the importance of information management in disaster risk reduction, the 2005 World Disaster Report (IFRC, 2005) has focused on information in Disaster. From the experience of response to Indian Ocean Tsunami in 2004, the report has stressed the fact that "aid organizations have focused on gathering information for their own needs and not enough on exchanging information with the people they aim to support… [T]he problem sometimes is not so much lack

of information as a lack of communication and understanding between scientists, governments concerned and donors (IFRC, 2005, pp. 12)." From hazard assessment and risk identification to emergency response and recovery, information plays instrumental role in reducing the impacts of disasters to people and the economy. With the advancement of information and communication systems in the past decades, lack of information is no more a major issue. The major issue, rather, is managing the information, translating it into a comprehensive knowledge for decision making and disseminating it to the communities at risk for action.

The importance and usefulness of information lies in its accessibility to the communities in need. "Information must not rest in the hands of a few officials with evacuation plans, but be spread throughout communities at risk before and after the event" (IFRC, 2005). In order to avoid duplication, to make efficient use of information and to ensure better accessibility, it is necessary to gather information using standard procedures and to make available the information. It will be a major step for transparent decision making. The 2005 World Disaster Report mentions that agencies did not follow standardized procedures and reports were not made available. Co-ordination was undermined by competition and information sharing has been one of the most challenging and least successful aspects of tsunami response (IFRC, 2005).

Therefore, the problems with optimum use of ICT in disaster risk reduction can be divided in two categories: First, focus of ICT has so far been only in responding to or managing emergency situations and second, there has been lack of proper communication and dissemination of information to the communities at risk. Although advancement in IT has transformed disaster communications (Marincioni, 2007), the focus has primarily on managing emergency situations (as in Stephensen & Anderson, 1997), mapping the risk (as in Kumar & Bhagavanulu, 2007) and

Figure 1. Data vs. information

early warning system (as in Nakamura & Saita, 2007). Web-based information management system is the most common use of ICT in disaster management. They have been used in all areas of disaster risk reduction: resource mapping (Troy, Carson, Vanderbeek & Hutton, 2008), resource management (Montells, Montero, Diaz & Aedo, 2006), public alert system (Seddigh, Nandy & Lambadaris, 2006), logistic coordination (USGS, 2009), post-disaster damage evaluation (USGS, 2009; Wyss, 2005; WAPMERR, 2009) and integration of user created information (Fahland, Gläßer, Quilitz, Weißleder & Leser, 2007). However, all of these systems are developed with ultimate aim of responding to emergencies (Dilekli & Rashed, 2007) or for effective management of emergencies rather than encompassing the overarching goal of preventing or reducing impacts of disasters in the communities and increasing resiliency of the communities. Emergency response aims at reducing the impacts of disasters and increasing efficiency of relief operation, which, is only a part of the overall goal of building resilient communities to disasters.

Effective and efficient disaster management requires collection and storage of information, management and efficient flow of the information and extraction of essential knowledge and wisdom from the information and providing access of the knowledge and wisdom to the communities at risk. As the paradigm of disaster risk reduction is evolving from emergency relief to building resilient communities and growing emphasis has been put on effectiveness of community based disaster risk reduction, the ICT solution must address and suffice to these developments (Srivastava, Hedge

& Jayaraman, 2007; Troy et al., 2007). Both in general applications of ICT and web-based technologies, information management for integrated disaster risk reduction to build resilient communities is lacking which has hindered the effort for efficient use of IT in disaster management.

INFORMATION DISSEMINATION

Hierarchical organization of disaster communication has been suggested by Ben Wisner in the 2005 World Disaster Report. The hierarchy includes data (basic unorganized fact), information (organized data), knowledge (understanding of information) and wisdom (choices based on understanding, experience and principle) (IFRC, 2005; Marincioni, 2007). Information is something which is not only logical, rational and reproducible but also something which is accessible to the end users. Data set which is not communicable and is not accessible to users is not information. In order to qualify for a data set as information it has to be collected, processed into something meaningful and made accessible to the users through some interface (Figure 1).

Data can be anything from measurements done using equipments (such as rainfall intensity, catchment area, quantity of water in a river and earthquake time-histories) to personal experiences (such as experience of previous flood and highest water level mark in the river bed from previous disaster and earthquake intensity at a particular space and time). However, in order for it to qualify as information, it has to be received by user as a meaningful content.

Figure 2. Interaction of information and individual's decision making for risk reduction

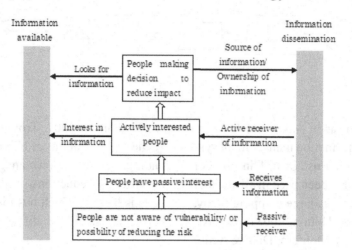

Access to information is the fundamental requirement for effective use of information in disaster risk reduction. Access to information can be either availability of information or dissemination of information (Figure 2). People are not interested in available information unless awareness is created through information dissemination. Only after people are actively interested, they start showing interest to the available information. Interest in information is critical factor for making individual decision to work towards reducing vulnerability which will be a source of information in itself. This also provides ownership of information to the communities which is essential for information dissemination.

Availability of information does not guarantee access to information if the users are unaware of the risk they are facing or of their vulnerability to hazards. Therefore, access to information requires creating awareness as a first step and then building ownership of the communities in the information. Information dissemination to communities at risk is not limited to providing a package of 'rational ideas' about risk reduction but it is also a dynamic process where building ownership of the communities is essential (Figure 2).

INFORMATION MANAGEMENT AND WEB

Advancement in the technology has laid open the possibilities of application of many communication tools for collecting the information, storing it efficiently, retrieving it at the time of need and disseminating it at grass-roots. Furthermore, and all of these can be completed just in one click. Many information and communication platforms - ranging from printed media to electronics media such radio, mobile phone and Internet – are available for managing the information. Among all these platforms, Internet has growing popularity and its reach is increasing in a faster way. Because of its potential for information management, Internet is going to dominate the basic source of disaster information as well in the future. Radio and mobile phones are other mediums with huge potential because of their widespread availability at grass-root level even in the remote areas without access to electricity and computer. The advancement of technology and reach of communication tools at grass-root levels has created an opportunity to increase effectiveness of disaster risk management with the optimum use of disaster informatics.

Table 1. An example of layered approach in the web for providing information to target groups about earthquakes

Information layer	Information content	Target group
Layer 1	Information video on how earthquake damages structural and non-structural members	General public School students
Layer 2	Provision of open space in communities	School teachers Policy makers
Layer 3	Hazard map	Practitioners and professionals
Layer 4	Time histories	Researchers

Although information is important, it has to be realized that information required for different purposes and for different users are different. This important consideration is often overlooked in information dissemination which has not only distanced the desired positive impacts from information sharing but also has increased the adverse impacts on effort to minimize the disaster risks. One case in point may be considered the Tsunami early warning system. The content of information required for scientific community, for disaster management authorities, for policy makers and for general public is different. At the time of Tsunami, the information on fault and fracture, time history data of earthquake and location of earthquake may be important for scientific community; information on depth of tsunami wave, its intensity distribution and time of arrival may be important to disaster management authorities; information on communities at risk and access to them may be important information for decision makers and warning through alert messages or sirens may be important to the communities at risk.

For effective management of information, it is necessary to understand the fact that information needs to be organized according to the target group and end users (Rajbhandari & Subedi, 2005). Mixing of information in the web is a hindrance for effective dissemination and optimum utilization of information. For example, time history data of earthquake event is not of interest to general public; whereas, information of Ricther Scale magnitude and location of epicenter is of interest

to the public. For a researcher, however, magnitude and location data is insufficient and whole time history and other relevant information is required. If information is not layered according to target group and their level of expertise, optimization of the information available in the internet for disaster risk management cannot be achieved. One of the important issues that disaster informatics has to confront with is not the availability of information but with the management of the information so that it can reach to the target group with less effort. Today, in many cases the information available in the web is buried so deep in the pool of information that it never reaches the target group. Following table gives an example of layered approach for information presentation for an internet website dedicated to provide information on earthquakes (Table 1). The approach as explained in Figure 3 is to ensure accessibility to information in layer 1 conveniently to the end users compared to that of layer 2 or 3 (Figure 3).

The different levels of accessibility, visual significance and amount of data are shown schematically in Figure 3. The information in layer 4 can be relatively difficult to access as the data will be used by persons of specific interests or by machines (such as crawlers). The layer has very large amount of unprocessed (or slightly processed) data and has almost no visual significance. As the layer 4 moves to layer 3, further processed information will be available which has more visual significance but contains less amount of data compared to layer 3.

Figure 3. Layered approach of information management in the web

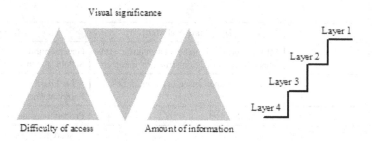

Application Examples

The next generation of information providing service in internet will be tailored information to suit the need of individual end-users, self-evolving website (e.g. self processing mechanism in the web-site to generate specific and required information from a pool of data such as from Layer 4 to Layer 3 and so in the Table 1). Few examples of how web-based technologies are using raw data to produce logical information are discussed in the following sections.

One of the examples of processing raw data from different sources in the internet (Layer 4) to generate spatial and temporal location of seismicity and volcanic activity (Layer 3, 2) is Seismic eruption program (Jones, 2008). This tool uses earthquake occurrence data from different sources in the internet which updates automatically within half-an-hour to one-hour as the source information are updated. The visual output is location of earthquakes according to their time of occurrence and size in the screen. The visual output can be used as a tool to disseminate information about plate tectonics, global earthquake phenomenon and pattern of earthquake generation.

Another examples is that of Prompt Assessment of Global Earthquake for Response (PAGER) which is an internet site launched by the U.S. Geological Survey (USGS) to help international agencies to make quick decision on scale of emergency response needed after an earthquake

anywhere in the world (USGS, 2009). As has been witnessed in the past, the initial information of extent of loss in an earthquake is underestimated as the information comes from peripheral areas where impact is much less and the communication system is still working. This underestimation is one of the major obstacles to mobilize emergency resources to reach to the affected communities in the early hour of disasters which is crucial to save lives. Realizing the need for estimating the impact of an earthquake at the earliest, the PAGER system was initiated. The schematic flow of information in PAGER system is given in Figure 4.

WAPMERR which stands for World Agency of Planetary Monitoring and Earthquake Risk Reduction is another system which "is currently offering a service of loss estimates for earthquakes with M 6 anywhere in the world, to anyone who wishes to subscribe (Wyss, 2005, pp. 298)." The system uses QLARM which is a computer tool to estimate building damage and human losses due to earthquakes anywhere in the world. The program estimates damages to houses, numbers of fatalities and injured based upon information on earthquake origin hour, earthquake features such as coordinates of epicenter, depth and magnitude of the earthquake and building fragility profile of target settlements. WAPMERR disseminates the information through three services which include "(1) an estimate of the number of casualties sent by email, (2) a telephone call with discussion of the estimate and its error limits, if desired, and

Figure 4. Process of PAGER system (From USGS, 2009; Earle & Wald, 2006)

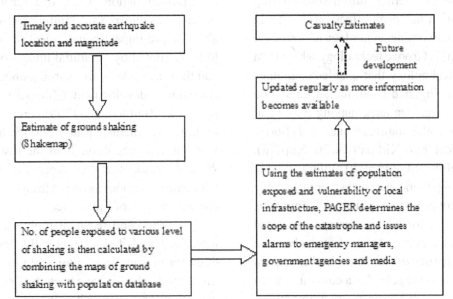

(3) a password-protected web site with a map showing the calculated average destruction in all settlements affected, as well as a list of those settlements (Wyss, 2005, pp. 298)."

One of the richest contributions of Internet in the realm of communication is its flexibility for participation of end-users. Internet sites such as Flickr.com, Twitter, Facebook and YouTube provide end-users with unlimited opportunity to provide information which has proved to be valuable resources in many areas – disaster risk reduction being one of them. Realizing the importance and scope of information from end-users, television networks such as CNN and BBC have also started programs where viewers themselves can provide news. In disaster information management also the potential of gathering data from end users themselves has been effectively initiated. USGS has launched a site by name "Did you feel it?" which is mapping of earthquake intensity by the people themselves experiencing the shaking. The system "is an automatic Web-based system for rapidly generating seismic intensity maps based on shaking and damage reports collected from

Internet users immediately following earthquakes (Earle & Wald, 2006). The system is also integrated with PAGER where the information is used to update estimate of ground shaking.

Another such initiative of collecting data from end users is HOUDINI (Fahland et al., 2007) which stands for HUmbOldt DIsaster MaNagement Interface. The system has been developed by a team of researchers from Humboldt University which is "a prototype system for the flexible integration and visu-alization of heterogeneous data sources for disaster management" (HOUDINI, 2009). HUODINI is based on Semantic Web technologies "which collects information from a number of freely available data sources on the web, such as news feeds, personal blogs, tagged images, and seismo-graphic information" (Fahland et al., 2007, pp. 255). The system envisages supporting disaster management by gathering information published in the internet sites by affected people which is turning out to be a huge and useful resources as observed in the aftermath of disasters like Hurricane Katrina and 2004 Indian Ocean Tsunami.

In addition to Internet, information sharing and coordination has been effectively tested in peer-to-peer networking technology as well. One of such system is Groove technology which is "a peer-to-peer technology that enables communication, file sharing, and coordination for ad hoc groups across organizations, making it an ideal application for collaboration around relief efforts" (Farnham, Pedersen & Kirkpatrick, 2006, pp. 46).

Unlike web-based services it (a) allows people to work in group workspaces whether or not they are currently connected to the Internet, (b) provides status information for those in workspaces, and (c) securely shares any content in workspaces because it is stored on participant machines and encrypted for transport. Groove provides a communication and data synchronization infrastructure without any centralized point of failure, enabling people to reliably share and integrate information across locations, while mobile, and with intermittent Internet access. Groove also enables ad hoc access to colleagues across organizational boundaries, providing a conduit for information sharing. (Farnham et al., 2006, pp. 46).

One of the major advantages of this peer-to-peer networking is that it can be used effectively for horizontal information exchange among agencies which can minimize duplication of data, promote better coordination and save resources.

OTHER DISSEMINATION TOOLS

Although internet has potential for storage and processing of data, internet as information delivery medium to end-users has limited application compared to mobile phone and radio. Combination of potential of internet as information resources with dissemination tools such as radio and mobile phone has the potential to produce tremendous results in disaster risk reduction. Integration of communication system is necessary not only for effective information management but also to establish redundancy for post-disaster situa-

tion. Dissemination in post-disaster situation is different as the physical infrastructures available for pre-impact communication are "likely to be disrupted by the initial impact of an event and thus unavailable for warning about possible subsequent development (Zimmermann, 2005 pp. 328)." Zimmermann (2005) provides insight on how the Amateur Radio Service has been a valuable resource throughout the history of radio and stresses that "the support of emergency telecommunications as one of the most important characteristics" of the service.

The scope and reach of mobile phones are increasing and it is currently the leading medium for personal communication in day to day life. Mobile phones have been used successfully even in the emergency situation and it has shown demonstrated potentiality as tool for emergency communication. The infrastructure damage will have minimal effect in mobile phones as they can receive signals from towers in peripheral areas and disturbed towers can be repaired quickly compared to other infrastructures. Farhham et al. (2006) point that "many cell towers were operating within a week, and it was striking to observe the extent to which personal cell phones were the primary form of communication both at the personal level and in the context of relief work" (Farnham et al., 2006, pp. 41) after hurricane Katrina. Second advantage of mobile phone is its potential for informing the public especially through the use of text-based features (Sillem & Wiersma, 2006). Mobile phone services are now available with internet services or they can be used as mobile modem to connect to internet. This feature has added advantages for its wider application in pre and post disaster situation as information management tool.

INTEGRATED SYSTEM

Information management in disaster informatics can be divided into three nodes: collection,

Figure 5. Three nodes of information management as applied in disaster informatics

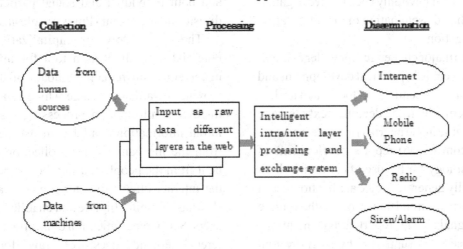

processing and dissemination. Integrating the different tools which are efficient in data collection (such as satellites), in storage and processing (such as internet) and in dissemination (such as TV, Radio and Mobile phones) is essential for capitalizing the potential of ICT in disaster risk reduction. Such an integration system with layered approach in Internet is shown in Figure 5.

Source of data can be human experience or machine reading. Internet and mobile phone have provided opportunities for automatic data collection from human experience in a very short time as explained above in "Did you feel it" website of USGS. The data collected in such way can be channeled as input to different layers in the web. Layers are portal for information dissemination (and collection as well) in the web which are organized according to target audience. The system can be elaborated by considering an example of early warning for flood in a community. The early warning needs data collection (such as weather forecasting, rainfall intensity etc.), processing the data in advance to identify risk and automatic transmission of warning to communities at risk through communication tools such as TV (automatically triggered to 'on' position in case of an warning), Mobile phones (text message or alarm through a central system) and Radio.

One of the successful applications of this model of integrated system is earthquake early warning (EEW) in Japan by Japan Meteorological Agency (JMA). JMA provides residents in Japan with earthquake early warnings that issue prompt alerts just as an earthquake starts, providing valuable seconds for people to protect themselves before strong tremors arrive. The EEW "system provides advance announcement of the estimated seismic intensities and expected arrival time of principal motion. These estimations are based on prompt analysis of the focus and magnitude of the earthquake using wave form data observed by seismographs near the epicenter. The EEW is aimed at mitigating earthquake-related damages by allowing countermeasures such as promptly slowing down trains, controlling elevators to avoid danger and enabling people to quickly protect themselves in various environments such as factories, offices and houses (JMA, 2008)".

Although successful applications of integrated system of ICT in disaster risk reduction are yet to be seen at wider level, partial application of combination of such system do exist showing promising scope for future use. The PAGER and "Did you feel it?" are examples of data collection system from automated machines and from human experiences, respectively. Their practical applications so far have only been limited to gather data rather than dissemination of the information for risk mitigation.

One of the major challenges for wider application of the system is in practical development and advancement in dissemination tools such as data crawler system which can decipher text message (e.g. from a table in the internet) to voice message. PHYSorg. com (2006) reports development of such a technology for the use of car radio which "automatically generates voice applications from Internet information and transmits it to the vehicle via radio signals." The report further mentions that the process "is supported by software that is capable of combining individual items of information into sentences and of breaking down spoken sentences into semantic units in order to understand them. The result is natural-language interactions between humans and computers where voice input no longer has to be restricted to a preset menu PHYSorg.com (2006)." The next steps in the system involve transmission of the information via radio signals and receiving the signal at the driver's end.

In the light these developing practices and tools, the integrated system of use of ICT for disaster risk reduction has a wider possibility in the recent future. This not only will contribute to the mitigation of disaster risks in the future but also will be a step towards developing ownership of the process of risk mitigation to the communities.

FUTURE RESEARCH DIRECTIONS

The evolution of internet has reached to Web2.0 (Puras & Iglesias, 2009) where role of end-users has evolved from mere passive receiver of the content to the active developer of the content. This evolution has already provided huge scope for application of the web technology for disaster risk reduction and researchers are proposing it as a "valuation tool to contribute to address both lessons, achieving public knowledge of new disasters

and both individual and social participation in disaster management (Puras & Iglesias, 2009)."

The efforts towards capitalization of the potential of Web2.0 as a tool for information management in reducing impact of disaster is moving towards next stage which can be called intelligent information processing system. The initial information available in the web – be it from the mechanized data collection center or be it from the people sharing their experience in the internet after a disaster such as earthquake shaking – is unintelligible and chaotic for general users. Self-Communication of information in different layers and processing the raw information to translate it into wisdom or knowledge to support decision making in a vulnerable community will be the one landmark step towards efficient use of disaster informatics.

However, as witnessed in the last decades, advancement in information technology in itself is not going to increase effectiveness of disaster risk reduction effort. "An important challenge lies with the wrong belief that the technology of information systems itself can solve all the problems.... [There also exist] complications of assuming that a well-operational IS [information system] in a particular context, whether this context is a given application domain (e.g., emergency management, natural resources), a country, or scale of implementation (e.g., local, regional) can be adopted into another different context in a straightforward fashion (Dilekli & Rashed, 2007, pp. 59)." Another important aspect of application of IT in disaster management is the growing realization of the fact that disaster risk reduction can be best achieved through participation of communities, empowering the communities and building their ownership to reduce the risk.

Therefore application of ICT for disaster risk reduction requires ownership of the communities in the different nodes of the integrated system. The approach so far in information communication in ICT is usually from one to many which required end-users to approach information collected at one point (without ownership of the end-users). The evolution of web2.0 is changing this towards more many to many which maintains certain level of ownership for the end-users such as in Wikipedia, Blogs, Facebook, Twitter and Youtube. Further ground-work is necessary on how the experience in information sharing in such sites can be used effectively for reducing disaster risk.

One of the major issues in the use of integrated system for disaster risk reduction is dissemination. The Web 2.0 has opened the possibility of information collection through a large community. However, disseminating the required information from the collected sources to the communities in need is still conventional. The recent development of smart-web radio devices which can translate the text message in internet to voice message is a landmark achievement. This system can be developed further and has very wide possibility for delivery of early warning of hazards like typhoon, tidal waves, flash floods and tsunamis to the coastal communities through radio and mobile phone.

The future research work should be directed towards realizing the theoretical concept of integrated system into practice of disaster risk reduction. It requires exploring the application of Web 2.0 in disaster information collection and dissemination, improving the intelligent information processing system further so that raw data in the internet can be processed to make comprehensive information for different stakeholders and developing smart-web tools so that the processed information in the internet can be disseminated to communities through media such as mobile phones, radios and TV.

CONCLUSION

Disaster informatics is information management for increasing the efficiency of disaster risk reduction efforts. The risk reduction efforts include building resilient communities i.e. a community well prepared to reduce the impact of disasters, a community where response and relief operation in case of a disaster is efficient because of better preparedness and a community which can return to normalcy and build back better efficiently in the post disaster situation. Information plays vital role for building resilient communities. So far the focus has been to use ICT as a tool for information management for emergency situations. With the evolution and progress of disaster management to encompass the overall aspect of disaster risk reduction (from preparedness to recovery) and gradual shift from centrally managed emergency operations to community based disaster management, the focus of information management also needs to shift from emergency management to overall disaster risk reduction. Information management is not only collection of information, processing it and storing the wisdom locked in the technological space. The domain of information management requires access of the information to the required end-users. One important aspect of managing information is to ensure access to information of communities at risk. Disaster informatics needs to be defined in this broader context and further research and development should ensure on how overall aspect of disaster management is covered, how the information is disseminated to the communities at risk and how communities build ownership for sustainability and efficient use of the system.

Next generation of web technologies will be based on intelligent processing of disaster information at different layers which will be communicated to the communities at risk through an integrated approach on use of IT for disaster risk reduction. This approach is sufficed by the potential of web-based technology to store and process huge amount of data and wide reach of dissemination tools such as Mobile phone and Radio. Integration of the collection, storage and processing capacities of web technologies and dissemination and feedback potentialities of communication tools like mobiles phones combined with their availability even in remote areas is the key for optimum use of information technology in disaster risk reduction.

REFERENCES

Alexander, D. (1997). The study of natural disasters, 1977-1997: Some reflections on a changing field of knowledge. *Disasters, 21*(4), 284–304. doi:10.1111/1467-7717.00064

Dilekli, N., & Rashed, T. (2007). Towards a GIS data model for improving the emergency response in the least developing countries: Challenges and opportunities. In Van de Walle, B., & Carlé, B. (Eds.), *Proceedings ISCRAM2007* (pp. 57–62).

Earle, P. S., & Wald, S. J. (2006). Rapid post-earthquake information and assessment tools from the U.S. Geological Survey National Earthquake Information Center. In B. Van de Walle & M. Turoff (Eds.), *Proceedings of the 3rd International ISCRAM Conference Newark, NJ (USA), May 2006*

Fahland, D., Gläßer, T. M., Quilitz, B., Weißleder, S., & Leser, U. (2007). HUODINI – Flexible information integration for disaster management. In Van de Walle, B., Burghardt, P., & Nieuwenhuis, C. (Eds.), *Proceedings ISCRAM2007* (pp. 255–262).

Farnham, S., Pedersen, E. R., & Kirkpatrick, R. (2006). Observation of Katrina/Rita Groove deployment: Addressing social and communication challenges of ephemeral groups. In B. Van de Walle & M. Turoff (Eds.) *Proceedings of the 3rd International ISCRAM Conference, Newark, NJ (USA), May 2006* (pp. 39-49).

Glossary of Risk Management. (2009). Retrieved June, 20, 2009, from www.merrea.org

IFRC (International Federation of Red Cross and Red Crescent Societies). (2005). *World Disasters Report 2005: Focus on Information in Disasters*. Bloomfield, CT: Kumarian Press Inc.

JMA. (Japan Meteorological Agency (JMA). (2008). What is an earthquake early warning? Retrieved September, 30, 2009 from http://www.jma.go.jp/jma/en/Activities/eew1.html

Jones, A. L. (2008). Seismic eruption. Retrieved September, 30, 2009 from http://www.smate.wwu.edu/ slibrary/SeismicEruption/main.html

Kumar, S. V., & Bhagavanulu, D. V. S. (2007). Flood simulation and inundation mapping of Adyar river: A case study using GIS. *Disaster and Development: Journal of the National Institute of Disaster Management, 1*(2), 155–168.

Marincioni, F. (2007). Information technologies and the sharing of disaster knowledge: The critical role of professional culture. *Disasters, 31*(4), 459–476. doi:10.1111/j.1467-7717.2007.01019.x

Montells, L., Montero, S., Díaz, P., & Aedo, I. (2006). SIGAME: Web-based system for resources management on emergencies. In B. Van de Walle & M. Turoff (Eds.), *Proceedings of the 3rd International ISCRAM Conference (B.), Newark, NJ (USA), May 2006* (pp. 1-5).

Nakamura, Y., & Saita, J. (2007). UrEDAS, the earthquake early warning system: Today and tomorrow. In Gasparini, P., Manfredi, G., & Zschau, J. (Eds.), *Earthquake early warning systems* (pp. 249–281). Berlin: Springer-Verlag. doi:10.1007/978-3-540-72241-0_13

PHYSorg.com. (2006, March 3). Ask your car radio! Retrieved September, 30, 2009 from www.physorg.com/news11399.html

Puras, J. C., & Iglesias, C. A. (2009). Disasters2.0. Application of Web2.0 technologies in emergency situations. In J. Landgren & S. Jul (Eds.), *Proceedings of the 6th International ISCRAM Conference – Gothenburg, Sweden, May 2009.*

Rajbhandari, R., & Subedi, J. (2005). Disaster informatics: Issues and future of information management. *Proceedings of International Conference on Disaster Management: Achievements and Challenges* (pp. 146-150). Nepal Engineering College, Kathmandu

Seddigh, N., Nandy, B., & Lambadaris, J. (2006). An Internet public alerting system: A Canadian experience. In B. Van de Walle & M. Turoff (Eds.), *Proceedings of the 3rd International ISCRAM Conference, Newark, NJ (USA), May 2006* (pp. 141-146).

Sillem, S., & Wiersma, E. (J.W.F). (2006). Comparing Cell Broadcast and Text Messaging for Citizens Warning. In B. Van de Walle & M. Turoff (Eds.) *Proceedings of the 3rd International ISCRAM Conference, Newark, NJ (USA), May 2006* (pp. 147-153).

Srivastava, S. K., Hedge, V. S., & Jayaraman, V. (2007). Integrating technological interventions and a community-centric approach for disaster-risk reduction. *Disaster and Development: Journal of the National Institute of Disaster Management*, *1*(2), 111–118.

Stephenson, R., & Anderson, P. S. (1997). Disasters and the Information Technology Revolution. *Disasters*, *21*(4), 305–334. doi:10.1111/1467-7717.00065

Troy, D. A., Carson, A., Vanderbeek, J., & Hutton, A. (2008). Enhancing community-based disaster preparedness with information technology. *Disasters*, *32*(1), 149–165. doi:10.1111/j.1467-7717.2007.01032.x

UNISDR (UN International Strategy for Disaster Reduction). (2005). Building the resilience of nations and communities to disasters. In *Proceedings of the World Conference on Disaster Reduction* (pp. 18-22). January 2005, Kobe, Hyogo, Japan. United Nations Inter-Agency Secretariat of the International Strategy for Disaster Reduction, Geneva.

United Nations Development Programme (UNDP). (2004). *Reducing Disaster Risk: A Challenge for Development* (p. 146). New York: United Nations Development Programme, Bureau for Crisis Prevention and Recovery.

United Nations International Strategy for Disaster Reduction (UNISDR). (2004). *Living With Risk: A Global Review of Disaster Reduction Initiatives.*

USGS (United States Geological Survey). (2009). Prompt Assessment of Global Earthquake for Response (PAGER). Retrieved on June 20, 2009, from http://earthquake.usgs.gov/eqcenter/pager/background.php

WAPMERR (World Agency of Planetary Monitoring and Earthquake Risk Reduction). 2009. Retrieved on June 20, 2009, from http://www.wapmerr.org/qlarm.asp

Wyss, M. (2005). Earthquake loss estimates applied in real time and to megacity risk assessment. In B. Van de Walle and B. Carlé (Eds.) *Proceedings of the 2nd International ISCRAM Conference, Brussels, Belgium, April 2005* (pp. 297-299).

Zimmermann, H. (2005). Recent developments in emergency telecommunications. In B. Van de Walle & B. Carle (Eds.), *Proceedings of the 2nd International ISCRAM Conference Brussels, Belgium* (pp. 327-334).

Section 2
Advanced Collaborative Technologies for Disaster Management

Chapter 7
Efficient Deployment of ICT Tools in Disaster Management Process

Aysu Sagun
Anglia Ruskin University, UK

ABSTRACT

This chapter will emphasize that efficient integration of Information and Communication Technology (ICT) in disaster management process can help mitigation of impacts of disasters on people and the environment, minimizing the failures and maximizing the collaboration. It summarizes the nature of information flow and management processes during disasters and the potential of recent ICT at three stages of disaster management. The requirements and problems faced during their deployment at different stages of disaster management process are stated. The solutions for common constraints are discussed as well as the critical factors that should be considered in efficient deployment of ICT in the disaster management process.

INTRODUCTION

Millions of people suffer from natural or man-induced disasters around the world. As unexpected and sudden events, disasters make serious impacts on people, causing death and damage to the built and the natural environment. There is a significant need to focus on mitigation of disasters because the increase in population and built assets make the needs and requirements of the response and relief activities more expen-

sive and complex (Oberoi and Thakur, 2005). Mitigation refers to those activities conducted to decrease the vulnerability of society by reducing the residual destructive effects of disasters. This can be achieved by a systematic disaster management approach that uses sufficient resources and technologies at preparedness, response and recovery stages. Disaster management is a cyclic and collaborative process in which the gathering, organization and dissemination of information and data are critical. Information and Communication Technology (ICT) has a vital role in management and mitigation of disasters by supporting data

DOI: 10.4018/978-1-61520-987-3.ch007

collection, decision making, communication and collaboration. The challenges of ICT are more related to the effective management of technology and its appropriate application than their capacity. Therefore, a focus on emerging ICT is necessary to employ the appropriate technology and tools in the disaster management process.

This chapter focuses on deployment of ICT in the disaster management process to mitigate the impacts of disasters on people and their environment. The nature of information flow during disaster management process will be explained in more detail giving examples from various natural and man induced disasters. The chapter will clarify how the role of advanced ICT at preparedness, response and recovery phases changes and explain the capabilities of recent ICT tools integrated in disaster management process. The chapter emphasize that the experts, collaboration patterns and appropriate ICT tools used vary regarding the nature of the disaster. Similar ICT tools are used in most types of disasters, but they may be used for different purposes. The constraints in efficient deployment of ICT in disaster management are also discussed by proposing solutions to common problems.

BACKGROUND

Disaster management can be defined as the collaborative process and the set of actions of the relevant organizations/agencies and government to minimize and inhibit the crushing effects of disasters (Scalem, et. al., 2005; Rajabifard, et. al., 2004). This collaboration process involves the sharing of decision making as well as data and resources (Popp, et.al., 2004). Various bodies are involved in the process such as governments, disaster management organizations, responders, the construction sector and the general public. ICT plays an important role in management and mitigation of disasters by facilitating information flow as well as enhancing data collection and

decision making in disaster planning, mitigation and management. ICT products include any product that can receive, disseminate, communicate, edit, recover, manipulate and store information electronically in a digital form. The deployment of advanced ICT at the pre-, during and post-disaster stages of disaster management enhances communication and collaboration. However, being prepared for a potential threat of a disaster does not mean to provide high technology information and communication tools at response stage of a disaster. The important point is to provide the appropriate technologies which are reliable, resilient and flexible to adapt to changes caused by the impacts of the extreme event. Use of sophisticated means of communication systems can cause failure in communication and collaboration.

It is important to reduce the problems in organization of disaster information in order to minimize the challenges in disaster management. McEntire (2002) claimed that the information challenges and lack of communication between the field and the operation centre during the post disaster operations of Fort Worth Tornado were as a result of inaccurate, incomplete or too much information that caused delays in decision making as well as insufficient amount of information in some cases. Therefore, providing the information flow is not enough in disaster management but there is the need for the systems that can help evaluating, filtering and integrating information for the responders so that the rapid decision making process is enhanced and expedited. Moreover, there is a need to set up common semantics for clear information flow because different expressions or different definitions used by different collaborators at various levels can cause chaos and delays in collaboration. Generating standards can solve this problem but they should be flexible to be changed with the new emerging situations. As Midkiff and Bostian (2002) stated, the communication infrastructure should be flexible to respond to different situations in different types of disasters. Michalowski et. al. (1991) also stated that the disaster management process requires flexibility in decision support and the ability to respond to varying situations because the scale of the residual effects varies according to the type and scale of the disaster.

Current ICT can help to achieve the flexibility needed in collaboration during the disaster management process. However, as it is stated in the CSTB Report (1999), misplaced reliance on technology could itself trigger a crisis. Therefore relevant integration of various ICT systems with each other (such as the World Wide Web, simulations or decision support systems) and with other technologies is necessary to generate an effective setting for disaster management (Asimakopoulou, et. al., 2006; Sagun, et. al., 2006a; Sagun, et. al., 2006b; Sagun, et. al., 2009). The problems in ICT are more dependent on misuse or wrong choice of technology for implementation and their insufficient management (Scalem, et. al., 2005).

THE POTENTIAL TO USE INFORMATION AND COMMUNICATION TECHNOLOGIES (ICT) IN DISASTER MANAGEMENT

Disaster management is an information and communication intensive activity. Adequate management of information is the key for a successful disaster management. All of the stakeholders team up physically or virtually to collaborate in case of a disaster. Information is managed in a centralized or decentralized way. In both of the approaches, it is important to have or provide an alternative for a common form of communication among individuals and groups such as using the same language or a means of translation of information. The organization of data is critical for representing, accumulating and dissemination of information because the disaster site is a chaotic work area with rapid changes. Efforts must be taken to avoid misunderstandings and delays in communication during disaster management

because any problem in communication may be lead to loss of more lives and further dangers. Within this context, another vital issue is the targeting of information. It is very important to pass the right information, to the right person(s) or organization(s) at the right time.

Nature of Information Flow in Disaster Management

There is a significant amount of information flow that has to be handled to reveal the entire picture of the impacts and consequences of an extreme event. It is a complex and time-consuming process that includes the collection, organization, integration and dissemination of a huge amount for information. Effective systems are needed to access, retrieve and filter the relevant information. The scope of "relevant information" cannot be standardized because it can be a general picture, or an abstraction, or a detail. Besides, it can vary at different stages of disaster management for different people or organizations within a wide range from disaster impacts to resources or medical records to family contacts. Therefore, flexibility is required in presentation of data to configure, extract and integrate information with high understanding of user requirements and information architectures. The shared information should also support making choices and decisions (De Marchi, 1990). The scope of information needed during disaster management includes a clear statement of the uncertainties and available alternative solutions to specific problems in addition to disaster notification and any practical actions that institutions/individuals can take using a simple and clear language.

In addition to the scope of information, 'when' and 'how' the information is released are important (Covello, 1995). Time is a very important factor in disaster management because rapid decision making is necessary and the tolerance for a delay in the collection, organization and dissemination of data is very low. Moreover, it is critical to provide reliable, secure and continuous communication between response teams and the public (Caribbean Information and Communications Technology Community, 2004). Especially at the response stage, robust, mobile and flexible communications systems are the first requirements because it is a very dynamic state where the circumstances change rapidly over time and there is the risk of unpredicted consecutive events. Repetition of the disaster impacts, such as simultaneous explosions and aftershocks, or new impacts caused by another affected environmental element, such as fire and building collapse, may cause worse impacts after a disaster. Therefore, up-to-date information is essential for the stakeholders to take proper actions to accelerate rescue as well as to avoid panic in public to prevent any consequential crisis. There is an immediate need to gather information on critical areas, buildings and road networks at the response phase to inform the stakeholders. The communication between the stakeholders are maximized at this stage for a speedy decision making and to eliminate the overlaps, conflicts, duplications or gaps in information exchange. The priorities can change at any time and the load of traffic can increase as a result of consecutive events. A heavy traffic of information flow decreases the performance of the networks, bandwidth or speed. For this reason, development and testing of advanced information infrastructures are essential to cope with an overload of information.

Voice communication is mostly superior for emergency situations because input and flow of visual and textual data is time-consuming. Waiting to access loaded visual and/or textual data can be a waste of time in some situations where time is very critical. However, we can not underestimate the support of data communication (such as such as textual and visual data including video, pictures and maps) for communicating with the disaster field. If the existing communication systems and tools are damaged or destroyed during the disaster, it becomes even more difficult to keep collaborators informed about the current and changing circumstances. Different groups of people require different kinds of urgency at differing degrees. A human-centered approach is essential in communication where human behavior and ergonomics are considered for the whole population. For instance, different types of warning systems are needed for people with different abilities. Use of a visual warning system would not be suitable for blind people, so it should be supported by audio warning systems. Also, different kinds of disaster cases enable different ways of information flow. The experts and advanced ICT tools involved in disaster management may change according to the nature of the disaster. These differences are discussed below in more detail.

Advanced ICT in Three Basic Stages of Disaster Management

There is a huge variety of ICT tools that can be used for information exchange and collaboration in disaster management and they can be categorized as telecommunication technologies, space technologies, and other computer-based technologies (Sagun, et al., 2006a, Sagun, et al., 2006a and Sagun, et al., 2009). Use of an inefficient technology could obstruct the communication and collaboration processes and increase the impacts of disasters. Therefore, it is necessary to find out the most suitable ICT solutions based on the stage of the disaster management process as well as the type of the disaster. Disaster management process has three basic stages: preparedness, response and recovery (Mansourian, et. al. 2006). Identification of the actions taken at each stage of disaster management can help to choose and deploy the appropriate ICTs. Table 1 summarizes the actions taken at each stage of disaster management process in detail.

There are various ICT systems used at different stages of disaster management for natural and man-induced disasters. Developments in telecommunication systems aim to enable more flexibility and mobility with lower costs and higher

Table 1. Actions taken in disaster management

Preparedness	Response	Recovery
Analysis of data from previous disaster cases *Raising public awareness* *Monitoring of potential risks* *Early disaster predictions* *Structural strengthening of buildings and infrastructure* *Possible damage assessments* *Reviewing and updating strategies, tools and resources* *Training of stakeholders* *Establishment of communication and collaboration systems* *Response and recovery planning with collaborators* *Establishment of disaster warning systems*	*Notification of stakeholders* *Transportation of response personnel (police, fire-fighters, rescue teams, health personnel, etc.)* *Transportation of resources (food, medicine, etc.)* *Communication of rapidly changing disaster data* *Collaboration with stakeholders* *Management of available resources* *Warnings for a possible consequent disaster/crisis/chaos*	*Sharing disaster specific information* *Damage assessments* *Archiving of resources and information* *Registration of claims for disaster relief funds* *Repair of infrastructures and buildings* *Reconstruction of infrastructures and buildings*

bandwidths. Radio communication services can be helpful especially when the wired communications are destroyed or overloaded (ITU, 2005). Developments in wireless technology are in rapid increase. They are more reliable than mobile and portable systems that use utility mains power because they are powered by independent batteries and they enable connection in moving vehicles such as automobiles and ships. Therefore, they can be emergency backups for fixed wire line and mobile systems in case of a power cut caused by a disaster. Wireless networks can be constructed at various scales. Recent developments in space technologies also enhance disaster management process saving time and effort in collecting disaster relevant data or location detection. Besides, web-based computer technologies have high potentials for disaster communication and collaboration because interactive websites for collaboration, online web forums for discussions, weblogs and wikis for exchange of information enable rapid dissemination of both the up-to-date and archive information to the whole world. The capabilities of web tools are developing rapidly to improve exchange of different types of data (such as 2D, 3D redlining and textual data exchange) and to provide personal/institutional virtual spaces connected to a common database to form a disaster network.

Flexibility of web channels facilitates publishing and updating information on the internet. Katrina help blog (http://katrinahelp.blogspot. com) and South East Asia Earthquake and Tsunami (SEA-EAT) blog (http://tsunamihelp.blog.spot. com) are examples which were set up to provide an information channel for disaster relief right after the disasters stroke. There are also various protocols being developed to support exchange of information during emergency response activities such as Emergency Data eXchange Language (EDXL) and CAP. EDXL is specifically designed for emergency data exchange by the U.S. Department of Homeland Security (DHS), Federal Emergency Management Agency (FEMA), and industry members of the Emergency Interoperability Consortium (EIC) to include resource queries and requests, situation status and forecasts, financial and personnel data, and message routing instructions. EDXL and CAP also enable disaster warnings to be sent over various types of networks (Gustavsson, et. al., 2006). It is also possible to use more than one protocol and language to set up a system for information distribution. For instance, as Raymond (2004) stated, alerts can be sent to cellular phones, PDAs and computer screens by using the CAP and XACML (eXtensible Access Control Markup Language) together, where En-

Figure 1. ICT in disaster management process

		Examples	Preparedness	Respose	Recovery
Telecommunication Systems	Radio communications	General Packet Radio Service (GPRS)	Collaboration, information exchange, data collection	Disaster warning, collaboration, information exchange, data collection	Collaboration, information exchange, data collection
	Wire-line Mobile Technologies	high speed internet connections	Collaboration, information exchange, data collection	Disaster warning, collaboration, information exchange, data collection	Collaboration, information exchange, data collection
	Wireless Mobile Technologies	Wireless Application Protocol (WAP), Universal Mobile Telecommunications System (UMTS), 3G, Wi-Fi, WiMAX, and and Bluetooth	Collaboration, information exchange, data collection	Disaster warning, collaboration, information exchange, data collection	Collaboration, information exchange, data collection
	Network Systems	Wireless Local Area Networks (WLAN), Wireless Fidelity (Wi-Fi), Wireless Wide Area Networks (WWANs), Mobile Ad-Hoc Network (MANET), Wireless Mesh Networking	Collaboration, information exchange, training of stakeholders	Disaster warning, collaboration, information exchange, data collection	Collaboration, information exchange, data collection
Space Technologies	Satellite Systems	Geographical Information System (GIS), Global Positioning System (GPS), Information, Communication and Space Technologies (ICST)	Collaboration, information exchange, data collection,	data collection, location tracking, damage assessment	data collection, damage assessment
	Remote Systems	Laser Scanning, Remote Sensors	early disaster predictions, data collection	data collection, location tracking, damage assessment	data collection, damage assessment
Computer-based Technologies	Modelling and Simulations	2D/3D Modelling and simulations, Agent-based systems	raising public awareness, training of stakeholders, predictions of possible damage, risk identification	initial damage estimates,	performance evaluation for reconstruction
	Protocols and Standards	Extensible Markup Language (XML), UML eXchange Format (UXF), Common Alert Protocol (CAP), Emergency Data Exchange Language (EDXL), Specific Area Message Encoding (SAME), and IEEE 1512	Collaboration, information exchange, training of stakeholders	Collaboration, information exchange, data collection	Collaboration, information exchange, data collection
	Databases and Networks	information fusion, SDI, Grid technology, Evolutionary computing	Collaboration, information exchange, data collection, training of stakeholders	Collaboration, information exchange, data collection	Collaboration, information exchange, data collection
	Web-based technologies	smartphones for Web browsing i-Mode), Electronic Commerce (EC), Websites, online web forums, weblogs and wikis	Collaboration, information exchange, data collection, training of stakeholders, raising public awareness	Disaster warning, collaboration, information exchange, data collection	Collaboration, information exchange, data collection

coding of needed information such as contact data and alert interests is done using the language CAP Markup and the access policies for the protection of the data are written in the XACML. Although flexibility of network and web tools and channels is an advantage for collaboration, they may be hazardous in some cases. Physical infrastructural systems may be in danger of being damaged because most are connected to the Internet and ICT systems in various ways (Hellstrom, 2006). Breakdown or cyber attacks through the web may lead to technological disasters.

Figure 1 summarizes the use of three categories of ICT at preparedness, response and recovery stages of disaster management process including examples from advanced technologies and tools.

Some of the ICT systems can be applied at all stages of disaster management for the same purpose such as telecommunication systems and tools used for communication of collaborators. In case of damage on a wire-line communication channel, an alternative wireless communication channel can be provided. Some of the ICTs are used at different stages for different purposes. For instance, the laser scanning technologies are used for damage detection at the recovery stage as well as used to collect disaster information as an input to predictive modeling and analysis.

ICTs used in disaster management also vary according to the type of threat or nature of the disaster. Different types of sensors can be used to build sensor networks for different purposes such as the Data Sensors used for monitoring and detection of fire and waves, Microwave Sensors used for detection of soil moisture or smoke, Video Surveillance Systems, multi-resolution cameras and Response Sensors used for tracking people of interest. Another example can be the variances in sensor Networks which consist of groups of heterogeneous sensors such as cameras, strain gauges, accelerometers, etc. to provide real-time data streams (Elgamal, et.al., 2005).

Special ICT tools have been developed based on the nature of the disaster for a particular reason. For instance, in case of terrorism, special technologies are developed to provide private communications, data analysis and management such as pattern analysis, deception detection, identity management and intelligent surveillance systems.

The scale of the disaster does not change the type of ICT employed, but it may result in slight variations in location, targets and number of people affected. It is also possible to observe a chain in the use of some of the ICT tools during disaster management where the information gathered with one type of ICT is used to initiate or support the use of another ICT tool for analysis or predictions. For instance, the data collected in laser scanning can be used for 3D computer modeling of buildings to identify their changes (Steinle and Vögtle, 2001). Integrating ICT with each other can also improve the potential of technologies in disaster management. Information, Communication and Space Technologies (ICST) is an example for this approach which integrates space technologies that enable immediate data collection via satellite systems with the ICTs to enhance natural disaster management. Another recent approach is the use of GIS with the laser technology. GIS can provide support to see the potential interactions of various elements by combining layers of information (Hill, 2005). Its integration with laser technology support 3D data gathering in space to represent the surfaces. Moreover, Raheja, et. al. (1999) claim that integration of GIS and the WWW would increase of the use and accessibility of spatial data, facilitating the gathering, analysis and use

of a huge quantity of data related to disasters with its user friendly interface (Raheja, et. al., 1999).

Constraints in Use of ICT in Disaster Management

The most common problem is the inefficient management of resources as a result of poor communication and coordination in relief operations (Scalem, et. al., 2005). Inefficient management can be due to non-technical and technical constraints in information flow and deployment of ICT. Non-technical constraints are more related to people and organizations. Variation in capabilities and needs of people is an important challenge for information flow and use of ICT during disaster management process. There is an additional need to support disabled people or those who cannot use new technologies (such as handicapped, elderly or young children). Moreover, the performance of people changes with stress under a real-time disastrous event and this can also influence the information flow, communication and collaboration considerably.

Problems also arise as a result of lack of awareness of the nature of the extreme events and possible consecutive events; and awareness of the requirements for efficient communication and available technologies that can enhance the disaster collaboration. Besides, organizational resistance to sharing information due to particular reasons such as security or willingness of people/ organizations to get organized and cooperate before or after the disastrous event affects the communication and collaboration process. There can also be financial constraints of the organizations in deployment of relevant ICT as highlighted by Telecommunications Regulatory Commission of Sri Lanka (1998).

Technical constraints include the lack of overall system architectures or relevant standards and misuse or failures of available technologies. It is important to be aware that technological disasters can also occur as a consecutive event because of destructive attacks or failures of technology products. Failures of ICT tools may be due to poor or lack of regular maintenance as well as misuse. Some of the common management challenges in use of ICT for disaster management are summarized by CSTB (1999) as:

- Resistance to change which is a result of acceptance of ICT as a new task by the some organizations rather than a useful tool.
- Outdated Technology base and equipment which can cause failures and/or delays in communication and exchange of information.
- Lack of training and education which can help collaborators get the most out of the ICT tools/systems performance and to minimize failures during their use.
- Limited resources which include the need for operational, maintenance and training costs.
- Lack of funding which is essential to develop effective communication channels and collaboration patterns, especially to cover all residential and public areas in developing countries.
- Coping with multiple standards which obstruct the coordination of communication and exchange of information.

SOLUTIONS AND RECOMMENDATIONS

A reliable and disaster-resistant communication infrastructure is essential to minimize the interruption of the relevant information flow for an effective collaboration. Provision of reliable power supplies such as generators, batteries or appropriate management software and tools with capabilities of back up and archiving is helpful in supporting the operability of the communication system during extreme events. However, the communication infrastructure is not the only issue to

be considered in efficient deployment of ICT as it was discussed above. The challenges of use of ICT for disaster management are mostly related to the effective management of technology and its appropriate application rather than their capacity. Therefore, all organizations and collaborators involved in the process of disaster management should react efficiently both in their individual responsibilities and coordinated activities. Actions are taken according to past, present and future circumstances during disaster management process. Lessons can be learned from the records of previous disaster cases. Analysis of failures in management and implementations can help to define a reliable strategy that minimizes possible risks in disaster management process. Audit trails are helpful in testing the effectiveness of various communication and information management strategies. They are useful for capturing the actual state of events (when and where they happened) which enables the tracking of where the data is lost or delayed as well as the location of the bottleneck in the flow of information.

Web-based systems can support rapid sharing of disaster information in databases. Databases provide systematic storage of information to be used as a reference to observe past experiences and applications as well as to follow the progress of recorded issues by exchanging information between the levels of disaster managers in future (See Figure 2).

For instance, a pilot study conducted by cooperation of 12 organizations from local and national levels in Tehran for a possible earthquake case showed that employment of a web-based system based on SDI during disaster response can facilitate communication by reducing on third of the response time (Mansourian, et. al., 2004). There is also growing interest in multi-agent systems and information fusion in recent researches, where the concept of information fusion is highlighted. Information fusion is the process in which information gathered from different sources (such as different data bases and sensors) is integrated

Figure 2. Database levels

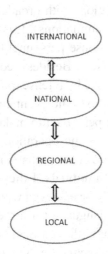

into a single structure to be used by a computer system in decision making process. D'Agostino, et. al. (2002) stated basically three approaches to information fusion which involve the process of gathering, integrating, interpreting information and decision making, and acting based on the decisions. Centralized information fusion approach involves the gathering of data into a single central unit; whereas hierarchical fusion architecture is like a tree structure and it based on different layers of fusion nodes through which the data flows from sensors through low fusion nodes to higher fusion nodes. In distributed architectures, each node performs locally and it has a different topology of fusion nodes than hierarchical ones. D'Agostino, et al (2002) stated that agent approaches to information fusion can be as a system component that allows for a modular software approach to both system design and modelling concepts. Grid technology also enables users to share, select, and cluster large amounts and varieties of geographically distributed computational resources and to present them as a unique resource. This approach is effective in dealing with large-scale computer and data intensive computing applications allowing collaborative works and large-scale data analysis. The UK e-Science Program has employed the Grid

Operations Support Centre (GOSC), which is a distributed virtual environment built to support UK services by providing deployment and operation for user authentication, user authorization and project organization, resource discovery, etc.

The complex structures of the recent rapidly developing telecommunication infrastructures and applications can also make use of evolutionary algorithms to find efficient solutions to complex problems. Evolutionary computing, which is a system development for computers formed by combining several computational techniques to be used as an automatic optimization and design tool for complex systems (Kicinger, 2005), reflects the ideas of natural selection, mutation and the Darwinian principle of survival of the fittest found in nature. It can be applied in critical network infrastructure and scenario-based studies in the DM planning process.

Provision of a budget is essential to set up these systems as well as for planning, construction, equipments and organization of disaster communications, by organizations, institutions and the government. Funds are needed for research on recent technologies, provision of equipment, training of the personnel, construction of the communication system, and construction of a database to store disaster information. ICT alternatives that will be used in all stages of disaster management need to be observed in terms of cost. For instance, wireless technology products can be preferred because they enable fast and safe data transmissions at lower costs if other organizational, technical and environmental circumstances permit its deployment. Additionally, consideration of the following factors can enhance the efficient use of ICT for disaster management:

- **Potentials of recently available technologies.** It is essential to follow the developments on ICT technologies to take advantage of them during the disaster management process.

- **Flexibility and ease of use of ICT.** This approach can help sufficient use and flow of available assets, resources and information, if they are designed and employed by considering all people at different literacy levels or people with various physical characteristics, such as elderly, children, or people with various types of disabilities. For instance, effective Graphical User Interface techniques that can be easily understood and used by all people with minimal training can save time during response stage. The design for all approach also enables the adaptability and continuity of information flow and collaboration in case of changing circumstances during the extreme event.

- **Pros and cons of the available technologies.** The advantages and disadvantages of ICT need to be compared to decide on the most efficient ICT that will be used in every stage of disaster management process. The decision is also based on the type and scale of the disaster. For instance, Wire-line mobile technologies require a lengthy planning and construction process and cannot be employed in urgent situations immediately (Midkiff and Bostian, 2002). Mobile technologies eliminate the place constraints in communication and information exchange but they have a low memory, limited capacity and power for interaction.

- **Training.** The stress and panic in using the time efficiently draws people to use the communication tools that they are familiar with. For this reason, it is a key issue to familiarize people with the recent ICTs developed and employed in disaster information flow and communication. ICT can play a vital role for induction process within the society by engaging people with the possible impacts of disasters (with representations of disaster virtual environments and simulations) and introducing recently available tools and procedures for disaster communication (with broadcastings, computers and simulations). It is also used for training response personnel by simulating disaster cases. Besides, communication failures caused by intense traffic during the chaos of an extreme event can be reduced by training personnel to make them aware of procedures and operations and alternative solutions at the preparedness level.

- **Common Standards.** Efficient collaboration can only be achieved with provision of common a language (or a means of translation among the parties) and common standards of technology that enable the exchange of information with a common data format.

- **Improved real-time data access.** Real-time communication based on conditions, users, personnel accountability, medical information, etc. requires additional characteristics in ICT such as the ability to transmit signals through/around obstacles, integration of functions found in multiple pieces of equipments into a single piece of equipment, and standardization of equipment.

- **Update and maintenance of the current equipments.** The control and recovery of telecommunication systems and electronic tools is essential at preparedness and recovery stage as well as the physical recovery of the built environment.

FUTURE RESEARCH DIRECTIONS

Most research approaches in the use of ICT for disaster management refer to the design and improvement of capacities of information and communication technologies and tools to support the collaboration process rather than the development of ICT based on collaboration patterns or nature of information flow. User-centred approach and focus on collaboration patterns and complex nature of information flow during extreme events would mitigate the destructive impacts on people and the built environment with more effective communication and efficient and relevant information flow. Moreover, it would enhance time management during disaster collaboration increasing accessibility of information. Such an approach would enable the accomplishment of critical tasks on time and prevent waste of time caused by information flow to wrong targets.

In order to focus more on the nature of collaboration and information flow, the future studies need to consider the constraints in deployment of ICT for better communication and collaboration solutions in more detail by identifying the ICT problems at various stages of disaster management process. Scenario-based studies can help the identification of the problems and development and testing of alternative solutions. The importance of continuity of efficient information flow needs to be emphasized as the most important issue during the development of ICT to cope with constantly changing information and to minimize the problems in information flow such as information overload, dissemination of incorrect, incomplete or conflicting information.

Use of ICT technologies cannot be limited to communication technologies so further research may also contribute to the training of personnel and public as well as recovery of the built environment in design or redesign of the built environment via simulation technologies. All of these future research studies need to be based on a user-centred approach where people's interaction with their immediate environment is also investigated in addition to their interaction with each other.

CONCLUSION

A systematic disaster management approach with efficient use of resources is required in order to cope with any type of disaster and to mitigate the impacts of disasters on people and the environment. This is a continuous, highly collaborative process and it involves the use of various ICT for various purposes based on the type and scale of the extreme event. This chapter focused on deployment of relevant ICT for an enhanced and advanced collaboration process in disaster management. The following points were highlighted for efficient use of ICT and their integration with each other at different stages of the disaster management process.

- the experts, collaboration patterns and appropriate ICT tools used vary according to the nature of the disaster;
- some of the ICT tools can be used in the management process of all types of disasters for similar or different purposes at the same or different stages of DM;
- there are specific ICT tools that enhance the information collection, information flow and collaboration process depending on the nature of the disaster;
- changes in the scale of the disaster do not alter the type of ICT employed, but they may result in slight variations in location, targets and number; and
- it is important to take into consideration any consecutive event that may arise after a disaster.

The efficient integration of advanced ICT to support the management and mitigation of impacts of extreme events was discussed by pointing out the constraints as well as the advantages in their deployment. This study can be conceptual and practical guidance for the researchers, managers, practitioners, from the industry and the academics in the areas of disaster management, threat detection, and ICT.

REFERENCES

Asimakopoulou, E., Sagun, A., Anumba, C. J., & Bouchlaghem, N. M. (2006). Use of ICT during the Response Phase in Emergency Management in Greece and the United Kingdom. *International Disaster Reduction Conference*, Davos, Switzerland.

Caribbean Information and Communications Technology Community. (2004). *Facilitating Effective Disaster Management in the Caribbean.* Retrieved May, 2009, from http://www.devnet.org.gy /documents/Caribbeandisasterbrief-Final.pdf

Covello, V. T., McCallum, D. B., & Pavlova, M. T. (1988). *Effective Risk Communication.* New York: Plenum Press.

D'Agostino, F., Farinelli, A., Grisetti, G., Iocchi, L., & Nardi, D. (2002). Monitoring and Information Fusion for Search and Rescue Operations in Large-scale Disasters. In *Proceedings of the Fifth International Conference on Information Fusion.* Annapolis, MD: Omnipress, USA.

De Marchi, B. (1990). Assessing People's Information Needs about Major Accident Hazards: Improving Knowledge for a Better Response. In Gow, H. B. F., & Otway, H. (Eds.), *Communicating with the Public about Major Accident Hazards.* New York: Elsevier Science Pub.

Elgamal, A., Yan, L., Fraser, M., Lu, J., & Conte, J. P. (2005). Large- Scale Simulation and Data Analysis. In *Proceedings of the 2005 ASCE International Conference on Computing in Civil Engineering,* Cancun, Mexico.

Gustavsson, P. M., Wemmergård, J., Garcia, J. J., & Larsson, M. N. (2006). (CML). In *Simulation Interoperability Workshop 2006.* Stockholm, Sweden: Expanding the Management Language Smorgasbord Towards Standardization of Crisis Management Language.

Hellstrom, T. (2006). Critical infrastructure and systemic vulnerability: Towards a planning framework. *Safety Science, 45*(33), 415–430.

Hill, A. (2005). Information Technology & GIS Employment During an Emergency Event. ESRI Professional Papers. Retrieved November 2005, from http://gis.esri.com/library/userconf/ proc05/papers/pap1024.pdf

ITU. (2005) *International Telecommunication Union Web Site.* Retrieved May 2009, from http://www.itu. int/ITU-R/information/emergency/index.asp

Kicinger, R., Arciszewski, T., & De Jong, K. (2005). Evolutionary Computation and Structural Design: A Survey of the State-of-the-art. *Computers & Structures, 83*, 1943–1973. doi:10.1016/j.compstruc.2005.03.002

Mansourian, A., Rajabifard, A., Valadan Zoej, M. J., & Williamson, I. (2004). Facilitating Disaster Management Using SDI. *Journal of Geospatial Engineering, 6*(1), 39–44.

Mansourian, A., Rajabifard, A., Zoeja, M. J. V., & Williamson, I. (2006). Using SDI and web-based system to facilitate disaster management. *Journal of Computers & Geosciences, 32*, 303–315. doi:10.1016/j.cageo.2005.06.017

McEntire, D. E. (2002). Coordinating Multi-organizational Responses to Disasters: Lessons from the March 28, 2000, Forth, Worth Tornado. *Disaster Prevention and Management, 11*(5), 369–379. doi:10.1108/09653560210453416

Michalowski, W., Kersten, G., Koperczak, Z., & Szpakowicz, S. (1991). Disaster management with NEGOPLA. *Expert Systems with Applications, 2*, 107–120. doi:10.1016/0957-4174(91)90108-Q

Midkiff, S. F., & Bostian, C. (2002). Rapidly-Deployable Broadband Wireless Networks for Disaster and Emergency Response. In *Proceedings of the First IEEE Workshop on Disaster Recovery Networks.* New York City, USA.

Oberoi, S. V., & Thakur, N. K. (2005). Disaster Preparedness in The Hills: Natural Hazard Modeling Using GIS and Remote Sensing. In *Proceedings of the 2005 ASCE International Conference on Computing in Civil Engineering,* Cancun, Mexico.

Popp, R., Armour, T., Senator, T., & Numrych, K. (2004). Countering Terrorism Through Information Technology. *Communications of the ACM, 47*(3), 36–43. doi:10.1145/971617.971642

Raheja, N., Ojha, R., & Mallik, S. R. (1999). Role of internet-based GIS in Effective Natural Disaster Management. Retrieved June, 2009, from http://www.gisdevelopment.net/technology/gis/techgi0030.htm

Rajabifard, A., Mansourian, A., Williamson, I., & Valadan Zoej, M. J. (2004). Developing Spatial Data Infrastructure to Facilitate Disaster Management. In *Proceedings GEOMATICS 83 Conference,* Tehran, Iran.

Raymond, M. (2004). Personalized Emergency Alerting System. In *Extreme Mark-Up Languages 2004.* Montreal, Quebec: WIRLED PEAS-World Information Resources, Localized Environment Distribution.

Report, C. S. T. B. Computer Science and Telecommunications Board (1999). *Summary of a Workshop on Information Technology Research for Crisis Management. National Research Council, Committee on Computing and Communications research to Enable Better Use of Information Technology in Government.* Washington, DC: National Academy Press. Retrieved November 2005, from http://books.nap.edu/html/itr_crisis_mgmt

Sagun, A., Anumba, C. J., & Bouchlaghem, D. (2006a). Coping with Extreme Events in the Built Environment: ICT for Disaster Mitigation and Collaboration. In *Proceedings of APSEC 2006 Asia Pacific Structural Engineering and Construction Conference, Challenges Toward Sustainable Construction,* Kuala Lumpur, Malaysia.

Sagun, A., Bouchlaghem, D., & Anumba, C. J. (2006b). Improving Safety and Security Through the Deployment of ICT in Disaster Management and Mitigation. In *INCITE-ITCSED 2006 World Conference on IT in Design and Construction,* New Delhi, India.

Sagun, A., Bouchlaghem, D., & Anumba, C. J. (2009). A Scenario-Based Study on Information Flow and Collaboration Patterns in Disaster Management. *Disasters: The Journal of Disaster Studies. Policy and Management, 33*(2), 214–238.

Scalem, M., Sincar, A. K., Bandyopadhyay, S., & Sinha, S. (2005). A Decentralized Disaster Management Information Network (DDMIN) for Coordinated Relief Operations. In *9th World Multiconference on Systemics, Cybernetics and Informatics (WMSCI 2005),* Orlando, USA.

Steinle, E., & Vögtle, T. (2001). Automated Extraction and Reconstruction of Buildings in Laserscanning Data for Disaster Management. In E. P. Baltsavias, A. Gruen, & L. Van Gool (Eds.), *Proceedings of the Workshop Automatic Extraction of Man-Made Objects from Aerial and Space Images (III),* Swets & Zeitlinger, Lisse, The Netherlands.

Telecommunications Regulatory Commission of Sri Lanka. (1998). *Pilot Study on the Use of Telecommunications in Disaster and Emergency Situations in Sri Lanka.* In association with United Nations Office for the Coordination of Humanitarian Affairs, Working Group on Emergency Telecommunications and ICO Global Communications Interim Report. Retrieved November, 2009, from http://www.reliefweb.int/telecoms/tampere/slcs.html#EXECUTIVE%20SUMMARY

Chapter 8

Current State and Solutions for Future Challenges in Early Warning Systems and Alerting Technologies

Ulrich Meissen
Fraunhofer Institute for Software and Systems Engineering (ISST), Germany

Agnès Voisard
Fraunhofer Institute for Software and Systems Engineering (ISST), Germany

ABSTRACT

The deployment of Early Warning Systems (EWS) and Alerting Technologies (AT) is one of the best measures for improved disaster prevention and mitigation. With the evolution of Information and Communication Technologies (ICT), we face new opportunities as well as new challenges for improving classical warning processes. This chapter concentrates on the main aspects of existing early warning systems and alerting technologies. Beginning with the definition and classifications in this field, we describe general approaches, representative systems, and interoperability aspects of EWS. Furthermore, we introduce a list of criteria for evaluating and comparing existing systems. It is worth noting that the deployment of an operational EWS is a complex challenge and remains a young field of research. This is due to many reasons, ranging from the political to the technical. The most critical issues regarding efficient alerting are described in this chapter, along with areas for future research.

INTRODUCTION

Evident in recent natural disasters, such as the 2004 Tsunami in Indonesia, the key challenges are to detect the threat and alert the public within a short timeframe, especially to achieve the most important objective: preventing the loss of human life. In the last decade we have witnessed considerable effort and progress towards improved risk detection, monitoring and prediction, not only for long term threats (such as El Niño) but also for short term threats. In this context the technologies for the monitoring and prediction of threats that are meteorological (e.g., hurricanes, local tornados or thunderstorms) or hydrological (e.g., coastal and

DOI: 10.4018/978-1-61520-987-3.ch008

river flooding) have advanced significantly and are increasingly important due to the effects of climate change. Recent developments in tsunami prediction are promising, and several monitoring systems for wild fires have improved. Even for short term threats, the volcano or earthquake warnings arrive within seconds (based on p-waves), and meteorite alerts provide promising approaches for future warning systems. Since 2001, man-made threats to our infrastructure have also become a major issue in the development of new early warning technologies.

However, the best warnings are ineffective if they cannot be distributed in a timely way and targeted to people at risk. The dissemination of information in this case implies (1) elaborate processing of the available information, (2) the understanding of people's current need for information, and (3) efficient use of the available communication channels. The new generation of early warning and alerting systems, when implemented as an integral part of disaster prevention strategies in industrialized and developing countries, will be an effective answer for dealing with increasing worldwide threats from natural (hydro-meteorological, volcanic, earthquake, tsunamis, landslides, etc.), human-caused (wars, terrorist attacks, etc.), technological (infrastructure failures, etc.) or biological (diseases, etc.) risks. Several ICT challenges are emerging with the evolving blueprint of new early warning and alerting infrastructures. This chapter describes these challenges and possible solutions, based on an overview of the state-of-the-art early warning and alerting technologies. We present the current status of early warning and disaster alert systems with a strong focus on ICT in general and alerting technologies in particular. The aim is to give the reader an insight into existing ICT challenges and various solutions in this field. By compiling the upcoming challenges and possible solutions, we give a short preview to future trends and developments over the next several years. Note that the

monitoring and prediction technologies of EWS are not the major focus of this chapter since these are highly domain dependent and therefore should be described in the context of the addressed threat.

The chapter is organized as follows. Section 2 gives background information on EWS.

Section 3 focuses on representative systems and interoperability issues in EWS. Section 4 presents common alerting technologies and compares them systematically. Section 5 summarizes future challenges in EWS. Finally, Section 6 draws conclusions in this domain.

BACKGROUND ON EARLY WARNING SYSTEMS

This section introduces the basic principles of early warning systems. It starts with a common definition (2.1) before presenting various categories of EWS using task-oriented, functional, process-oriented, and architectural views (2.2). We discuss the increasing importance of the Internet in EWS (2.3). The main ICT challenges in this context as well as a general evaluation for EWS are presented (2.4). Finally, we describe a general evaluation methodology for EWS (2.5).

Definition of Early Warning

A first problem in the field of EWS is the existence of a variety of different notions of what defines and constitutes an early warning system. The general UN definition describes the term early warning as follows: "The provision of timely and effective information, through identifying institutions, that allow individuals exposed to hazard to take action to avoid or reduce their risk and prepare for effective response" (UN/ISDR, 2006). Despite this quite common definition, the views and the understanding of early warning are often considerably different and depend on the domain of a user or developer. This is partly due to the fact that the components of an early warning system are complying heterogeneous and interdisciplinary tasks. We find a bandwidth of notions of early warning in diverse contexts stretching from long term sociological, environmental or economic broad threats to short term individual risks (e.g., car driving).

Accordingly, we find very different definitions of early warning and hence many possible underlying technologies to implement the corresponding systems. In its effort to support a common understanding of early warning, the UN Inter-Agency Secretariat of the International Strategy for Disaster Reduction (UN/ISDR) has defined four key elements of an EWS: risk knowledge, monitoring and warning service, dissemination and communication, and response capability (UN/ISDR, 2006). Looking at existing EWS shows that the focus is often set on one or two of these elements only and rarely on the complete set. Because long term warnings do not face the problems of rapid alerting, their focus is naturally more on monitoring and prediction rather than on information dissemination. Usually, even if this formal definition is incomplete, one can still speak of an EWS. Thus, researchers developing a system with a focus on risk knowledge, for instance, and another researcher focusing on dissemination and communication might both call their solution an EWS even though they are tackling completely different tasks and challenges. It is therefore important to define the specific field and aim of one's contribution to EWS research.

Categorization and Views on Early Warning Systems

In order to categorize developments and solutions in EWS we can consider several views. A common set of criteria is given by the overall task and system boundaries of an EWS.

1. Criteria, such as the hazard type:
   ○ natural (e.g., hydro-meteorological, volcanic, earthquake, tsunamis, landslides),
   ○ human-caused (e.g., wars, terrorist attacks),
   ○ technological (e.g., infrastructure failures),
   ○ biological: (e.g., diseases), or
   ○ multi-hazard
2. Time scale
   ○ long-term,
   ○ short-term, or
   ○ real-time
3. Geographical or political scale:
   ○ global (i.e., the entire world),
   ○ international (e.g., the Indian Ocean),
   ○ regional,
   ○ national, or
   ○ local
4. Organizational scale:
   ○ governmental,
   ○ non-governmental, or
   ○ private sector

In addition, the maturity state of an EWS should not be neglected and can be considered as pre-science, ad-hoc science-based, systematic end-to-end, or integrated (Basher, 2006). We can identify these stages often as improvement stages of existing EWS, though rarely has the final stage been reached. Aside from these classification criteria, other model views can be applied to structure a EWS: task and functional views (as above), process views, and architectural views. Structural and process views are proposed by Glanz (2004).

From a system engineering and ICT point of view, we propose an additional architectural view that consists of four layers, which are: (1) monitoring and data collection layer, (2) information processing layer, (3) warning generation layer,

Figure 1. Layers of an early warning system

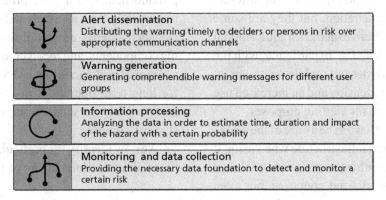

and (4) alert dissemination layer (see Figure 1). It is important to note that this view is strongly system-oriented and leaves out organizational aspects.

The Role of Internet in EWS

In the last 10 years, a technological breakthrough has occurred for a wide range of precise, fast and effective early warning systems, all of which involve the intensive use of the Internet. A main factor in this progress was that the pervasive emergence of the Internet provided cost-effective solutions for fast, high-capacity and multi-modal communication infrastructures necessary for complex monitoring and data collection layers as well as effective alert dissemination. Furthermore, the Internet facilitated several interoperability solutions, from the protocol layer (TCP/IP) to the application layer (web services), which enable the integration of several heterogeneous information sources and the distribution of functionalities within an EWS architecture based on a worldwide common data exchange infrastructure. The inevitable shift towards multi-hazard and multi-channel EWS is unthinkable without the infrastructure provided by the Internet as a basis for these architectures. Following, we discuss the role of the Internet in EWS particularly for the monitoring and data collection layer, and the alert dissemination layer.

Based on a common and almost ubiquitous data exchange infrastructure, the data collection of an EWS can integrate almost any worldwide information source connected to the Internet. A prominent example for the effective use of this infrastructure is the provision of meteorological sensor data from weather stations or seismic networks with several hundred sensors providing worldwide data in almost real-time. As complex simulation and prediction components are increasingly deployed in EWS to enable warnings for complex hazards or to increase the accuracy of the forecasts, the infrastructure for data exchange provided by the Internet is becoming an increasingly essential foundation for the data infrastructures of EWS. Even the behavior of Internet users becomes an interesting information source for forecasting, as demonstrated by Google in introducing a flu warning system based on monitoring search entries for increased occurrence of certain keywords. Further, the new micro-blog technologies used by services such as Twitter might offer possibilities to integrate human sensors in EWS in a scale that has been unthinkable before the global success of Internet technologies.

For alert dissemination, the Internet also offers a wide range of new potentials. The possibility to spread information worldwide at a minimum cost is one of the major advantages of this infrastructure. Alert channels provided by the Internet such as web sites, e-mail, RSS-feeds and – currently evolving

XMPP-messages are clearly imperfect in terms of coverage and penetration, but they are some of the most cost-effective media for multi-modal information dissemination. In particular, with the convergence of Internet and mobile communication, we expect an extension and an increased use of this infrastructure for alerting purposes.

Nevertheless we have to be aware of the weaknesses of the Internet infrastructure when using it in an EWS. First, the infrastructure might be almost ubiquitous in industrialized countries, but even in these countries we find large rural areas with only limited access. In less developed countries, especially in rural areas, Internet access is only rarely available or completely absent. There are promising approaches to overcome the lack of coverage by Internet units installed on vehicles or the implementation of municipal Internet access, especially in the context of enabling better preparedness and response to disasters. But still we have to consider that this problem will not be easily solved in the short term.

Second, and most important, the Internet infrastructure is highly vulnerable. With EWS as part of critical applications for disaster management, the consequences of a likely failure of infrastructures before or during a disaster must be carefully considered. In recent years, we witnessed an increased use of the Internet infrastructure in almost every newly developed EWS. Even when failures of the main infrastructures are considered in the system design, secondary side effects are often not taken into account. Besides the potential for the implementation of powerful EWS, we should be aware of the risks for system availability by using the Internet as a foundation for these systems.

ICT Challenges in Early Warning Systems

EWS are generally highly demanding in terms of ICT requirements. Particularly in the field of short term EWS, we refer to ideal systems that must work reliably way in a 24*7*365 timeframe,

provide nearly real time processing, and have minimum failures in distributed ICT environments (sensors systems, communication links, processing components, and receiver infrastructures). An additional problem is that EWS must be designed for long term operation in a changing environment, which is a challenging requirement for a systems engineer.

We now identify the particular ICT challenges to the above defined layers of EWS:

1. **Monitoring and data collection:** The necessary monitoring and data collection infrastructure for risk detection has to be secure, reliable, timely, scalable and standardized. Security in this context means that the data sources such as sensors or databases can not be easily manipulated or modified externally. The data sources and the collection process should ensure a certain overall reliability in terms of data correctness, consistency and data provision. In an optimal EWS design the data collection is based on several sources that can substitute others in case of failure or inconsistencies. Especially in the case of short term EWS, the question of timely data supply is a crucial issue. For example, in meteorological casting systems using radar or satellite data the provision has to be provided within real time (including pre-processing and transmission) in order to provide adequate pre-warning times. In this case, these requirements have ruled out the use of satellite data until there is no higher measuring and transmission frequency available. Furthermore, the data sources should be exchangeable and extensible in order to ensure an evolutionary adaptation to changing environments and advancements in technology. Progress in this direction is evident in the increasing use of the Sensor Web Enablement Standard (SWE) for connecting single sensors or sensor systems to early warning systems (Botts, 2008). This

standard enables an open system design to new sensor technologies in the future and is highly useful for operation with a changing number of connected sensors (e.g., due to maintenance or extension of the sensor sources).

2. **Information processing:** The analysis, prognosis, and decision support processes within an EWS have to be quick and accurate in terms of time, location, and impact estimation. These requirements often force a compromise between speed and accuracy. For example, in the prognosis of a tsunami impact, the necessary processing and simulation of data - in particular GPS and seismic refinement - is taking considerable time within the warning timeframe. The solution is to use on demand clustering of computing resources and provide algorithms with iterative results with increasing accuracy and accompanying heuristic methods to define a quality seed from where a warning can be issued. In most fields, there are mainly domain specific restrictions to accuracy in terms of time, location and impact that cannot simply be solved by computing power. It is important to determine the minimal accepted accuracy, which can be also called the break-even-point of early warning, before implementing an EWS. This point defines the necessary quality requirements expressed in a false-negative and false-positive parameters for a defined spatial-temporal frame from which the provision of early warning starts to be useful. For example, the provision of meteorological alerts in Germany started to be accepted by the broader public when warnings were provided on the location accuracy of postal codes and the time accuracy of +/-5 minutes with over 90% impact probability based on now-casting alerts. Anything below these accuracy parameters would lead to a completely failed investment in implementing such an EWS.

3. **Generation of warnings:** The generated warnings should be comprehensive and supporting individual response capabilities of the individual receivers and receiver groups. As classical early warning is often understood as the broadcast of one message to deciders, response forces, and people at risk, the challenge is to provide individual warning messages according to the needs of each receiver. The reason is that the effectiveness of a warning is strongly dependent on the content in relation to the profile of a user and his/her current situation. This means that the user has to comprehend the content of the warning in the right way and should be provided with appropriate response measures that best fit his profile and current response capabilities. Only in this case it can be expected that the receiver takes the right response measures. The ICT problem of these requirements is far more complex than it can be expected at first glance. It comprises issues such as cognitive and behavioral requirement analysis, dynamic ontology-based and multi-lingual text generation and dynamic (context-aware) profile representations of single receivers and receiver groups. In terms of available alerting technologies, it has to be considered that the warning content is highly restricted to available channel technologies and bandwidth. Nevertheless, the system should be able to provide scalable warning messages to both individual receivers (multiple message approach) and receiver groups (broadcast approach).

4. **Alert dissemination:** The alerts and warnings issued by an EWS have to be highly accessible, targeted, secure, reliable, timely, and standardized. "Highly accessible" in this context means that the alerts should be ideally ubiquitous to all affected people. Even in highly industrialized countries with extensive communication infrastructures this

task is ambitious. And even with simple and effective solutions, such as sirens or radio, the ubiquity of alerts can be in conflict with the second requirement: targeted alerts. Targeting in this context is understood as the ability to (i) restrict alerts to a certain area (e.g., to prevent panic) and (ii) specifically address and convey message content to single receivers or receiver groups. To this end, more sophisticated alerting technologies than the classical siren and radio alerts are required. Analogous to the data collection requirements, alerting has to be reliable and timely in a distributed infrastructure, which also is a highly demanding task. The infrastructure of the alerting layer is usually highly distributed due to the need of supporting multiple channels, and the communication technology environment will change over time. The emergence of the Common Alerting Protocol (CAP) as a general standard provides the basis for distributed and expandable infrastructures in this field (Bottorell, 2006). Since this chapter focuses on alerting technologies, the challenges and solutions will be discussed later in more detail.

Evaluation Methods for Early Warning Systems

Given the actual state of early warning, more – and most likely the major part of - investment has to be made in the next decade. Despite the necessity to invest in better monitoring technologies it will be a major challenge to ensure the targeted and effective distribution of warnings to the people in danger, especially in the area of short-term risks. These improvements will be costly, and furthermore, the long-term operational costs of EWS in the future will become a major issue. In this context, the question of cost-effectiveness is becoming increasingly important. Decision mak-

ers will ask more and more about the utility and effectiveness of investments in EWS. Accordingly, the UN-ISDR recommendations for EWS research mention the importance of the development for cost-benefit models (UN/ISDR, 2008).

The challenge in this field is not so much to calculate the costs of an EWS, but more importantly to find appropriate evaluation criteria for the benefit of an EWS. Some work (Bayrak, 2007; De Groeve, 2005; Held, 2001, Sillem 2006) already has been done on calculating the possible benefit of an EWS on a macro-economic scale by comparing total economic damage and estimating possible mitigation percentages using early warnings. These models do not detail down to EWS performance and its benefit for certain scenarios and therefore are only partially helpful for decision-making. The aim is to find detailed evaluation criteria that take the EWS performance into account in order to calculate possible benefits.

Based on the work of (Bayrak, 2007), (De Groeve, 2005), (Held, 2001), and (Sillem 2006) who proposed performance measures for alerting systems, we developed a hierarchical set of parameters which enable us to measure the overall performance of an EWS but also to identify the performance of single aspects of the system. The five main parameters are:

1. **Frequency:** The average frequency of the occurrence of the specific disaster for a certain location.
2. **Accuracy:** The accuracy of correctly predicting a certain disaster in given response time corridors.
3. **Response:** The aggregated probability that warnings reach the receiver, are understood, and response action has been taken correctly.
4. **Prevention:** The aggregated probability that a certain preventable damage occurs during a disaster.
5. **Damage cost:** The average cost for a single damage scenario that can be prevented

Figure 2. Sub-parameters of accuracy and response

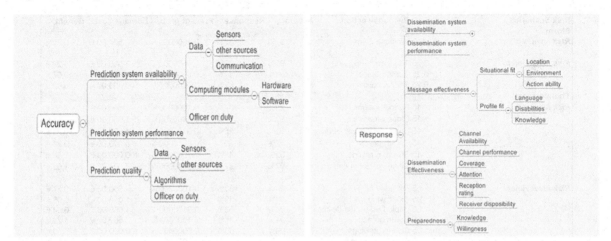

Especially the accuracy and response parameters depend on a variety of sub-parameters. Figure 2 shows the sub-parameters that influence the performance of a EWS in terms of accuracy and response.

Example:

We applied these criteria as a basis for a cost-benefit evaluation of two meteorological EWS WIND and SAFE. WIND is a weather warning service in Germany, Austria and Switzerland with 470,000 subscribed users. SAFE is a research pilot funded by the German Ministry of Education and Research (BMBF) that aims at improving meteorological EWS for better climate adaptation (Meissen, 2007). We used our criteria in order to compare the expected benefits of both systems.

Figure 3 shows an excerpt of the benefit calculation for storm events and response actions in SAFE. It calculates a yearly benefit for a single user per year. The estimations are based on single threat scenarios, statistics and surveys from insurance companies that take part in both projects. Our findings show that in SAFE the average response performance with around 40% is expected to be considerably higher than the one in WIND with measured 20%-30%. The accuracy of WIND is already between 85%-90% and can

only be increased slightly in SAFE. Looking at the necessary IT investments, namely better prognosis (accuracy) or warning dissemination (response) we suggest that the investment in the latter has a much stronger influence on the benefit in this case.

We believe that the presented evaluation framework can serve as a basis for general cost-benefit evaluation for EWS. Still work has to be done for the adaptation and generalization for other types of EWS. We see this as a first step towards a solid basis for decision making in the case of either the implementation of a new or the improvement of an existing EWS infrastructure. The model provides parameters for evaluating single aspects of EWS and can distinguish between user groups. With the increasing need for economic efficiency of EWS and the integration of private stake holders such an evaluation procedure for EWS will become necessary if not inevitable in the future.

REPRESENTATIVE SYSTEMS AND INTEROPERABILITY

In this section we present examples for international, national, regional, and local EWS that are representative for the different classes of EWS introduced above (3.1). The interoperability as-

Figure 3. Excerpt from the benefit calculation of SAFE

Risk Scenarios	Frequency	Response action	Accuracy	Response	Prevent prob.	Damage cost	Benefit
Storm							
Risk level red	2	Fix loose items	87%	37%	0,1000000%	500,00 €	0,32 €
Risk level violett	0,25	Fix loose items	93%	39%	0,5000000%	500,00 €	0,23 €
	0,25	Car in garage	93%	30%	0,0100000%	10.000,00 €	0,07 €
	0,25	Stay in buildings	93%	42%	0,0000005%	500.000,00 €	0,00 €
Thunderstorm							
Risk level red	15	Fix loose items	92%	37%	0,1000000%	500,00 €	2,55 €
	15	Close windows	92%	37%	0,0100000%	1.000,00 €	0,51 €
	15	Unplug electronic devices	92%	25%	0,0100000%	1.000,00 €	0,35 €
	15	Car in garage	92%	30%	0,0010000%	5.000,00 €	0,21 €
	15	Stay in buildings	92%	45%	0,0000005%	100.000,00 €	0,00 €
	15	Close valvets in cellar	92%	37%	0,0010000%	20.000,00 €	1,02 €
Risk level violet	1,5	Fix loose items	94%	39%	0,1000000%	500,00 €	0,27 €
	1,5	Close windows	94%	39%	0,0100000%	1.000,00 €	0,05 €
	1,5	Unplug electronic devices	94%	30%	0,0100000%	1.000,00 €	0,04 €
	1,5	Car in garage	94%	35%	0,1000000%	2.500,00 €	1,23 €
	1,5	Stay in buildings	94%	50%	0,0000005%	100.000,00 €	0,00 €
	1,5	Close valvets in cellar	94%	42%	0,0050000%	20.000,00 €	0,59 €
Local flooding							
Risk level red	0,5	Close valvets in cellar	82%	37%	0,0050000%	20.000,00 €	0,15 €
Risk level violet	0,2	Close valvets in cellar	88%	39%	0,0100000%	20.000,00 €	0,14 €
	0,2	Install barriers	88%	10%	0,0050000%	20.000,00 €	0,02 €
	0,2	Secure oil and gas	88%	15%	0,0050000%	250.000,00 €	0,33 €
Sum:							**8,09 €**

pects of EWS are then discussed along with these examples (3.2).

Selected EWS

Listing and categorizing existing EWS alone would probably fill a complete book on early warning systems. The UN/ISDR has generated a compendium of existing EWS (United Nations, 2006). Although this collection is by far incomplete, as it is mainly missing local and private sector EWS, it is nonetheless one of the best common overviews.

Global Level

On the global level the most prominent System is the Global Disaster Alert and Coordination System (GDACS) (GDACS, 2009). This multi-hazard EWS provides warnings on the global level for natural disasters, in particular floods, volcanoes, tsunamis, earthquakes, and cyclones. The system can be described as a meta-EWS that collects warnings from national warning systems and combines them in a common global system. The main dissemination platform is Internet-based, but it also offers subscription services for e-mail, fax, SMS, and RSS-feeds. The selection does provide localization on the global level (e.g., in Asia, North America, Europe), multi-lingual messages (e.g., in English, Spanish, French), hazard types and warning levels, thus providing a typical subscription-based alert dissemination facility. The system is provided by a joint initiative of the United Nations and the European Commission.

The system is connected to existing detection and alerting systems and automatically incorporates their warnings. Hence, it does not provide its own prediction components, but rather it has an impact estimation component that assigns the incoming events to a common alerting level model. The core system is based on ASGARD, a component-based monitoring and alert notification system that has been implemented by the Joint Research Center in 2005, providing a general tool set for early warning (Jacobsen, 2005). Additional components provide direct alerts issued from authorized national organizations.

The GDACS approach shows the major advantages and disadvantages of global EWS. The system provides a general, complete and – most importantly - standardized overview on global disaster for the covered hazard type, thus providing an important basis for international coordination of response and relief forces. The system concept has its advantages since it does not try to build up its own monitoring and prediction infrastructures - which would be technical and financially unfeasible. However, it is restricted to the collection and common presentation of existing national detection systems. This concept reaches its limits when it comes to effective alert dissemination to the public. Analogous to the dependency on existing disaster detection and prediction facilities, the system needs local alert dissemination systems that can send alerts to the people at risk in a more targeted way. The GDACS already provides the Common Alerting Protocol (CAP) interface in order to enable multiplicator systems, however, the connected local dissemination systems are missing.

International and National Level

A prominent example here is the enforced cooperation of different states after the 2004 tsunami in Indonesia, which aims at setting up a tsunami warning infrastructure in the Indian Ocean. Although the developments are mainly nationally driven, we can see a strong movement towards the sharing and harmonizing of infrastructures and systems. One of the main new developments is the German Indonesian Tsunami Early Warning System (GITEWS) that became pre-operational in November 2008. The system concept set up as one of the first EWS using the new Sensor Web Enablement (SWE) standard, thus providing a better basis for integrating heterogeneous sensor sources. This concept enables an easier integration

of the sensor system infrastructures of the different states active in earthquake and tsunami detection in this region. It can be expected that the national tsunami detecting and monitoring infrastructures will increasingly be incorporated. A major reason for this enforced cooperation is the high cost of the necessary sensor systems (i.e., seismic sensor, deep ocean sensors, GPS-sensors, buoys, and tide gauges). In this context GITEWS can be seen both as a national and international system in the sense that infrastructures are shared between different states. This is likely to be the future model of early warning infrastructures. On the other hand, it has to be expected that the decision making and alert dissemination process be kept strictly on the national level and will not be shared.

Another illustration of EWS defined at the national and international level is the weather hazard warning system WIND (Jaksch, 2003), which was developed and operated by private stakeholders. The WIND system was introduced by insurance companies in 2003 and is based on a research prototype that implements an EWS based on information logistic principles. The system provides users with precisely localized weather hazard alerts based on subscribed geo-positions (zip codes). The monitoring and prediction components include weather station networks as well as radar and satellite images for precise now-casting in nearly real-time. The warning generation and dissemination component provides alerts for a variety of media adapted to the profile of the user. The system has become successful with approximately 470,000 current users and increasing numbers of subscribers in Germany, Austria, Switzerland, Hungary, Slovakia and Poland. It is the largest commercial EWS in Europe. WIND is solely financed by the insurance industry, which expects a benefit in terms of damage reduction. This commercial dimension enables the extensive use of mobile network infrastructures for individual alerting that would usually be too costly to be provided by public authorities. Also the ability to consider profiles and dynamic locations of receivers for better alert adaptation is a feature that can be better realized by commercial providers (as long as privacy issues are well managed).

Regional and Local Level

Below the national level, another class of EWS provides warnings for regions, towns, an industrial plant, or even just for a local event. An example for such systems is the weather and flood warning system SAFE, a research prototype funded by the national research ministry in Germany (BMBF), which is aiming at extending the limits of local accuracy of warnings in a cost-effective fashion and at alerting effectively on a local level (Meissen, 2007). Furthermore, it controls actuators (e.g., activating an awning if there is too much sun). The system integrates dense local sensor networks (for better detection and monitoring) and pervasive alerting both for persons and systems. In a test field around the town of Mering in Bavaria and an industrial plant of Wacker Chemicals, specialized sensor networks improve the local warnings which are then adapted and distributed over a variety of channels according to the individual needs of the receivers (users and telematic systems). The system directly controls systems in buildings (closing windows or sewage valves) and a variety of alert channels (light signals, message boards, TV-Set-Top-Boxes, mobile devices, fire-alert systems). The adaptation of warnings is based on detailed profiles and context information about the receivers, and thus the message content can be much more targeted to the current need of the user. The system concept focuses on cost effective solutions that incorporate existing and inexpensive infrastructures (e.g., integration of private weather stations of households in the sensor network or using existing DSL-Routers for controlling systems in buildings) thus offering a

realistic base for the development of such systems both for commercial and public use.

A further step towards localized EWS is the IMSK system that started as a large scale EU research project (European Commission, 2009). The system aims at providing a mobile security toolkit for large scale events (e.g., sports events or political rallies). Its major tasks are to monitor, detect, and alert the organizers and the public in the event of any kind of threat. Hence, the system can partly be seen as an EWS. The system will set the basis for a new paradigm of a rapid, deployable, and component-based EWS that could be deployed anywhere where the demand of early warning is restricted to the location and duration of an event.

Although both systems are still in research phase, it is likely that SAFE and IMSK will both be the first representatives of a new generation of regional and local EWS that make full use of state-of-the-art and upcoming ICTs to provide effective local detection, monitoring, and alert capabilities.

Interoperability Issues

A large variety of approaches for implementing effective early warning infrastructure exists. The major questions are whether centralized or de-centralized warning infrastructures should be implemented, whether multi-hazard systems are more effective then single-hazard systems, whether private stakeholders can provide warning systems, or whether the challenges can only be solved based on international approaches. These questions have a considerable impact on the design and technology used for the implementation of future EWS. At the current stage, we estimate a strong trend towards distributed systems with specific functionalities. Interoperability and functional orchestration will be the key issues for the successful implementation of such infrastructures.

Figure 4. Proposal for alert dissemination

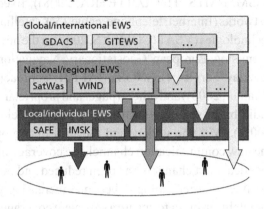

As seen in the examples above, the main synergy potentials for distributed EWS infrastructures are:

1. Shared sensor and monitoring infrastructures
2. Shared decision support modules
3. Shared alert dissemination capacities

In the first field, current EWS developments such as GITEWS are providing the basis for better interoperability based on the SWE standard. We expect that this standard will play an important role in the future for the implementation of shared sensor infrastructures. A successfully established standard in this field can open the door to new dimensions of detection and monitoring possibilities: thinking alone about all the existing sensors in our daily life (e.g., in cars, buildings, mobile devices, and infrastructures) gives us just a glimpse of the potential use of this data for the next generation of EWS.

In the second field we can identify high synergy potentials when it comes to shared decision support modules. The example of IMSK shows a demand of a new generation of highly flexible and modular EWS. Also in other fields the simulation, prediction, and impact estimation modules have to support an adaptive decision support process for different areas, infrastructures and user groups (e.g., flexible urban flood warning systems for traffic control authorities, civil services, industries and public). It would be highly beneficial if these

system concepts could be based on the flexible configuration and orchestration of shared modules especially, in the field of risk and impact estimation. Yet only in the field of GIS on the OGC and web service standards we find solutions that are flexible and integrated in EWS. For the future, we see a demand for common data models for simulation, prediction, risk, and impact estimation in order to have a better interoperability of EWS in this field. The upcoming Emergency Data Exchange Language (EDXL) standard might be able to provide a possible basis (Pack, 2008).

In the third field, we are just at the beginning of an inevitable and beneficial trend towards sharing alerting infrastructures. With the example of GDACS, reaching efficiently and targeting the persons at risk is a challenging and cost-intensive task. In general, regional, local and commercial systems can solve this problem better than systems on a higher level that are provided by authorities. The later systems can mainly provide broadcast-based alerting, whereas the former systems can easier provide targeted and individually addressed warnings to persons and systems (seen in the example of SAFE) that are much more effective for better comprehension and response (Meissen, 2008).

Figure 2 shows possible interoperability model for systems on different scale levels. In this model, alerts can be passed to the alerting components of systems on the next level that can provide better and more targeted alerting capacities on the last mile. Further, these systems can offer receiver group representations and dynamic service level parameters (coverage, availability, and so on) that can be used to orchestrate the warning process. A necessary prerequisite for this sharing of alerting capacities is the existence of an established standard for alerting. The Common Alerting Protocol (CAP) introduced in 2005 is filling this gap, enabling and easing the sharing of alert information. Nevertheless, the standard has its limits when it comes to the representation of detailed emergency information. We expect that the upcoming EDXL standard will provide a solution to this problem of missing ontologies that can be used for standardized, precise and multi-lingual alerting.

ALERTING TECHNOLOGIES

This section first presents the main challenges in alerting the public (4.1) before presenting the current alerting channels (4.2). It then gives a set of criteria that serves as a basis for comparing the technologies (4.3). Finally, we compare existing technologies in a systematic manner (4.4).

Challenges in Alerting the Public

For most of existing EWS, the efforts in the implementation were mainly focusing on the detection, monitoring and prediction of risks. The produced warnings of these systems are usually available for governmental authorities who are then responsible for the dissemination of alerts to the public. Often we witness major problems in the effective dissemination of these warnings to persons at risk. In less developed countries this is due mainly to missing infrastructures, but even in developed countries this task has been underestimated. Several projects have been initiated to implement better alerting mechanisms for the public (Improvement of EAS in the US (Moore, 2006), Cell-Broadcast-Alerting in the Netherlands (Sillem, 2006), the SATWAS in Germany (BBK, 2005), MyRescue in Japan (Government of Japan, 2006), IPAS in Canada (Seddigh, 2006), and others). In this context, newly available information communication technology (ICT) is offering greater potential as well as new challenges to effective warning. While classical public warning until the nineties was transmitted via sirens, loud speakers, mass media and partly radio-based receiver-specific solutions, new channels are now using digital broadcast technologies (digital radio and TV), mobile network technologies

(GSM, UMTS, TETRA (TETRA, 2008)), fixed networks (Internet, telephone, cable TV), satellite technologies (VSAT (COMSYS, 2007) or the new promising European Geostationary Navigation Overlay Service (EGNOS)[1] -ALIVE approach (Mathur, 2006)) and others (pager and proprietary radio-based solutions such as DCF77, EFR (Held, 2001)). With the variety and the shift of use of the new communication channel, the coverage of classical alert channels has been reduced and the use these new technologies becomes increasingly inevitable in order to ensure a sufficient coverage. Furthermore, it is now common sense between experts that a variety of channels should be used in parallel, also to make the warning process more resilient (FCC, 2007). With these technologies, other options – such as particularly addressing specific areas and user groups, or locating the receiver – can now be used to increase the efficiency of warning.

Aside from this existing potential, new challenges occur in providing systems that make full use of these heterogeneous channels in an intelligent, sustainable and cost-effective way:

- **Intelligent use of alerting channels:** First, as it is stated by the UN/ISDR (UN/ISDR, 2006), warning messages are often not sufficiently targeted to the users and therefore inefficient. The new communication technologies offer means to disseminate specified warnings to certain user groups down to the individual level. The main challenge will be to provide intelligent mechanisms for warning adaptation to the needs of the users. Second, a further problem arises with the possible - and in the case of disasters, most likely - failure of communication infrastructures (physical, congestion, failure of supporting infrastructure) (Townsend, 2005). In this context intelligent mechanisms for the efficient selection and resource allocations of available alerting channels have to be provided.

- **Sustainability:** A major part of new communication channels comprises the integration of private stakeholders. This fact makes public-private partnership (PPP) models inevitable and can be strongly beneficial for the sustainability of EWS, especially in terms of long-term operational and maintenance costs that are often underestimated in this area.
- **Cost-effectiveness:** The necessary infrastructure should be interoperable with existing warning systems and the synergy of using the same dissemination infrastructure among EWS should be exploited. A major milestone in this direction has been reached through the development of the Common Alerting Protocol (CAP), which offers a good basis for interoperability and is now adopted by several EWS. The task now is to identify synergies and realize interoperability between isolated EWS solutions (in the US alone we count eight different warning systems on the federal level (FCC, 2007)) in the area of alert dissemination. Furthermore, the cost-benefit ratio for the realization and maintenance of an alerting infrastructure has to be estimated. As it applies to all elements of EWS, not everything that is desirable is practical or cost-efficient in the area of alerting infrastructures.

Available Alerting Technologies

This section gives an overview of applicable alerting technologies (AT) for public alerts and warnings. The technologies can be generally divided into terrestrial-based (audio, cable network and radio network) and satellite-based communication. Furthermore, an orthogonal categorization considers the functional aspects such as alert systems (e.g., sirens, alarm systems in buildings), broadcast/information systems (e.g., digital/analogue radio/TV), telecommunication systems (e.g., telephone, GSM, TETRA, Internet, etc.), or telecontrol systems (e.g., DCF-77, EFR, Powerline, EGNOS).

Classical Alert Systems (AT1)

Classical alert systems, usually sirens, provide wide distance audio (public sirens) or local audio/light (building/industry alert systems) signals. Wide range alarm systems are only of interest for certain kinds of major alerts addressing the whole area, thus complementing other alert technologies. General disadvantages of sirens are the lack of coverage in major cities and densely populated areas, the lack of system conformity (local control only), and the dependency on electric power supply. Local information and alert systems at event sites (e.g., loud speakers of emergency forces, stadium loud speakers), in buildings (fire alarm systems) also exist. A major problem is then the possible lack of accessibility of such systems for external use. One of the major obstacles in terms of content is that they can usually provide only little comprehensible information (except for loud speakers). One major advantage of these basic alerting technologies is that they are usually less vulnerable to failure in comparison to more advanced and complex communication methods.

Radio and TV Broadcast (AT2)

Here we cover all kinds of radio and TV broadcast technologies: analogue, digital, terrestrial radio, cable or satellite. In the last decades, radio and TV have substituted classical siren alerts and are the media most often used to disseminate warnings and information. However, without a prior alert, these can reach the population only if receivers are turned on and tuned in to a station broadcasting the alert messages. In recent years, the coverage of the classical terrestrial radio and TV broadcasting decreased and is only partly substituted by digital broadcast or internet radio/TV. Further developments in the area of digital broadcasting are DVB-T, DVB-C and DVB-S. These channels provide general alerting possibilities to the public in two ways:

1. Classical alerting over the television program, like the classical alerting over analogue channels.
2. Using additional data/information channels. A major problem here is that effective alerting would require specific client software on the receiving devices which handle alerts. Technically, it even would be possible to provide a wake-up solution for alerts, thus overcoming the problem that the TV has to be switched on in order to receive alerts. In this case, with an alert client on the DVB-T, DVB-C or DVB-S receiver, incoming alerts could be processed and lead to a certain action (sound or switching on the TV). Note, however, that these technologies are not provided on the market.

Telecommunication Systems (AT3)

Under this category we cover a number of different technologies (radio terrestrial, cable, satellite) with the major aim of providing telecommunication services to the public, industries and authorities. We focus here on public accessible technologies.

Mobile Radio Networks

In this field GSM and UMTS are the leading standards for providing mobile communication with an almost full coverage of the population. Both network density and number of end user devices are continuously increasing.[2] In normal operation, cellular phones use individual point-to-point connections, but this is unsuitable for population alerts. One method used to distribute alerts and warnings in several warning and emergency notifications systems is to distribute SMS to end users. The main obstacle of this possibility is to prioritize SMS over voice traffic. However, GSM systems provide a

"cell broadcast" function: a broadcasting mode for the transmission of text over the data channel. This function makes it possible for all mobile radio units that are active in a cell of the mobile radio network to receive information simultaneously in the form of text shown on the display with a maximum of 93 symbols, similar to SMS (Short Message Service). The cells in which messages are broadcast are selected at the Cell Broadcast Centers (CBC) or gateways. The feasibility of Cell Broadcast for public alerts was extensively tested in the Netherlands (Sillem, 2006). An important aspect of GSM- and UMTS-based networks is the decentralized structure of the network and the possibility to imitate sender units for building an ad-hoc communication infrastructure. There are attempts in disaster management to substitute the failing mobile network (e.g., due to destruction or power failure) with local mobile senders addressing the still functional phone devices of the citizens in a defined area. This opportunity at the same time shows the vulnerability of the mobile network to destruction and malicious attacks. One method is that it requires a priori subscriptions of end users or the use of customer information by the telecommunication provider which is forbidden due to privacy regulations.

Another mobile communication technology applicable for public alerting are paging systems provided in some countries in Europe. These broadcast based systems offer also solutions for location-based alerting where the receiving device picks the relevant alert according to its location. Paging devices are in a larger scale used in Israel for alerting purposes (rocket and terrorist attacks), whereas fixed devices are installed in private and public buildings. Other states are considering similar systems for emergency notification. The main obstacle is the lack of coverage due to sparse end user devices.

In the future the increasing availability of smart phones with GPS in combination with UMTS and following standards with higher data transmission capacities or W-LAN-based Internet access can offer new dimensions for alert services based on client application accessed via the Common Alerting Protocol (CAP) over Push-Methods (XMTPP, SMS) or constant requesting. These alert services could offer the full range of precise targeted alerting, warning and information based on tracking information (GPS, Cell, W-LAN) and including the potential for rich situational feedback for command and control units.

Fixed Cable Networks (AT4)

Today's fixed cable networks mainly provide three services: telephone, cable TV and Internet. Nearly every household and every organization has an analog or digital telephone connection and this infrastructure has been proven to be robust in disaster cases (although this robustness is decreasing with the increasing use of power supply dependent devices such as digital or wireless phones). One of the major advantages of telephone devices is their always-on status and their wake-up ability especially during the night when almost no other communication technology can be used for alerting (most mobile phones are switched off). It can thus be assumed that the telephone would be suitable for use as an alert system. Unfortunately, however, the fixed network is only designed for point-to point connections; this does not allow rapid and simultaneous alerts. An exception to this may be the bell signals of analog connections. These could be generated almost in parallel by the telephone exchanges. The US invested extensively in the use of the public telephone networks for alerting (FCC, 2007). Today we witness a shift towards mobile network-based solutions. In countries where the telephone system infrastructure is not prepared for alert purposes it cannot be considered as a feasible alert channel.

Cable-based internet provides a large range of possible alert services from e-mail, RSS, instant messaging or voice over IP services. Although still mainly restricted to fixed devices such as PCs, the coverage should not be neglected, especially in working and organizational environments. Furthermore, the Internet increasingly provides the infrastructure for telematic applications (Internet of things). With the emergence of the Common Alerting Protocol (CAP) the foundations are set for common alert services for persons and systems (e.g., domotics, industries) in mobile and fixed environments. Furthermore, cable-based digital broadcast services such as DVB-C are, to a major extent, are accessible to the public and can provide an additional alerting technology discussed in the previous section. The general advantage of cable-based technologies lies in their protection against jamming attacks, but they are more vulnerable to physical destruction (caused by earthquakes or other incidents) and power failure.

Satellite Communication (AT5)

The advantage of satellite based digital broadcasting is its large coverage and its invulnerability in case of physical terrestrial incidents or power failures. The main problem is that for effective alerting certain client applications should be available on the receivers in order to make full use of the possibilities of this technology. In particular, feedback possibilities, which will be provided in the future (DVB-S2, RCS), would provide useful functions.

Other satellite communication technologies are not discussed here in detail since they are not accessible for the public to a great extent. Satellite communication systems play a crucial role as a backbone for national alerting infrastructures.

Telecontrol Systems (AT6)

Additionally, several domain specific communications systems for control purposes exist, such as the DCF-77 - a low-frequency radio transmitter operated by the German Federal Standards Laboratory, which provides standard civil time for a large number of radio controlled clocks used in everyday life – and the EFR a low-frequency radio system designed for electric power consumption control or power line systems. These systems can be used for alerting but they are dispersed on a national level and do not provide a relevant number of end user devices. In the field of telecontrol systems, an evolving satellite-based approach shows a significant potential for future alerting solution: The EGNOS-ALIVE approach. The availability of free bandwidth enables systems like EGNOS to broadcast additional communication messages that can be received by any modern GPS device (e.g., smart phones) (Mathur, 2006). This system would provide an excellent common available alerting system by overcoming the problems of terrestrial communication (vulnerability, national and provider dependencies, coverage) and satellite communication (missing end user devices) with the additional feature of localized alerts through the GPS unit. Still, the ALIVE-System is still in a concept phase, but we think that such promising technology should be considered for potential use.

Another telecontrol communication systems that can be used for alerting is the TMC system for the transmission of traffic information for navigation devices. These systems can be used for virtual blocking of roads and indirectly preventing individual transport from entering danger areas. Although yet not effectively applicable, the method provides great potential for the future.

Criteria for Comparing Alerting Technologies

Following, we list and briefly describe the criteria considered as relevant for estimating the applicability of alerting communication technologies for different aims and environments. They are derived from previous studies on alerting technologies (Held, 2001; McGinley, 2006, Ward 2000), our own studies, and the technical specification of the ETSI for Emergency Communications (ETSI, 2008).

Coverage

The ability of the communication technology to provide the same level of services to individuals depending on their location, mobility and environment.

Availability

The service level that the communication technology provides in terms of secured, long term, stable, 24/7 availability.

Capacity

The ability of the communication technology to disseminate alerts and warnings in time, given a high number of recipients.

Vulnerability

The exposability and the effectiveness of protection measures of the communication technologies to external threats, which can be human-caused (e.g., physical destruction by attacks, misuse, jamming, intrusion, or spoofing) or disaster-caused (e.g., direct physical destruction or secondary effects such as power failure).

Scalability

The ability of a communication technology to dynamically target alerts and warnings to a defined geographic area or to specific recipients and recipient groups.

Multi-Tasking

The ability of the communication technology to perform in parallel different alerting tasks.

Table 1. A comparison of alerting channels

	Coverage	Availability	Capacity	Vulnerability	Scalability	Multi-Tasking	Interoperability	Content	Cost& Feasibility
Classical alert systems									
Sirens	+	+	++	O[3]	-	--	-	-	O
Loud Speaker	O[4]	+	++	O[5]	-	--	O	+	O
Building alarm systems	-	+	++	O	+	--	-	-	O
Radio/TV broadcast									
Radio	+	++	++	+	--	--	O	+	+
TV	+	+	++	+	--	--	O	+	+
RDS/DAB	O	+	++	+	+[6]	O	+	+	++
DVB-T/C/S (program)	+	++	++	+	O	O	O	+	+
DVB-T/C/S (data)	+	++	++	+	+	+	+	++	-
Telecommunication Systems									
SMS	+	O	O	-	+[7]	+	+	+	+
Cell Broadcast	+	O	+	-	+[6]	O	+	+	O
Pager	-[8]	+	+	O	+	O	+	+	+
Hybrid (Smartphone, GPS, UMTS, WLAN Internet)	O[9]	O	+	-	++	++	+	++	O
Telephone (ETAS)	+	++	O	+	+	O	+	O	-[10]
Internet (Alert clients)	-	O	+	O	+	++	+	++	+
Telecontrol Systems									
EGNOS-ALIVE	+	++	++	+	+	O	+	O	O
TMC	O	+	+	O	+	O	+	O	+

Interoperability

The ability of the communication technology to be integrated in an early warning and alerting infrastructure in a flexible manner.

Content

The ability of the communication technology to provide (a) alerts, (b) warnings, (c) information, and (d) bi-directional communication.

Cost-Effectiveness and Feasibility

The costs of a communication technology. It should be free of charge for the recipient of alerts and warnings and low for the provision of service.

Comparison of Existing Alerting Technologies

In Table 1 we compare the discussed communication technologies in regard to the previously derived criteria for alerting.

The overview shows that none of the communication technologies can provide an optimal solution. Therefore, we have to consider multichannel approaches as the only feasible solution for optimal alerting the public.

FUTURE CHALLENGES

Early warning systems design and implementation is an evolving and highly challenging field for software engineering in the near future. These systems are extremely demanding in terms of design and operability. In particular, general EWS solutions have to be provided for the following requirements:

1. The necessary monitoring and data collection infrastructure for risk detection has to be secure, reliable, timely, scalable and standardized.
2. The analysis, prognosis, and decision support processes within a EWS have to be timely and accurate in terms of time, location, and impact estimation.
3. The generated warnings should be comprehensive and support individual response capabilities of the individual receivers and receiver groups.
4. The alerts and warnings issued in an EWS have to be highly accessible, targeted, secure, reliable, timely, and standardized.

In general, we witness a strong demand for effectiveness and hence better interoperability of the upcoming solutions. The main challenges of interoperability are:

1. shared sensor and monitoring infrastructures
2. shared decision support modules
3. shared alert dissemination capacities

Even though existing and upcoming standards will provide the basis for the necessary innovation, it will be a challenging task to offer practical, efficient and stable solutions in this domain. As far as alerting technologies are concerned, the challenge in the future will be to provide hybrid (multi-channel) solutions that integrate and orchestrate multiple alerting technologies in order to optimize alerting infrastructure along the nine criteria mentioned previously. In particular, new ICT in the field of mobile devices and telematics have to be integrated in new concepts for efficient alerting. Context-aware alerting based on these technologies will be a key innovation in this field, providing a promising approach for targeted and situation-adaptive alerting that increases response effectiveness.

Future research in the field of EWS has to be jointly performed in an interdisciplinary and cooperative approach. Domain experts for particular hazards, together with sensor specialists and computer scientists, must explore new monitoring technologies. A major task of this research will be to find cost-effective ways to collect necessary data, integrate additional information sources and process large amounts of data in almost real-time. New sensor and communication technologies have enabled considerable progress in this field in the last decade but there is still potential for groundbreaking innovation. In particular, micro-sensing, low energy consumption and near-field communication in sensor network technologies are promising approaches for cost-effective and pervasive data collection, especially for improving the regional forecast of natural hazards. GRID computing might offer solutions for real-time information processing of large amounts of data. Research in the field of distributed system architectures, complex event processing and information logistics will lead to new system design for better system stability, interoperability and sustainability. The overall task enabled by this research will be to provide common design principles and components for high performance and long-term EWS-infrastructures as well as the enabling of synergies between different EWS in interoperable infrastructures (e.g., sharing sensor or alert dissemination facilities). The largely unsolved problem of efficiently alerting the population will remain a challenging research task. On the technical side, the efficient use of new communication technologies, the orchestration of multi-channel alerts, and the use of multi-modal and context aware alert messages should be explored. This research must be interdisciplinary, accompanied with sociological and psychological studies on the comprehension of warning messages. Due to the vast technological and cultural differences in the world, this research has to be adaptable for different and heterogeneous environments. In particular, the combination of technological and sociological solutions, such as the role of human

alert multiplicators in different cultural environments, have to be examined. Finally, general evaluation methodologies for EWS have to be jointly established, based on interdisciplinary research, in order to have a common evaluation measure for the efficiency of future EWS and, hence, the joint practical application of all above mentioned research directions.

CONCLUSION

In this chapter, we presented a broad introduction to the ICT-related problems and challenges of early warning systems and alerting technologies. Starting from the definition and classifications in this field, we described general approaches, representative systems, and interoperability aspects of EWS. As one of the most challenging tasks – and still an insufficiently resolved issue – the problem of effectively alerting the public was discussed in detail. We have seen that there is and – most likely –will be no single alerting technology that optimally fulfills all requirements for effective alerting. Therefore, we have to aim at systems that integrate and orchestrate multiple alert channels. In the future, ICT can lead towards more effective alerting through context-aware alerting, but a significant amount of research still needs to be carried out in this area until practical solutions are accessible to the broader public.

Finally, we have to be aware that all mentioned challenges and solutions have to consider cost-benefit constraints. Not everything desirable is technically and financially feasible, especially in EWS. In order to be able to evaluate proposed solutions we need a common criteria model for the effectiveness of early warning systems. The foundations of this model were presented in this chapter, in particular a common view on domain-dependent, organizational, and technical issues.

REFERENCES

Basher, R. (2006). Global Early Warning Systems for Natural Hazards: Systematic and People-centred, *Philosophical Transactions of the Royal Society*, (364), 2167-2182.

Bayrak, T. (2007). Performance Metrics for Disaster Monitoring Systems. In B. Van de Walle, P. Burghardt & C. Nieuwenhuis (Eds.), *Proceedings of the 4th International Conference on Information Systems for Crisis Response and Management ISCRAM2007* (pp. 125-132). Newark, NJ.

BBK. (2005). *Opportunities for Public Safety in Germany*. Bonn, Germany: German Federal Agency for Public Safety and Disaster Preparedness BBK.

Bottorell, A. (2006). The Common Alerting Protocol: An Open Standard for Alerting, Warning and Notification. In B. Van de Walle & M. Turoff (Eds.), *Proceedings of the 3rd International IS-CRAM Conference* (pp. 497-503). Newark, NJ.

Botts, M., Percivall, G., Reed, C., & Davidson, J. (2008). OGC® Sensor Web Enablement: Overview and High Level Architecture (LNCS 4540, pp. 175-190). Berlin/Heidelberg: Springer.

COMSYS. (2007). *The COMSYS VSAT Report*. Retrieved January 2, 2008, from http://www.comsys.co.uk/vsat_rep.htm

De Groeve, T., & Eriksson, D. (2005). *An Evaluation of the performance of the JRC Earthquake Alert Tool*. DG Joint Research Centre of the European Commission. Retrieved June 30, 2009, from http://dma.jrc.it/services/gdas/Performance_of_Earthquake_Alert_Tool.pdf

ETSI. (2006). Emergency Communications (EM-TEL); Requirements for communications from authorities/organizations to individuals, groups or the general public during emergencies. *ETSI TS 102 182 V1.2.1*. Retrieved June 30, 2009, from http://portal.etsi.org.

European Commission. (2009). *IMSK –Integrated Mobile Security Kit*, European Research Framework Program 7. Retrieved June 30, 2009, from http://ec.europa.eu/enterprise/security/doc/fp7_project_flyers/imsk.pdf

FCC. (2007). *Review of the Emergency Alert System. Report of the Federal Communications Commission*. Washington, D.C.: Federal Communications Commission.

GDACS. (2008). *Global Disaster Alert and Coordination System GDACS*. Retrieved June 30, 2009, from http://www.gdacs.org

Glanz, M. H. (2004). Usable Science 8: Early Warning Systems: Do's and Don'ts. National Center for Atmospheric Research Report of workshop, 20–23 October 2003, Shanghai, China. Boulder, CO.

Government of Japan. (2006). *Japan's Natural Disaster Early Warning Systems and International Cooperative Efforts*. Tokyo, Japan: Technical Report - Early Warning Sub-Committee of the Inter-Ministerial Committee on International Cooperation for Disaster Reduction.

Häkkinen, M. T., & Sullivan, H. T. (2007). Effective Communication of Warnings and Critical Information: Application of Accessible Design Methods to Auditory Warnings. In B. Van de Walle, P. Burghardt & C. Nieuwenhuis (Ed.), *Proceedings of the 4th International Conference on Information Systems for Crisis Response and Management ISCRAM2007* (pp. 167-171). Newark, NJ.

Held, V. (2001). Technological Options for an Early Alert of the Population, *Zivilschutzforschung* [Bonn, Germany.]. *Zentralstelle für Zivilschutz BBK*, *45*, 64–130.

Jacobsen, M. (2005). ASGARD – System Description 1.1, EU Joint Research Centre. Retrieved June 30, 2009, from http://dma.jrc.it/new_site/documents/AsgardSystemDescription.pdf

Jaksch, S., Pfennigschmidt, S., Sandkuhl, K., & Thiel, C. (2003). Information Logistic applications for information-on-demand scenarios: concepts and experiences from the WIND project. In *Proceedings of the 29th Conference on EUROMICRO* (pp. 41-147). Belek, Turkey.

Mathur, A. R., Ventura-Traveset, J., Montefusco, C., Toran, F., Plag, H.-P., & Ruiz, L. (2006). Provision of emergency communciation messages through SBAS: the ESA ALIVE concept. In *ION GNSS 2005* (pp. 2969–2975). Long Beach, California: Proceedings.

McGinley, M., Turk, A., & Benet, D. Design (2006). Criteria for Public Emergency Warning Systems. In B. Van de Walle, P. Burghardt and C. Nieuwenhuis (Ed.), *Proceedings of the 3rd International Conference on Information Systems for Crisis Response and Management ISCRAM2006* (pp. 154-164). Newark, NJ.

Meissen, U., Auge, J., & Fengler, M. (2007). SAFE - Sensor-Actuator-based Early-Warning System for Hazard Protection in Extreme Weather Conditions. In *Proceedings of the 1st Conference on Research for Climate Protection and Protection from Climate Impacts klimazwei,* (pp. 56-57). Berlin, Germany.

Meissen, U., & Voisard, A. (2007). Situation-based Alerting Strategies in Early Warning Systems. In *Proceedings of the International Conference Wireless Applications and Computing IADIS 2007,* Lisbon, Portugal.

Meissen, U., & Voisard, A. (2008). Increasing the effectiveness of early warning via context-aware alerting. In F. Friedrich (Ed.), *ISCRAM 2008, 5th International Conference on Information Systems for Crisis Response and Management. Proceedings. CD-ROM: May 4-7, 2008* (pp. 431-440). The George Washington University. Washington, DC.

Moore, L. K. (2006). *Emergency Communications: The Emergency Alert System (EAS) and All-Hazard Warnings. Congressional Research Service Report.* Washington, DC: Library of Congress.

Pack, D., & Coleman, C. (2008). Assessing interoperability in emergency management standards. In *Southeastcon, 2008* (pp. 334–339). Huntsville, AL: IEEE.

Seddigh, N., Nandy, B., & Lambardis, J. (2006). An Internet Public Alerting System: A Canadian Experience. In B. Van de Walle, P. Burghardt & C. Nieuwenhuis (Eds.), *Proceedings of the 3rd International Conference on Information Systems for Crisis Response and Management ISCRAM2006* (pp. 141-146). Newark, NJ.

Sillem, S., & Wiersma, E. (2006). Comparing Cell Broadcast and Text Messaging for Citizen Warning. In B. Van de Walle, P. Burghardt & C. Nieuwenhuis (Ed.), *Proceedings of the 3rd International Conference on Information Systems for Crisis Response and Management ISCRAM2006* (pp. 147-153). Newark, NJ.

TETRA. (2008). Terrestrial Trunked Radio TETRA. Retrieved June 30, 2009, from http://www.tetra-association.com

Townsend, A. M., & Moss, M. L. (2005). *Telecommunications Infrastructure in Disasters: Preparing Cities for Crisis Communications.* Graduate School of Public Service, New York University.

UN/ISDR. (2006). Developing Early Warning Systems: A Checklist. *Third International Conference on Early Warning EWC III.* Bonn, Germany.

United Nations. (2006). *Global Survey of Early Warning Systems.* Bonn, Germany: ISDR Platform for the Promotion of Early Warning PPEW.

Ward, P. (2000). *Effective Disaster Warnings.* Washington, DC: Report of the Working Group on Natural Disaster Information Systems – Sub-comitee on Natural Disaster Risk Reduction.

ENDNOTES

[1]   The European Geostationary Navigation Overlay Service (EGNOS) is a satellite based augmentation system (SBAS) under development by the European Space Agency, the European Commission and EUROCON-TROL. It is intended to supplement existing systems, including Galileo when it will be operational.

[3]   Problem of power supply, except for mobile sirens and load speakers

[4]   Depending on the scenario: + in a stadium, - in a city

[5]   Problem of power supply

[6]   Distinction of geographic areas

[7]   Distinction of user groups

[8]   Missing end user devices

[9]   Missing devices and clients yet

[10]   Depending on the availability of ETAS in the telephone network infrastructure

Chapter 9
MedISys:
Medical Information System

Jens P. Linge
Joint Research Centre of the European Commission Institute for the Protection and Security of the Citizen Global Security and Crisis Management Unit, Italy

Ralf Steinberger
Joint Research Centre of the European Commission Institute for the Protection and Security of the Citizen Global Security and Crisis Management Unit, Italy

Flavio Fuart
Joint Research Centre of the European Commission Institute for the Protection and Security of the Citizen Global Security and Crisis Management Unit, Italy

Stefano Bucci
Joint Research Centre of the European Commission Institute for the Protection and Security of the Citizen Global Security and Crisis Management Unit, Italy

Jenya Belyaeva
Joint Research Centre of the European Commission Institute for the Protection and Security of the Citizen Global Security and Crisis Management Unit, Italy

Monica Gemo
Joint Research Centre of the European Commission Institute for the Protection and Security of the Citizen Global Security and Crisis Management Unit, Italy

Delilah Al-Khudhairy
Joint Research Centre of the European Commission Institute for the Protection and Security of the Citizen Global Security and Crisis Management Unit, Italy

Roman Yangarber
University of Helsinki, Department of Computer Science, Finland

Erik van der Goot
Joint Research Centre of the European Commission Institute for the Protection and Security of the Citizen Global Security and Crisis Management Unit, Italy

ABSTRACT

The Medical Information System (MedISys) is a fully automatic 24/7 public health surveillance system monitoring human and animal infectious diseases and chemical, biological, radiological and nuclear (CBRN) threats in open-source media. In this article, we explain the technology behind MedISys, de-

DOI: 10.4018/978-1-61520-987-3.ch009

scribing the processing chain from the definition of news sources, scraping and grabbing articles from the internet, text mining, event extraction with the Pattern-based Understanding and Learning System (PULS, developed by the University of Helsinki), news clustering and alerting, to the display of results. The web interface and service applications are shown from a user's perspective. Users can display world maps in which event locations are highlighted as well as statistics on the reporting about diseases, countries and combinations thereof and can apply filters for language, disease or location or filters with orthogonal categories, e.g. outbreaks, via their browser. Specific entities such as persons, organizations and locations are identified automatically.

INTRODUCTION

In many fields, professionals need to scan vast quantities of information from multiple sources on a daily basis, e.g. journalists have to keep up with incoming news stories, press officers need to react quickly to evolving stories, and investors follow the latest developments affecting the stock markets.

In the area of Public Health (PH), national and international authorities continually monitor the widest possible set of available sources of information. In their daily surveillance routine, public health authorities use indicator-based and event-based surveillance tools to identify evolving public health threats and to track ongoing incidents [Paquet et al. 2006]. A broad range of threats needs to be covered, from outbreaks of communicable diseases, terrorism cases such as the deliberate release of biological or chemical agents, contaminations of food and feed to chemical or nuclear incidents. Public health authorities employ experts in these domains who monitor all available sources of information. Timely monitoring is critical for the risk assessment.

Indicator-based surveillance systems collect structured data from health-care centres, clinicians, etc., and propagate it through official channels to the top-level authorities. Event-based surveillance refers to the activity of monitoring a much wider range of unstructured sources to detect actual or perceived threats to public health. Event-based monitoring within public health authorities is becoming increasingly focused on event detection using informal media sources (news wires, online newspapers, specialist blogs, etc.) on the internet.

Event-based media-monitoring systems can greatly facilitate the work of the analysts by identifying potentially relevant news items [Steinberger et al. 2008]. These systems can be broadly classified in news aggregators, automatic systems and moderated systems [Linge et al. 2009]. News aggregators collect articles from several sources, usually filtered by language or country. Most news aggregators simply retrieve RSS (Really Simple Syndication) feeds from news providers, e.g. online versions of newspapers. Automatic systems such as MedISys, HealthMap [Freifeld et al. 2008] and BioCaster Global Health Monitor [Collier et al. 2008] collect and automatically analyse articles to facilitate the work of the analysts. Automatic systems differ in the range of information sources, their language coverage, the speed of delivering information, level of analysis and visualization.

Moderated systems rely on dedicated teams of human analysts who manually scan and analyse the retrieved documents. GPHIN (Canada) [Mykhalovskiy et al. 2006], ProMED-Mail [Madoff 2004], and Argus [Wilson et al. 2008] (USA) are examples of moderated systems.

The trade-offs between automatic processing and human moderation lie in speed, accuracy of analysis, and cost. Human moderation is resource-intensive; GPHIN and Argus employ dozens of highly-qualified (and highly-paid) analysts to

Figure 1. A complex news web page with menus and advertising (www.usatoday.com). MedISys needs to extract the news article before being able to process its text.

process news retrieved by aggregators or in-house custom tools. ProMED-Mail avoids the cost of maintaining an analyst base by relying on volunteer moderators around the world. For this reason, it typically lags behind the original appearance of a story in the news media, usually by several days.

Automated systems are much faster and cheaper to operate, but are more prone to false positives. This is a burden for the human analysts who are required to separate signal from noise. All automated systems strive to improve the signal-to-noise ratio, developing increasingly sophisticated processing techniques.

EUROPE MEDIA MONITOR

MedISys is one of the publicly accessible systems of the Europe Media Monitor (EMM) [Steinberger et al. 2009] family of applications. MedISys came online in August 2004 and has since been continuously extended. NewsBrief and MedISys detect breaking news and short-term trends for early alerting and display up-to-date category-specific

news. While MedISys focuses on the public health domain, NewsBrief and the other EMM systems cover a wider variety of domains, with much less detail in public health. NewsExplorer focuses on daily overviews, long-term trends, in-depth analysis and extraction of information about people and organizations. EMM-Labs is a collection of more recent developments and includes various tools to visualize the extracted news data. For NewsBrief and MedISys, there are different access levels, distinguishing the entirely public web sites from an internal website for the European Commission. The public websites do not contain commercial sources and may have slightly reduced functionality. The public web pages of MedISys and NewsBrief are updated every ten minutes.

All EMM applications have in common that they receive the news articles from the EMM news engine, which gathers a daily average of approx. 90,000 news articles in 50 languages (figures of July 2009), from about 2,200 hand-selected news sources, several hundred specialist and government websites, and twenty commercial news providers. EMM visits the news web sites up to every five minutes to detect the latest articles, depending on the defined update frequency for each news website. EMM reads RSS feeds or HTML pages, depending on availability. Extracting full text from HTML pages is not a trivial task; the news text is usually embedded in a complex HTML page that may contain menus, links to other articles or pages, and advertising (see Figure 1). All news items are converted to Unicode and fed into a processing chain in which each module adds additional information. Internally, news articles are stored in UTF-8-encoded RSS format.

The system uses scraping rather than crawling news sites. This has several reasons: (a) scraping allows the robot to visit only the relevant pages of the newspaper and to ignore pages that are typically not relevant for the EMM user groups (e.g. celebrities, gossip); (b) it ensures that only the most recent articles are taken and that the robot does not enter the archives that contain old articles; (c) the burden on the web service providers is reduced drastically; EMM does not represent a burden for the servers of news providers and is thus usually not excluded by the providers from gathering news articles.

STATISTICS AND TEXT PROCESSING IN MEDISYS

All EMM news items are fed into MedISys instantly. Every ten minutes and in each of the languages, the application clusters the latest news items considering each time a window of four hours (or more, depending on the number of recent articles) and presents the largest clusters as the current top-ranking media themes, referred to as Top Stories. Documents are represented as frequency lists of words from which stop words are automatically removed. Each document that arrived within the last four hours is compared to each of the others by using the cosine similarity

measure. Documents are clustered using a hierarchical bottom-up group-averaging clustering algorithm where all clusters with certain homogeneity are retained. The title of the cluster's medoid (the article closest to the cluster centroid) is selected as the most representative title and thus as the title for the cluster. The top stories section of MedISys thus shows at any given moment which public health-related themes are being discussed most often across the abundant media sources world-wide. Unlike the news categories, which contain news about a specific disease or health threat, the clusters are formed automatically and in an unbiased manner. They may talk about a disease outbreak, a health threat relevant for a public event or new medical insights, or cover any other possible health subject that may be discussed in the news. All current clusters are automatically compared to the clusters produced in the previous round (i.e. produced 10 minutes earlier). If at least 10% of the articles overlap between a new cluster and any of the previous ones, the clusters get linked to form stories (different clusters linked over time) and those articles that have fallen out of the current four hour window are attached to the current cluster. When new articles on the same subject arrive, all related articles are cumulatively added to the cluster. A graph (see Figure 2) visualizes the development of a story over time, showing both the number of articles accumulated over time (the black rising line) and the amount of articles coming in at given times of the day (each represented by a blue bar). The red line shows the number of articles in a sliding four-hour window. The graph on the top stories page in MedISys shows the development of the ten biggest stories at the moment and thus gives an overview of the current state of (media) affairs per language around the world.

Larger new clusters (without overlap to previous clusters) and clusters of a rapidly increasing size are automatically classified as breaking news, which are automatically emailed to subscribed

Figure 2. Graph showing the development of a news story (about a new flu virus in Canada) over a period of a few hours (left), and of the major location mentioned in the articles (right).

Story Edition - Yet another new flu virus emerges in Canada

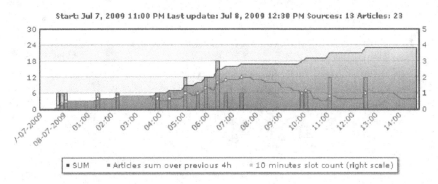

users. The statistical breaking news detection algorithm uses information on the number of articles and the number of different news sources, comparing the news of the last 30 minutes with longer periods of time.

Each article is geo-tagged, i.e. potential location place names are identified by comparing each uppercase word in the text with the entries in a multilingual gazetteer (also considering morphological variants to ensure proper recognition in highly inflected languages), and ambiguities are resolved. Resolving ambiguities is necessary because place names are often homographic (same spelling) with common words (e.g. there is a place called And in Iran and a city called Split in Croatia), with people's names, and even with other locations. For instance, world-wide there are 15 places called Paris, 102 places called San Francisco and 195 places called Victoria. An algorithm that considers the place hierarchy (city is part of a region, which is part of a country) and how often it is mentioned determines the major location in each cluster. While the system cannot know whether locations mentioned are the place of the event or the place of publication, the most frequently mentioned location in the news cluster typically is the event location, as this is the only location that is the same across all news sources.

For details about the geo-tagging algorithm, see [Pouliquen et al. 2006].

The geo-tagging result is used to visualize the location of the current news items on a geographical map (see Figure 3 and the map on the right side of Figure 2). The MedISys screenshot in Figure 3 shows that languages have a regional bias so that they differ quite a lot in the world regions they report about. The regional complementarity of the different reporting languages shows how important it is to monitor the media in different languages.

All news items are additionally categorized into hundreds of pre-defined categories. Categories include geographic locations such as each country of the world, organizations, and diseases such as swine flu, anthrax or thyroid cancer. Articles are classified in a category, if they satisfy the category definition which consists of Boolean operators, proximity operators and wild cards. Cumulative positive or negative weights can be used with a threshold. Many categories are defined with the help from institutional users.

The system keeps statistics on the 14-day average number of articles in any given country-category combination (e.g. Poland-tuberculosis). If the number of articles for this combination found in the last 24 hours (normalized by weekday fluctuations) is significantly higher than this aver-

Figure 3. Map showing the event locations mentioned in the biggest live news clusters in different languages. Colors and language codes indicate the reporting language of the news, making it clear that incidents in some geographic areas of the globe are reported in only one or two languages. In this snapshot, Latin America is only covered by Spanish and the Middle East is almost exclusively covered by Arabic news.

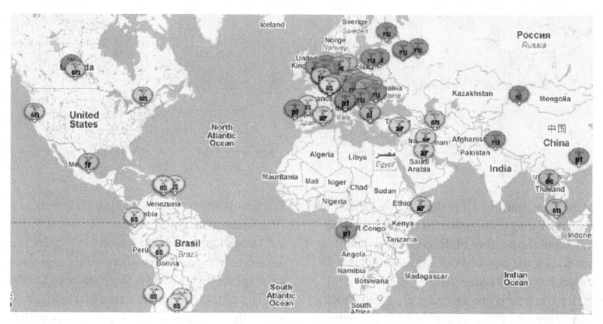

age, a country-category-specific alert is triggered and users are notified using ranking graphs (see Figure 4) or email notifications. Ranking graphs exist for the whole globe and for selected regions (e.g. European Union or South America).

As categories are defined in several languages (depending on user interest, some are defined in all, others in only a few languages), the statistics are language-independent. Since the system will detect a sudden rise in any of the languages, users may see any change in a category even before the event is reported in their own language. For humanitarian and public health institutions with a strong interest in early warning, this aspect is particularly important.

EXTRACTING DETAILED INFORMATION ABOUT MEDICAL OUTBREAK EVENTS USING PULS

The PULS system (http://puls.cs.helsinki.fi/medical/) is being developed at the University of Helsinki in collaboration with MedISys. The PULS system traces its origins to the IFE-BIO Project, which initially aimed at analysing events reported in ProMED-¬Mail [Grishman et al. 2003]. PULS tracks the occurrence of communicable human, animal and plant diseases, at present covering over 1,500 base terms related to infectious diseases and disease agents, with a total of 2,500 variants. The focus in PULS is on the analysis of news texts for information extraction, aggregation, and visualization. PULS is fully automatic and does not rely on human intervention. MedISys is its main source, and PULS employs natural language processing (NLP) methods for analyzing the news stream to

Figure 4. Live alert statistics for Asia, going from the left (high alert) to the right (lower alert level). Each pair of bars shows the expected number of articles and confronts it with the observed number of articles in the past 24 hours. The higher the differences between these two values are, the higher the alert level.

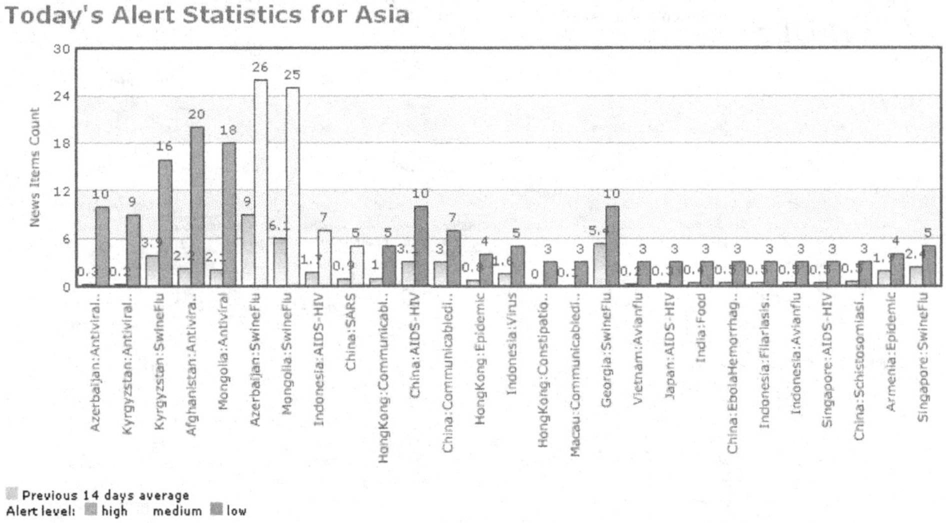

Today's Alert Statistics for Asia

build a database of cases about epidemiological events.

A single case describes an incident involving an occurrence of a certain disease at a certain location. The system tries to extract the disease and location name from the text, as well as other descriptive information about the case: the number of victims, whether they were human or animal, whether they died from the disease, etc. The rationale behind using linguistic analysis is that it can provide more detailed information than category-based searches, but this comes at a higher cost, as it requires a greater effort to customize the linguistic analysis system.

The output of PULS is a spreadsheet-like view of the fact base (see Figure 5). Facts are updated every 15 minutes based on new incoming information from MedISys via an RSS tunnel, and are sent back for integration with MedISys via the same tunnel. Linguistic coverage is primarily English (also covering ProMED-mail as source); French has recently been introduced.

PULS' average daily extraction rate varies from 300 entries during average periods to over 1000 per day during important events, e.g. the 2009 Swine Flu pandemic.

Further objectives include expanded multi-lingual support (with additional extensions to Spanish, and possibly Russian, Chinese), trend analysis, and data visualization. An important consideration will be the inclusion of analysis of indirect indicators, in conjunction with MedISys.

BLOG MONITORING

Blogs have become a mainstream distribution channel for various contents. Several free sites exist that allow everyone to publish a blog with a few mouse clicks. In the area of public health, blogs from scientists, experts, journalists, etc. are of interest since they often report new incidents quickly. Blog entries are usually displayed in reverse-chronological order and mostly contain text, but may also include pictures, video, audio (e.g. podcasts). An important feature of blogs is the ability for readers to leave comments in an interactive format.

Figure 5. A screenshot of the PULS interface, showing extracted cases, in reverse chronological order of publication. The base can be queried by specific fields, e.g., for certain countries, diseases, time intervals, etc.

		Published	Source	Disease	Country	Begin	End	Total	↑	Descriptor
[20]	+	2009.07.10	bbc_en	Influenza	Serbia	2009.07.09	2009.07.09	34		34 cases
[1047]	+	2009.07.10	bbc_en	Influenza	UK	2009.07.09	2009.07.09	--		The girl
[4963]	+	2009.07.10	icWales	Swine Flu	UK		2009.07.10	2009.07.10	279		about 279 cases
[4963]	+	2009.07.10	guardian	Swine Flu	UK		2009.07.10	2009.07.10	14	↑	..
[4963]	+	2009.07.10	YorkshirePost	Swine Flu	UK		2009.07.09	2009.07.09	14	↑	FOURTEEN people
[4963]	+	2009.07.10	YorkshirePost	Swine Flu	UK	2009.07.09	2009.07.09			,000 confir...
[82]	+	2009.07.10	MaltaToday	Swine Flu	Malta					ases
[82]	+	2009.07.10	MaltaToday	Swine Flu	Malta	2009.07.07	2009.07.07	10		Ten new cases
[82]	+	2009.07.10	MaltaToday	Swine Flu	Malta	2009.07.06	2009.07.06	3		three people
[82]	+	2009.07.10	MaltaToday	Swine Flu	Malta	2009.07.06	2009.07.06	2		two men
[82]	+	2009.07.10	MaltaToday	Swine Flu	Malta	2009.07.06	2009.07.06	1		a 24-year-old woman
[1047]	+	2009.07.10	channel4news	Influenza	UK	2009.07.10	2009.07.10	43	↑	10 July 2009 Source ..
[4963]	+	2009.07.10	channel4news	Swine Flu	UK	2009.07.10	2009.07.10	9 718		at least 9,718 confi...
		2009.07.10	googlenewshealth	Escherichia Coli	USA/Minnesota	2009.07.09	2009.07.09	2		two people
		2009.07.10	trinidadexpress	Swine Flu	Trinidad and Tobago	_	2009.07.10	65		65 laboratory confir...
[26]	+	2009.07.10	msnbchealth	Escherichia Coli	USA	2009.07.09	2009.07.09	69		at least 69 people
[11]	+	2009.07.10	moreoverhealth	Swine Flu	European Union	2009.07.09	2009.07.09	870		confirmed cases
[84]	+	2009.07.10	moreoverhealth	Influenza	European Union	2009.07.09	2009.07.09	--		the H1N1 flu cases
[4963]	+	2009.07.10	examiner	Swine Flu	UK	2009.07.10	2009.07.10	14	↑	FOURTEEN people
[4963]	+	2009.07.10	topixhealth	Swine Flu	UK	2009.07.10	2009.07.10	14	↑	Fourteen people

1 2 3 4 5 6 ... 99 100 101 >>

Viewing 2000 events in 691749 documents

In terms of media monitoring, blogs share some properties with common web sites, e.g. blogs can be scraped like any other web site and individual posts can be obtained and processed as single news items. However, there are also some peculiarities, e.g. it is quite common that authors modify entries several times. Also, the added value of a blog might lie in the discussion which often develops in the comments section of a particular post.

Usually, the blog provides an RSS feed with the last entries. For each entry, the title, publication date and some general information are provided; the actual content and the comments must be extracted from the HTML pages. This is not trivial since the entries must be re-scanned looking for modifications of the original post and new comments. In EMM and MedISys, blogs can be monitored, using a list of user-defined URLs. An enriched RSS file is produced and is inserted into the EMM processing chain. The blog entries and comments are reconstructed to make them more readable.

Blog monitoring offers some interesting challenges. The tone of the publication is usually less formal and more direct. Often, blogs spawn controversial debates in the comments section. This can be studied with the help of sentiment and tonality analysis.

USER INTERFACE AND CUSTOMIZATION

MedISys distinguishes five main domains: diseases, bioterrorism, nuclear, chemical and other (see Figure 6), although the domains nuclear and chemical threats are only visible to institutional users using the password-protected site. The domain other includes categories such as Public Health organizations (e.g. WHO and ECDC), medicinal products, multi-drug resistance, vaccinations and more. For each domain, the individual categories can be accessed via the menu structure, either by groups (e.g. blood-borne or hemorrhagic diseases) or alphabetically. Natural disasters are monitored

Figure 6. MedISys (http://medusa.jrc.it) screenshot: In this example, the results for the Influenza (H1N1) are shown. Hot spots are identified on the world map (right); statistics are given for the last 14 days (centre).

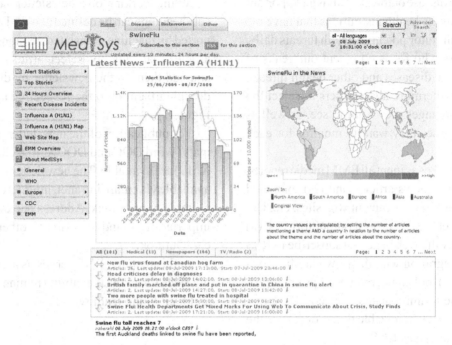

within EMM; often disaster events, e.g. hurricanes, earthquakes or floods, not only lead to direct casualties, but also affect the population indirectly, e.g. via contaminated water or exposure.

For each category, users see 14-day statistics and a world map showing the countries mentioned in the context of this category in the last 24 hours, as well as a list of articles in any of the 50 MedISys languages, sub-divided by source type (medical publications, newspapers, TV/radio, news wires and other commercial sources – the latter two are only visible to EC-internal users). For each news report, the title is shown with the article's description (the first few words of the article), the name of the source and the time of publication. Users can choose to view all articles displayed on one page or select those articles in a specific language. By clicking on the title, the original website is shown from which the article was retrieved. This enables the user to read the original news item, if the original link still exists.

On demand (by clicking on the '?' symbol), users can see the category definition.

When clicking on the green information button 'i', additional information for each article will be displayed (see the top right corner of Figure 6). This can include the category definition words found, the names of persons or organizations mentioned in the article, other categories to which the article belongs, as well as links to the Google translation engine. This latter feature allows users to read the full article in their own language. For Arabic articles, title and description are translated in-house so that users can immediately see the English translation without relying on Google Translate.

An important MedISys feature is the filtering function, which can be activated by clicking on the filter symbol at the top of the webpage. As MedISys has a large user group with widely varying user interests, it will collect and show articles on disease outbreaks, treatment, legislation, scientific response, and more. By using filters, users can, for instance, choose to view only those articles concerning a disease outbreak. This is an important feature for public health institutions that have the mandate to monitor potential health threats daily, i.e., for users who focus on recent outbreaks of communicable diseases and other health threats. The filter furthermore allows users to select the subset of languages they want to see, as well as the news sources they want to monitor for each country.

MedISys takes as input all EMM media reports, which includes news articles and text extracted from specialist or government web sites, etc. Users have the choice of several output formats: (a) they can view the web page, (b) subscribe to RSS feeds for integration with their own applications or viewers, (c) look at interactive maps or graphs, (d) subscribe to immediate or once-a-day email notifications at customizable times, or (e) use the moderation interface RNS.

RAPID NEWS SERVICE

The Rapid News Service (RNS) tool allows users to monitor categories in a customized view, to produce newsletters by selecting and grouping news items, to send the produced newsletters to user-defined groups of recipients via email or publish them on a web site. For urgent events, SMS messages can be sent from within RNS with a simple drag and drop operation. Users can also search for certain keywords, using Boolean expressions, and save these searches for future work.

The setup allows for different account types: users can edit and publish multiple newsletters, working on sets of categories of interest. Users are organized in user groups allowing a team to work on the same newsletter and to share an address book for storing email addresses and mobile phone numbers of newsletter recipients. User administrators set the capabilities for each user in a group. Finally, the RNS administrator is able to create and manage groups and to define the roles of each user.

While working on a newsletter, headings and sub-headings can be defined at will. The user, supported by machine-translated versions of original articles, can quickly select articles of interest and add them to the newsletter by dragging and dropping them onto the respective heading (see Figure 7).

For publication of the newsletter, various formats (Word DOC and DOCX, PDF, RTF, HTML, XML) are available. The newsletters can be made available via email, intranet, internet or WAP with a simple mouse-click. Address books can also be imported from and exported to other programs, e.g. Outlook.

In this manner, RNS serves as a user-friendly interface for analysts enabling human moderation and facilitating team work.

CONCLUSIONS & FUTURE RESEARCH DIRECTIONS

Although open-source media monitoring has demonstrated its value for public health authorities e.g. during the early stages of the A(H1N1)v pandemic in March/April 2009, many challenges still lie ahead.

News coverage across national and language barriers needs to be improved; more news sources have to be added and continuously updated. Future developments will enable us to monitor TV, radio and internet broadcasts in real time and thus expand the monitored media spectrum tremendously.

More data from various sources could be added to complement surveillance, under the presumption that the data is made available in a timely fashion, e.g. Google and Yahoo [Polgreen et al. 2008, Ginsberg et al. 2009] have demonstrated that patterns of searches matched with official influenza surveillance data.

Figure 7. Producing a newsletter in Rapid News Service (RNS). On the right-hand panel, articles from a MedISys category or from saved searches can be displayed (in this example, articles on natural disasters were selected). For each article, title, link, date, source, and a short description are shown; links to machine-translated texts are available. The structure of the newsletter is shown in the left panel, with user-defined headings onto which individual articles can be dropped. Users can work on several newsletters in parallel by clicking on tabs (upper-left corner). Eventually, newsletters can be published or emailed with a single mouse click. In urgent cases, SMS alerts can be sent out immediately.

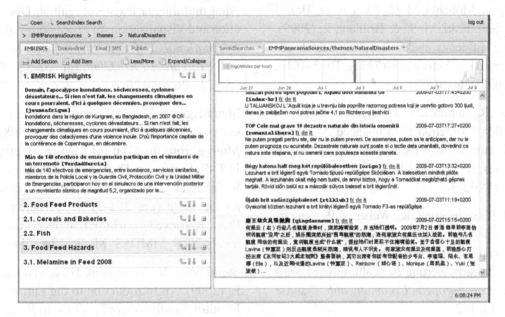

At the same time, the already existing sources need to be updated continuously. The multilingual category definitions need to be improved.

Advances in machine translation software will provide in-house translations of all incoming articles to facilitate the users' daily work. Ideally, machine translation is built into the processing chain so that original and translated text can be displayed together.

To increase specificity and speed up the analysis of incoming news articles, event extraction will identify important events in several languages. The same event can be reported at various time points in the media and through official channels. For this reason, events might be misclassified or overestimated. Better methods to reduce redundancy and avoid feedback loops between automatic systems and moderated systems need to be developed.

The specificity of automatic systems has to be improved without affecting the sensitivity. The analysts still need to screen a large amount of news items on a daily basis. Event extraction software helps the analysts to screen articles by identifying the nature of the incident and displaying the extracted information in a summarized form.

REFERENCES

Collier, N., Doan, S., Kawazoe, A., Goodwin, R. M., Conway, M., & Tateno, Y. (2008). BioCaster: detecting public health rumors with a Web-based text mining system. *Bioinformatics (Oxford, England), 24*, 2940–2941. doi:10.1093/bioinformatics/btn534

Freifeld, C. C., Mandl, K. D., Reis, B. Y., & Brownstein, J. S. (2008). HealthMap: global infectious disease monitoring through automated classification and visualization of Internet media reports. *Journal of the American Medical Informatics Association*, *15*, 150–157. doi:10.1197/jamia.M2544

Ginsberg, J., Mohebbi, M. H., Patel, R. D., Brammer, L., Smolinski, M. S., & Brilliant, L. (2009). Detecting influenza epidemics using search engine query data. *Nature*, *457*(7232), 1012–1014. doi:10.1038/nature07634

Grishman, R., Huttunen, S., & Yangarber, R. (2003). Information Extraction for Enhanced Access to Disease Outbreak Reports. *Journal of Biomedical Informatics*, *35*(4), 236–246. doi:10.1016/S1532-0464(03)00013-3

Linge, J. P., Steinberger, R., Weber, T. P., Yangarber, R., van der Goot, E., Al Khudhairy, D. H., & Stilianakis, N. I. (2009). Internet surveillance systems for early alerting of health threats. *Euro Surveillance : European Communicable Disease Bulletin*, *14*, 1–2.

Madoff, L. C. (2004). ProMED-mail: an early warning system for emerging diseases. *Clinical Infectious Diseases*, *39*(2), 227–232. doi:10.1086/422003

Mykhalovskiy, E., & Weir, L. (2006). The Global Public Health Intelligence Network and early warning outbreak detection: a Canadian contribution to global public health. *Canadian Journal of Public Health*, *97*(1), 42–44.

Paquet, C., Coulombier, D., Kaiser, R., & Ciotti, M. (2006). Epidemic intelligence: a new framework for strengthening disease surveillance in Europe. *Euro Surveillance : European Communicable Disease Bulletin*, *11*, 665. Retrieved from http://www.eurosurveillance.org/ViewArticle.aspx?ArticleId=665.

Polgreen, P. M., Chen, Y., Pennock, D. M., & Nelson, F. D. (2008). Using internet searches for influenza surveillance. *Clinical Infectious Diseases*, *47*(11), 1443–1448. doi:10.1086/593098

Pouliquen, B., Kimler, M., Steinberger, R., Ignat, C., Oellinger, T., Blackler, K., et al. (2006). Geocoding multilingual texts: Recognition, Disambiguation and Visualisation. In *Proceedings of the 5th International Conference on Language Resources and Evaluation LREC2006* (pp. 53-58). Genoa, Italy, 24-26 May 2006.

Steinberger, R., Fuart, F., van der Goot, E., Best, C., von Etter, P., & Yangarber, R. (2008). Text Mining from the Web for Medical Intelligence. In F.-S. Françoise, D. Perrotta, J.Piskorski & R. Steinberger (Eds.), *Mining Massive Data Sets for Security* (pp. 295-310). Amsterdam: IOS Press. Retrieved from http://langtech.jrc.it/Documents/2009_MMDSS_Medical-Intelligence.pdf

Steinberger, R., Pouliquen, B., & van der Goot, E. (2009). An Introduction to the Europe Media Monitor Family of Applications. In *Proceedings of the Workshop 'Information Access in a Multilingual World: Transitioning from Research to Real-World Applications SIGIR-CLIR2009*. Boston, USA. 19-23 July 2009.

Wilson, J. M., Polyak, M. G., Blake, J. W., & Collmann, J. (2008). A heuristic indication and warning staging model for detection and assessment of biological events. *Journal of the American Medical Informatics Association*, *15*(2), 158–171. doi:10.1197/jamia.M2558

Chapter 10
Social Media (Web 2.0) and Crisis Information:
Case Study Gaza 2008–09

Miranda Dandoulaki
National Centre of Public Administration and Local Government, Greece

Matina Halkia
European Commission, Joint Research Centre, Italy

ABSTRACT

Social media technologies such as blogs, social networking sites, microblogs, instant messaging, wikis, widgets, social bookmarking, image/video sharing, virtual worlds, and internet forums, have been identified to have played a role in crises. This chapter examines how social media technologies interact with formal and informal crises communication and information management. We first review the background and history of social media (Web 2.0) in crisis contexts. We then focus on the use of social media in the recent Gaza humanitarian crisis (12.2008-1.2009) in an effort to detect signs of a paradigm shift in crisis information management. Finally, we point to directions in the future development of collaborative intelligence systems for crisis management.

INTRODUCTION: ICTS FOR FORWARD-LOOKING CRISIS MANAGEMENT

Crises in the 21st century are expected to astonish both the experts and the lay people. Lagatec (2005) was succinct in portraying the current situation in his paper titled "Crisis management in the 21st century: 'unthinkable' events in 'inconceivable' contexts." Therefore, traits such as responsiveness, flexibility, self-organization, improvisation, resil-

ience, agility[1] seem pertinent for forward-looking risk and crisis management.

Crisis information management is no different. Top – down fixed information systems and tools cannot fully capture the spatial - temporal - social dynamics and respond to the uncertainties of future emergencies, disasters and crises. Numerous tools that have recently been developed to support crisis and emergency management[2] have a pre-defined structure and rely primarily on information collected and maintained in normal (i.e. non-emergency) conditions. Although the deficiencies of central, hierarchical architectures

DOI: 10.4018/978-1-61520-987-3.ch010

have been pointed out and alternative solutions have been proposed (Dandoulaki & Andritsos, 2007), emergency management tools remain habitually centrally administrated and still reflect a top-down approach.

Current information tools, either formal (e.g. local traffic surveillance cameras, global alert and monitoring systems) or informal (mobile phones taking videos on the spot), gradually change the landscape in emergency information (Moss & Townsend, 2006). Social media technologies (social networks, microblogging, blogs, wikis, annotatable maps, image and video sharing, instant messaging, internet forums and other web forms)[3] have also been acknowledged to have played a role in emergencies, disasters and crises (Palen, 2008; Palmer, 2008).

The emergence of such shifts and incursions summon and merit the clarification of the basic concepts in our discussion. The conceptualization of emergency, crisis and disaster[4] is central in the current discussion among scholars (Quarantelli, 1998; Perry & Quarantelli, 2005, Gundel, 2005) and has significant implications in policies and practices. The relationship between crisis and disaster is still to be defined. The two are inextricably linked (Boin, 2005, p.155); however crisis is a general concept encompassing disaster (Quarantelli, 1998 p.235). Boin (2005, p.164) suggests that disaster is a "crisis gone bad", thus including in the disaster category a spectrum of events and processes such as riots, epidemics, acts of terrorism and massacres.

This chapter examines how social media technologies affect crisis information management. We first review the background and history of social media (Web 2.0) in crisis management. We then focus on the use of social media during the recent Gaza humanitarian crisis. Finally, we point to directions in the future development of collaborative intelligence systems for crisis management.

Why focus on Gaza in order to grasp future trends? For one, social media played a major role in redefining the spatial and social locus of the crisis, thus the focus of crisis management, by triggering response all over the world. They also contributed to more resilience of the information system once the formal systems and tools could not or would not perform. Moreover, the Gaza case demonstrates the influence of social media technologies even when the disaster area has huge inadequacies in technological infrastructure. Finally, the vast Gaza mobilization on a worldwide scale exemplifies on one hand, the potential and strengths, as well as the perils and weaknesses of collaborative media, and begs on the other hand the question of a new structured platform for collaborative intelligence systems in decision-making for disaster management. It should be stated, however, that our objective is not to survey the use of social media in this case study; rather, we would like to discuss certain aspects that highlight a paradigm shift in crisis information management.

THEORY AND PRACTICE: SOCIAL MEDIA (WEB 2.0) IN CRISIS CONTEXTS

Crisis Management and the Media in the Risk Society

Crisis management is not separate from understanding the processes of risk and vulnerability that unfold both at micro- and macro- level. Yet, while much effort is put in dealing with the macro-level, especially as regards current information systems, less progress is made in grasping the situation and the dynamics at the micro level.

Being there in a crisis, having experienced the crisis and its multiple realities, having been part of the dynamics of a crisis, is by itself valuable knowledge (Hewitt, 1998, p.87; Barton, 2005 p.136, Buckle, 2005). The knowledge of experience is highly significant and on par with scholar and expert knowledge. Therefore, it is a challenge for crisis management to bring up also the micro-level (i.e. place) and with this, the knowledge and experience of people who "were there".

At the level of working practices, the importance of risk governance is rising. Risk governance encompasses risk assessment and risk management. It concerns how decision-making unfolds when a range of stakeholders, with different expectations, goals, activities and roles, are involved (Renn and Walker 2008). Amongst all of this, communication is at the centre (International Risk Governance Council, 2006).

Risk communication has transitioned to today's internet times (Krimsky, 2007). Risk communicators - a core element of crisis information according to Fischoff (1995) - are not unquestionably central in risk communication any more. There are evident signs of lay people gaining a central role in communication and information through the use of current technological means, among them social media technologies. Social media technologies have already been used for different purposes in crises. Everybody has or can have a voice in the future; this can be significant in times when the challenge is to democratize society's response to risk and disaster (Alexander, 2005, p.35).

Global communications have contributed to the symbolic significance of disaster (Alexander, 2006, p.7). In the present era, disasters are seen as a mindset, their interpretation and symbolic meaning constructed by the mass communication industry (Alexander. 2005, p.38; Saw, 1996). Social media potentially counterbalance mass media.

Today's society is a risk society (Beck, 1992; Adam at al., 2000); risks and society are interwoven and affect one another through a continual negotiation and renegotiation. The risk society is also a reflexive society. Reflexivity entails reflection (knowledge) on risks but also not being aware of risks (Beck, 2009, p.119). In the era of globalization risk reflexivity is shaped by the mass media and the instrumentation of anxiety by a range of political players (Beck, 2009, p.198). Social media offer an opportunity so that every-

body can voice a view in the current and future negotiation of risk.

Web 2.0 Technologies and Crisis Information

Although the modern world is technology-dominated, much of the world's population is excluded from the benefits of technology. It is estimated that about a half of the population have never used a telephone (Alexander, 2006, p.3). Then again, peer-to-peer information and communication technologies become increasingly pervasive and among them, social media technologies (Palmer, 2008, p.1; Shankar 2008).

Social media have been identified to have served in crises in many ways. They are attributed with social convergence (Palen, 2008; Hughes et al., 2008), they facilitate people's participation in emergency management (Palen, Hiltz et al., 2007; Vieweg, 2008), they support collective intelligence in a crisis (Palen, Vieweg et al., 2007), and they counterbalance or compliment mass media through citizen journalism (Liu at al., 2008). The emerging role of Web 2.0 technologies in crisis contexts is even a topic of discussion in social media; there are blogs, especially in the USA, where relevant research and experiences are shared.[5]

The topic of if and how social media could change crisis management is a burning one (see for example Tinker & Fouse, 2009); after all, social media have been exploited in one way or another in every recent crisis.

Social Media Use in Crises: Experiences and Trends

There is a mounting number of experiences in the use of social media in crises and a fast-growing quantity of related literature. Depending on the situation, social media has been used for different purposes and in various ways (Sutton et al. 2008). The following selected cases were either studied in-depth and now serve as reference cases in the field, or bring up new dimensions of the issue.

Our reading of these cases is purposeful; the intent is to bring up functionalities served by social media in a crisis. Then we make an effort to identify strong and weak points of social media tools from the point of view of formal crisis management.

9/11 (2001)

Four airplanes were hijacked; two of them hit the WTC towers, one hit the Pentagon and the other crashed in a field. Mass media were transmitting live the situation. E-mail was used by trapped people and by the affected population to communicate with their peers (Palen & Liu 2007, p.729). Harrald et al. (2002) report several uses of web based technologies; web-based technologies were used by corporations to account for and communicate with employees. FEMA, the American Red Cross, The US Army Corps of Engineers, and the Environmental Protection agency all used web communication to inform the public and to provide status reports internally and externally. There is evidence that the terrorists also used the internet to plan the attacks.

Even at this relatively early stage of web use in crisis contexts, the internet proved a significant, although not primary, source of information in the USA with 64% of adults in the USA using the Internet to find information on the situation (Bucher, 2002, p.2).

New Orleans Floods (2005)

Hurricane Katrina (category 3 storm) caused damaged in Mississipi and Louisiana. Areas of New Orleans were flooded. The city was evacuated and the population was temporarily relocated throughout USA. Citizen-led online sites were used for aid in the emergency. The site (www.katrina.com), previously used to advertise a small company, attracted people seeking information and it was readily converted to link to other useful sites and to serve as a message board to facilitate

tracking down missing people. Other sites such as Hurricane Information Maps (www.scipionus.com) were created to collect and share location specific information (Palen & Liu, 2007; Moss & Townsend, 2006; Currion, 2005). The disaster was captured in weblogs that could serve as crisis data sources (James & Rashed, 2006).

Virginia Tech Shootings (2007)

On April 16, a shooter killed 32 people at Virginia Polytechnic Institute and State University. Findings[6] record large-scale social interaction that occurred after this event over multiple sites of interaction; at first peer-to-peer communications and later on-line and off-line IC activities concerning larger sets of data (Palen, Vieweg et al. 2007; Vieweg et al., 2008). Facebook and Wikipedia were the main tools used for the latter. Facebook groups such as "I'm OK at VT" were set for members to post that they were not among the victims; others such as the "We support Chief Flinchum" expressed solidarity to emergency managers (Byrne & Whitmore, 2008, p.8). Online efforts correctly identified deceased victims even before the university released the information (Byrne & Whitmore, 2008). Nonetheless, problems similar to old-fashioned jamming of telephone lines were observed; eventually the bandwidth got overloaded (The Risk Communicator, 2008, p.4).

California Wildfires (2007)

The California wildfires began on October 20, 2007 and burnt over 500.000 acres of land. Sutton et al. (2008) show a wide spectrum of means used for communication and information from mobile phones to contact family and friends, to all available web-based tools. The informal notification process has been identified as one of two significant warning mechanisms in San Diego County, the other being a reverse call-down emergency warning system that San Diego County and the City of San Diego had put in place (Sorensen et al., 2009, p.22).

People were informed through traditional media (television, radio) and new alternative media mainly advertised in the former (websites, information portals). They also posted and shared information and participated in web forums (Palen, 2007, p.78). Websites were created with annotated geo-information of interest to the affected people (burnt areas, evacuation areas, shelters etc.) (Hughes et al., 2008). This website had more than 1.7millions hits (Sutton et al., 2008). Flickr served as a repository of photos both for personal use and group based interaction; As a result, houses were also inventoried for insurance purposes (Liu et al., 2008).

The Greek Riots (2008)

The riots started on 6th December 2008, after a teenager was fatally shot by a policeman at around 9:00 pm. His death triggered large protests and demonstrations, which escalated to widespread rioting. Demonstrations and rioting were propagated to several other cities. Outside Greece, solidarity demonstrations, riots and in some cases clashes with local police also took place in a number of European cities.

Thousands of people were on the streets protesting while established media had not even reported the event (P.Tsimas as referenced in Lam, 2008). SMS messaging and re-broadcasting on the internet played a significant role for the mobilization of people (Gavriilidis, 2009). Twitter was extensively used, with the first twitter message sent around 15 minutes after the event. About three hours later Pathfindernews and Skaigr were the first official media to announce the event.[7] At around 3:00am on December 7th, the hashtag #griots started to be used on Twitter.[8] Thousands of messages were sent over the next days, many of them directly from the streets.[9] Citizen reporting was intense especially in terms of images. Flickr featured more than 1.100 photos (Tziros, 2009). Facebook also played a role in the protests with several protest groups set-up, totalling some 187,000 members. The largest one had around 136,500 members and

shared messages about the upcoming protests and commemorating activity (Joyce, 2008).

Wikipedia started repoting on the issue on December 7 th at 21:00, and was updated more than 200 times afterwards (http://en.wikipedia. org/wiki/2008_Greek_riots). Collaborative citizen journalism projects, like Global Voices Online, NowPublic, allvoices and CNN's iReport, were used to publish original reports.[10]

L'Aquila Earthquake (2009)

On April 6th 2009, at 3:32am local time, an earthquake (Ml=5.8) struck central Italy in the vicinity of L'Aquila (capital of the Abruzzo region). The earthquake killed 308 people,[11] destroyed more than 10,000 buildings and left more than 24,000 homeless.[12] Approximately 65,000 people had to evacuate their homes and were lodged in tent camps, hotels and other structures located along the Adriatic coast.[13] This quake was the strongest of a sequence that started a few months earlier.

Seismic activity in the area was an issue in social media even before the main event.[14] After the event, information was disseminated immediately by people directly on site that kept feeding the web and the mass media.[15] Wikipedia (Italian edition) had the first entry at 11:52 on the same day, something which is still updated even today[16] and then at 12:03 it appeared also in the English edition. A website was created and offered annotated geo-referenced information on the earthquakes, their effects, emergency management and aid provision activities (information on victims and damages, blocked roads, hospitals, evacuation areas, shelters, coordination centres etc.) (Figure 1).

Bearing in mind that social media have been used in one form or another in every recent crisis, the aforementioned cases are only a few examples.

Formal crisis management cannot but take into consideration emergency management technology advances[17] and the mainstreaming of social media technologies. The response to this challenge dif-

Figure 1. Google maps mashup of the L'Aquila earthquake disaster. (Source: Google.maps-terremoto L'Aquila. Retrieved April 14, 2009, from http://www.google.it/landing/terremoto_abruzzo.html)

fers in different contexts but the signs of change are already there (Collins, 2009).

As early as 2002, FBI established a website after 9/11 to receive information and tips from citizens that might aid the law enforcement investigation (Harrald et al., 2002). Many USA government agencies use Twitter, among them the Department of Homeland Security (DHS), the Environmental Protection Agency (EPA)[18] and the US Geological Survey, to mention only a few. [19] The Los Angeles Fire Department and Philadelphia Emergency Management are among agencies using social networking to send breaking news and preparedness information. FEMA developed the Integrated Public Alert and Warning System (IPAWS) that expands upon traditional (radio and television) media and adds new media - internet and cell phones[20]. NGOs dealing with emergencies and crises have already put social media in everyday use, among them the American Red Cross.[21]

The Australian Government has announced plans to use social media websites such as Twitter and Facebook, alongside traditional warning mechanisms in order to improve the "quality and timeliness of bushfire warnings" ahead of the upcoming fire season.[22] Toronto's Police Service and Toronto Fire Service are both on Twitter.[23] Initiatives involving social media and government can be currently noticed in many countries all over the world.[24] Recently, a 20-page guide to using twitter was published targeting the UK government agencies.[25] These are initial but important steps that allow testing social media in formal crisis management, although only as peripheral tools for the time being.

Mass media (TV, radio and the press) also use social media to involve people in offering opinions and sharing information; see for example: the CNN user-generated site (http://www.ireport.com/) (Catone 2008), the Guardian blogposts (http://www.guardian.co.uk/tone/blog), The New York Times twitter (http://twitter.com/NyTimes).

Main Functionalities of Social Media in Crises

Studying a number of cases, we can identify three core functions of social media during crises each having strong and weak points as regards formal

Table 1. Strong and weak points of social media use for communication as regards emergency management

Strong points	Weak points
• Redundancy in communication • Vast territorial and social coverage • Able to communicate multi-type information • Facilitate social convergence • Potential for fast and broad communication of brief messages (such as alerts and warnings)	• Credibility of information not known • Information, information source and information flows not possible to control • No clear responsibility and no accountability • Risk of malicious spreading of information and easier dissemination of potentially problematic rumours • The bandwidth can get overloaded in the same way as telephone communications suffer in crises.

emergency management. It should be noted however, that these functions do not exclude one another since in many occasions they coexist and work in a complementary way.

Peer-to-Peer Communication

People tend to seek friends and family in a crisis in order to communicate personal information and get or give practical or psychological support; in doing so they also transmit information useful for emergency management. All means of communication are employed from leaving handwritten notes and signs, to mobile phones, e-mail and social networking tools such as Facebook and Twitter. As social media enable everyday networking, they also serve for peer-to-peer communications in a crisis.

Information

Information is vital in a crisis. People seek information. They also offer information demonstrating a socially convergent behavior; in so doing they also participate in emergency management. Social media tools facilitate searching, collecting, posting, sharing information by individuals and groups. Tools such as wikis, blogs, Facebook, Flickr, Twitter etc. are used. Previous networking of various types makes the activity easier.

Activism

Individuals and groups sharing similar goals in specific circumstances use social media tools for awareness raising and mobilization of others. Ad

hoc networks are created to channel messages and disseminate information. Clustering of nodes can usually be detected after some time. Twitter, blogs, wikis, Flickr have been used in different occasions, depending on the context and the situation.

Case Study Gaza 2008-2009

Between December 2008 and January 2009, Israeli military forces carried out a major offensive against the Gaza strip by air, land and sea. Although news of the attack was not readily available through the major media channels – due in part to a lack of foreign journalists in Gaza[26] – news about what was happening on the ground rapidly disseminated worldwide through social media, such as blogs, independent internet news broadcasting and social networking.

Although it might seem premature to assess objectively the impact of social media on the dissemination of information in the Gaza conflict, it can be said that this was probably the first time social media were used in absence of any other formal crisis management system,[27] unlike previous examples discussed. Information about the humanitarian aspects of the disaster were gathered by international organizations and observers on the ground and transmitted to western audiences at large through social media.

"Everyone has a Voice"

It was immediately apparent to westerners attempting to follow the conflict in its early days through social media, that there was a host of

Table 2. Strong and weak points of social media use for information as regards emergency management

Strong points	Weak points
• Multi-source information of various types with vast geographical and social coverage • Timely field information possible • Information can be disseminated easily • Potential for developing new forms of public involvement and attracting volunteers • Social convergence • Potential for raising global awareness and harvesting aid from all over the world with less dependency on mass media	• Credibility of information not known, information can be misleading, confusing or contradictory • Accountability very low, liability is obscure • Information stays on the internet long and might become out-of-date and be misleading • Huge bulk of information from different sources creates an information management nightmare • Authorities and official agencies are not accustomed to make use of possibly non-credible available information • Privacy protection issues. Low security of information • Possible manipulation and malicious use of information • Significant issues of transparency and trust

voices reporting, many of them without established credibility. Trust and identity had to be constructed on-the-fly, and very quickly, as news were coming in through blogs and websites. Many times during the three weeks of the assault news was conflicting, and the lack of institutional voices on the ground, in this highly controversial conflict, made news evaluation problematic.

The network of nodes in the social network that was reporting on the Gaza crisis was initially built on apparently strong, rather homogeneous ties, already in place before the conflict begun. In the USA, Gazasiege.org and FreeGaza.org and in Italy, Vittorio Arrigoni, an Italian volunteer contributing to the paramedic squads of Al-Quds hospital in Gaza city, were some of the most consistent providers of daily updates of the situation on the ground. Arrigoni's blogs were translated within hours in English and other languages and copied in a host of English-speaking sites including the

ones mentioned above. The Gaza social network expanded very rapidly to a worldwide network of now weaker, heterogeneous ties (for definitions and social network analysis see boyd & Ellison, 2007). Facebook groups decrying the Gaza assault gathered rapidly. As these lines are written, the number of groups returning the keyword "Gaza" in Facebook amount to 597, the largest of which counts short of 200,000 members. Of them, three supported or defended the assault. The others mainly in English (but also French, Italian, Spanish, and Norwegian) denounced the humanitarian catastrophe for Gazan children, disapproved the difficulty in access for humanitarian supplies or medical personnel, or praised Al-Jazeera for "providing the truth."[28]

Information spread through daily updates about the number of deaths, even though the actual numbers varied. Personal stories of family strife, of death and despair spread geographically

Table 3. Strong and weak points of social media use for activism as regards emergency management

Strong points	Weak points
• Fast notification of large numbers of people. • Independence from official information sources and mass media. • Social and political convergence in crisis • Participation and active involvement easier. • A good platform to democratize society's response to risk and disaster.	• Credibility of information not known. • Accountability very low. • Information stays on the internet long and might become outdated and misleading • Low security of information • Risk of manipulation of information and of malicious use of social media. • Difficult to control.

far away from the epicenter of the disaster, and became the theme for household small talk. For example, Facebook chats, wall posts, and comments were discussing Gaza events in the context of family or friend connections. The audience by the end of the three-week assault was international and heterogeneous. The news story on Gaza was constructed collaboratively in a distributed geography, dislocated from the crisis theatre, through opinions voiced in blogs and social network sites. Slowly and steadily the voices converged and the narrative became common ground. The Gaza disaster network was now forming a community.

Community Narratives, Distributed Collaborative Production of News: Multi-Author Narratives and the Construction of Meaning, Dissemination and Structure of the News Message

The Gaza network prompted the development of another community. Defenders of Israeli politics were quick to respond. While a Facebook group donated their status[29] in counting victims and wounded in the Gaza strip, another community was emphasizing the number of Qassam rockets fired at and dropping on Israel.

Different community narratives dictated different versions of news. The social media consumer had to navigate through these stories with caution. The construction of a coherent view of the crisis became relative to one's personal interest and motivation to follow another kind of story. A story that was multi-threaded, collaboratively constructed, fraught occasionally with conflicting or ambiguous news messages. The passive news consumer was called to active involvement in structuring the different elements of the news message. In practice, this involved navigating from blog to related blog, perusing blog comments, controversies, and criticism, consulting wikis, complimenting information by Google searches, and more often than not revert to "authority voices"[30] to balance, prioritize, and contextualize

reported stories. In so doing, the crisis reports created were as many as the reporters; every blogger or blog comment became potentially a report in itself, complimenting the edifice of the crisis from yet another point of view.

Although it is has been said that the Gaza crisis triggered a social media war,[31] this has not been war of winners and losers, as the news story was written by everyone and all: a conglomeration of eyewitness reports, comments, criticisms, background information and historical accounts brought to the international observer mainly through social media.

Establishing Credibility, Constructing Reliability, Accuracy in the Multifaceted Narrative, Discrediting and Slander

With a multitude of voices reporting their different realities of a crisis, trust to news sources is key. What is reported requires careful evaluation in relation to contextual and background information, but often verification is difficult, especially under the time pressures in a crisis situation. This was especially true in the Gaza conflict, where crisis reports were often antagonistic.[32] Credibility was built as more and more information was becoming available to confirm or contrast previous reports. Over a reasonable amount of time, certain blogs or news' sites risked losing credibility as military/ state propaganda,[33] and independent eyewitnesses on the ground emerged as dispassionate, objective reporters providing accurate crisis updates.

There was more than multifaceted reporting, which called for attention in the construction of the news message. Often, there was slander and deliberate discrediting of reporting voices through offensive sites. Offence included death threats against bloggers.[34] Although social media resist censorship and this remains one of their perceived advantages, in the Gaza crisis spontaneous informal efforts to control freedom of speech quickly became so serious as to instigate calls for official diplomatic action at state/national level.[35] While state control was exercised in other crises where social media were key,[36] in the Gaza conflict, censorship efforts were targeted towards exerting individual pressure, through offence or slander, in order to obtain self-censorship or at least curtail freedom of speech.

Roles: The Changing Role of the Audience; The Changing Role of the Journalist; The Expert

In social media, collection and dissemination of information transforms the passive crisis onlooker in an active news producer. By commenting on blogs, collecting reports and consulting search engines in order to piece the fragmented news landscape, she actively engages in an editorial effort even this does not materialize in written word. How can it be argued that a Facebook member who donates its status to a cause does not engage in activism for that cause? And how can it be claimed, that in the same act there is not implicit recruitment of supporters among its immediate social network for the said cause? By the same token, funds raised as a result of social media activism can be considered enabling financial instruments for crisis management. Undoubtedly, the case of Gaza resists any crisis management plan, as effective crisis mitigation was not possible. Among other ineffective crisis response mechanisms, humanitarian aid promised by the international community did not reach Gaza due to the practically closed border crossings.[37]

On the other hand, the professional journalist can then be seen as the key factor in filtering and editing the reported information, providing context and background and facilitating interpretation. In the Gaza example, due to access restrictions journalists played a secondary role in reporting from the crisis site. Still, CNN's i-report community provides a moderated social network in which eyewitnesses can provide information about events occurring real-time. Additionally, important texts providing background information and enabling

historical interpretation of the crisis climax were published in major western media. An impressive amount of international media coverage resulted, in order to elucidate the hows and whys of the humanitarian crisis. Indeed, academic experts and political analysts became popular reference figures[38] in this crisis, as there was a particular need to untangle complicated dependencies between cause and effect phenomena which would eventually lead and nurture a successful crisis management strategy.

Processes

We examined earlier in the chapter the three main uses of social media in crisis contexts: peer-to-peer communication; information dissemination; mobilization, awareness, and activism. In the Gaza example, these processes were particularly heightened because of the lack of any adequate crisis management system. The population used mobile phones and sms messages to notify the location of wounded to ambulance services. Sms notifications gave the latest information about whereabouts of Israeli forces, affected quarters and closed streets to family and friends. International awareness was raised through social media because of the access restrictions on international journalists. Activism through fundraising on social media platforms and organization of public support events and gatherings was also observed.

Social Media as a News Centre; The News Consumer as a News Producer and Peace Activist

In stark contrast to cases examined earlier in the chapter, in Gaza there was lack of formal crisis management systems. When the United Nations (UN) headquarters, two UN schools and the Al Quds Hospital,[39] were hit by Israeli forces it became apparent that any minimal organized crisis response that might have been in place, was no more. In this sense, the Gaza crisis is very different from other crises where social media played a role. Taken to extremes, this means that social media on Gaza can be likened to a crisis situation room. Because of the lack of international journalists to disseminate information and adequate civil protection mechanisms to respond to emergencies and support the population, social media provided the missing functions of a centralized emergency response system, as much as that was possible. For this reason, the Gaza example provides poignant lessons for system design of Web 2.0 technologies in crisis contexts: social media with all the weaknesses and defects that we have already examined, was a stand-alone support system in the Gaza disaster. Those participating in the Gaza crisis social network were perceived as emergency workers by the Gaza population.

The Two Faces of an Activist's Blog: A Forum, and a News Service

Vittorio Arrigoni's blog (http://guerrillaradio. iobloggo.com/) became, during the three weeks of the assault, the most visited blog in Italy with peaks of 20.000 visits per day[40]. Translations of his texts were read daily in USA and in South America. Arrigoni one of the few internationals remaining in Gaza during the assault, and by all accounts the only Italian citizen on the Gaza strip, provided eyewitness accounts from the ambulance service he helped to man in Al Quds hospital. He consistently updated his blog with disaster testimony, and a touching account of the ground experience. His audience expanded rapidly. He started receiving supportive emails from factory workers, university professors, students, housewives and air traffic controllers. He claims that the reason for the heterogeneity of his audience and the international impact of his texts was his independence; he does not belong to any interest group, political faction, or non-governmental organization. He says: "My language is clear, on level of the reader, and respects fundamental human values such as peace, justice, and freedom."[41]

Arrigoni, apart from providing ground reports, built overtime a knowledge of experience that indeed became as important as the scholar or expert knowledge on the Gaza crisis. His blog with its qualities of immediacy, lack of censorship, timeliness, and availability beyond national borders, became a reference node in the cluster of social media activity on the Gaza crisis. The comments following one of his daily blogs extended over several pages. Individual disputes, discussions and stories unfolded between his commentators. Arrigoni claims that, overtime, the blog was "infiltrated" by organized "trolls." He says "I believe in freedom of speech and for this reason I allowed trolls transformed into seemingly ordinary visitors to infiltrate and post racist remarks for months on end. When they surpassed any acceptable limits I ceded to my blogs' community pressure and banned the 'disturbing' comments". [42]

Indeed, during the days of the assault, the comments in Arrigoni's blog provided an impressively varied spectrum of different positions on the Gaza crisis; this demonstrated beyond doubt the heterogeneity of his readers. The passionate and occasionally inflammatory style of some comments supplied contextual information as rich as Arrigoni's eyewitness account itself; an indication of how important the blog had become to the Italian public interested in the Gaza crisis.

Summing Up Gaza: Does it Indicate a Paradigm Shift in Crisis Information Management?

In Gaza, social media were used to manage crisis information that was not readily available through state authorities or international journalists. In absence of any formal crisis management system, they replaced emergency response functions on the ground. They also provided a channel of information to the international community about the disaster. Because of the lack of established credibility, information verification was an intensive and elaborate process within the various Gaza crisis social networks. The cluster of network nodes was initially based on strong homogeneous ties, but rapidly grew to become a cluster of weak heterogeneous ties, crossing geographical borders, languages, social and ethno-religious backgrounds. While this heterogeneity was sometimes perceived, rightly so, as antagonistic -- so much as to be labeled a social media war -- it reflected an example of a multi-threaded narrative unprecedented so far in crisis and disaster contexts.

Over time, credibility was built as crisis reports converged. Credibility can be problematic in emergency response contexts; however, in Gaza we observed that if given enough time collective social network intelligence can be impressively accurate and coherent even if or because of, lack of any effective emergency response system. Although social networks do not easily facilitate application of censorship, in the Gaza crisis we have seen attempts to limit freedom of speech through self-censorship by exerting significant pressure at individual level.

Traditional roles were redefined. Eye-witnesses were reporting through blogs as international journalists could not. Experts and authority voices became key as they were uniquely positioned to explain conflicting information. Journalists had to provide context to all of this. More importantly though, the news consumer actively sought, produced and disseminated crisis updates, empowered by social media.

In Gaza, because social media functioned as a stand-alone support system, Web 2.0 technologies can be likened to a distributed, collaborative crisis situation centre. Online discussions were as informative about the crisis as were eyewitness reports. Momentous online activity and heated debate were both cause and effect of social media use. Ultimately, this demonstrates the importance of social media in managing and mitigating crises, albeit the effective benefits of the crisis response on the Gaza population are yet to be documented.

For all the above reasons, we sustain that the Gaza crisis may point to a paradigm shift in the use of social media in crisis contexts, the strengths and weaknesses of which we believe to have demonstrated above. We posit that social media use in Gaza can be employed as a poignant precedent for the development of systems, auxiliary to traditional crisis management technologies, which valorize collective distributed intelligence; provide for flexible role/actor definition in crisis management; further intrinsic resistance to censorship by supporting diversity; are complimented by adequate international legal instruments to protect individuals/entities against harmful material and internet crime, and support on-the-fly evaluation of network node connectivity and network cluster development to aid verification and confirmation of crisis information reports.

Towards Future Developments

The use of social media in crisis contexts has already been the focus of experts, academics and professionals. This interest seems compelled by the upsurge of social media use in recent crises: a reaction to unavoidable progress. To be sure, guidelines circulate on the internet on the integration of Web 2.0 technologies in existing risk communication strategies (Linder, 2006; Tinker & Fouse, 2009). We have not seen yet, design aspirations for integrated systems employing respective system strengths and addressing respective system weaknesses.

At the moment, social media technologies develop independently from formal emergency planning and management. What we observe today is facile usage of the former's capacities in the latter's service. For one, social media used for official alerts and warnings seems common sense: they decouple other technologies of communication thus offering redundancy in a crisis; some of them, Twitter for example, are believed to procure resilient means of communication; they offer rapid, timely and wide dissemination of messages, sent both by formal and informal sources.

However, it remains a challenge to see how the two could converge into more than the sum. It is time that the integration of social media in crisis management is considered a system design topic, bringing into focus traditional emergency management tools, together with computer-supported-collaborative-work (CSCW) know-how, human factors and interaction design experiences (HCI), as well as online communities' expertise.

Some experts call for a networked informed response to disaster, conflict and terrorism (Stephenson & Bonabeau, 2007; Linder, 2006). This cannot but include the successful integration of existing social media applications, to remain responsive to emergency situations where the latter seem to take precedence. Tools and methods developed in adjacent ICT fields are becoming available to make this possible.

Granted, one should acknowledge that social media in crisis contexts pose core issues concerning technology as discussed by Quarantelli (2007). Examples of potential hindrances are: inequalities in access to technology, further diminution of non-verbal communication, security of computer-based systems, social infrastructure and cultures that resist technology in a crisis. In the case of Gaza, these obstacles had a limited effect; on the contrary, issues such as trust and trustworthiness, identity management, reliability and accuracy, cooperation and transparency were key.

To address these concerns, one should think of the mapping of crisis events as a system of events that are interconnected and share similarities or feed into one another. The visualization and system design of such a mapping would provide a bird's eye view of an event when it happens in relation to a context of common or similar elements in space and time.

Given that current command and control systems would be unable to describe the complexity of such networks of events/actors/information, design requirements for such systemic approaches

would point to two directions for promoting system flexibility: a. cross-platform, cross-application systems, and super social networks, which visually map crisis information over space and time, and b. network cluster visualization, enabling different types of actors to digest, evaluate, and prioritize information, according to source, reliability, and trustworthiness. Tools and methods developed in adjacent ICT fields are becoming available to make this possible.

As a social network's accessibility and participation increases the more difficult it is to assess credibility. However, Donath (2007) has demonstrated how signals can be used in building trust and demonstrate trustworthiness. She has also demonstrated that the credibility cost can be balanced out by the richness of information obtained through increased connectivity. We argue that visually conveying the social network information flows, network node stability and behaviour, would aid the construction of trustworthiness in crisis contexts. In the case of Gaza the cost of trustworthiness was high because information required time for verification and confirmation. The design goal of an integrated "social network/crisis management" system would be to lower the costs of credibility assessment against time by increasing transparency of information flows, network connections, and actor definitions.

Kittur (2008) has demonstrated that users' trust of social media is directly proportional to revealing hidden information about elements, element stability and actor behaviour. This finding further supports system transparency to counterbalance credibility costs. Other tools are: visual signaling strategies, and *online fashions* (Donath, 2007) which can be used for identity definition of actors. Mapping information types, paths and provenance (clustering coefficients, the degree of bi-directionality in information flows, and types of media sharing) have also been found to play a role in the construction of trust in online communities (Donath, 2007).

Crisis management begs for something more than raw information. "Actionable knowledge" (von Lubitz et al., 2008) is required: usable, useful and immediately relevant information. Social media in crisis contexts have demonstrated the value of collective intelligence that can nourish actionable knowledge; they could effectively fertilize formal crisis information management. Acknowledging the importance of knowledge experience, integrating expert experience by system design on one hand, while providing metrics and peer-to-peer credibility evaluation, would be important milestones to achieve. Moreover, the convergence of Web 2.0 technologies with crisis management would need concerted effort towards an international legal framework that would provide security and accountability while respecting privacy and civil liberties.

Envisioning a global social network for crisis information, made of many separate interconnected emergency nets where individuals have a role to play, along with state authorities and emergency planners, may not be as far reaching as it seems, especially if the opportunity is seized now. Technological divergence might indeed prove to be a peril if openness is posited against security, and solutions to balance both, are not sought.

CONCLUSION

In the era of risk society the certainties of the past seem to fail us; what remains is a constant negotiation and renegotiation of our understanding of risk. In this, communication is core. Social media offer a platform of communication, sharing information and collaborative construction of meaning that cuts across geographical boundaries, social barriers and political limitations.

Especially in a crisis, social media offer a voice to many. They communicate the knowledge of extreme experience; they bring up the multiple realities of the situation; they counterbalance the dominant role of mass media in the construction of a symbolic meaning of the crisis.

Social media technologies have been used in many ways in different kinds of crisis contexts: disasters triggered by natural hazards, man-made and intended disasters, and conflicts. We identified three main functions of social media in crisis: communication, information and activism. These functions do not exclude one another but work in a complementary manner or in parallel. Each is of interest to formal crisis management, yet has weak and strong points that must be taken into account if more convergence between formal emergency management and social media is to be obtained.

A closer examination of social media use in the Gaza crisis (2008-09) reveals a paradigm shift in the use of social media in crisis contexts. Even in the case of strong technological inadequacies and weak formal crisis management in situ, they served as a substitute to traditional crisis management technologies; moreover they demonstrated the potential for collective distributed intelligence; they backed flexibility on the definition of roles and actors in crisis management; they resisted to censorship and supported expression of multiple realities in a crisis; finally, they enhanced social and geographic convergence in a crisis.

Academics, experts, governments and NGOs currently explore the present and future of Web 2.0 technologies in crisis management; moreover Web 2.0 use in a crisis has already been discussed in social media either occasionally or as a main field of interest. Current developments in the Computer Supported Collaborative Work (CSCW) and Human Computer Interaction (HCI) fields would address many of the technological issues at stake, and help to fertilize further integration of social media in crisis management. A design perspective is needed.

With Web 2.0 becoming more and more pervasive, formal crisis management is compelled to take it into account; but the way the two will be linked to one another is still to be defined. Developments both in crisis management and in social media will shape the future. These developments are not only technological, but are also mainly social and political. Re-examining the command and control structure of crisis management, which has been dominant over the last decade, is fundamental for better agreement between social media and crisis management.

Social media technologies are here to stay and set to change the landscape in crisis information management. They can be a way to democratize society's responses to risk and crisis, and at the same time boost formal crisis management. The question, however, remains whether the formal crisis communication and information community sees them as an uninvited guest or an intriguing newcomer.

ACKNOWLEDGMENT

We are indebted to the Italian blogger, writer and journalist, Vittorio Arrigoni who shared with us his eye-witness testimony and social media experience from the Gaza crisis ground-site.

DISCLAIMER

1. The views expressed here represent the individual views of the authors only. They do not necessarily express official positions of the organizations they represent.
2. The authors contributed equally and their names are placed in alphabetical order. Matina Halkia is the corresponding author.

REFERENCES

Adam, B., Beck, U., & van Loon, J. (Eds.). (2000). *The risk society and beyond: Critical issues in social theory*. London: Sage.

Alexander, D. (2005). An interpretation of disaster in terms of changes in culture, society and international relations. In R.W. Perry & E.L. Quarantelli (Eds.), What is a disaster? New answers to old questions (pp. 25-38). Philadelphia: XLibris.

Alexander, D. E. (2006). Globalization of disaster: trends, problems and dilemmas. *Journal of International Affairs, 59*(2), 1-22. Retrieved August 3, 2009, from http://www.policyinnovations.org/ideas/policy_library/data/01330/_res/id=sa_File1/alexander_globofdisaster.pdf

Barton, A. H. (2005). Disaster and collective stress. In R.W. Perry & E.L. Quarantelli (Eds.), What is a disaster? New answers to old questions (pp.125-152). Philadelphia: XLibris.

Beck, U. (1992). *Risk Society: Towards a new modernity*. London: Sage.

Beck, U. (2009). *World at risk*. Cambridge: Polity.

Boin, A. (2005). From crisis to disaster. In R.W. Perry & E.L. Quarantelli (Eds.), What is a disaster? New answers to old questions (pp. 153-172). Philadelphia: XLibris.

Boyd, D. M., & Ellison, N. B. (2007). Social network sites: Definition, history, and scholarship. *Journal of Computer-Mediated Communication, 13*(1), article 11. Retrieved August 3, 2009 from http://jcmc.indiana.edu/vol13/issue1/boyd.ellison.html

Bucher, H. G. (2002). Crisis communication and the internet: Risk and trust in a global media. *First Monday, 7*(4). Retrieved May 15, 2009, from http://outreach.lib.uic.edu/www/issues/issue7_4/bucher/index.html

Buckle, P. (2005). Disaster: Mandated definitions, local knowledge and complexity. In R.W. Perry & E.L. Quarantelli (Eds.), What is a disaster? New answers to old questions (pp. 173-200). Philadelphia: XLibris.

Byrne, M., & Whitmore, C. (2008). Crisis informatics. *IAEM Bulletin, February 2008*, 8.

Collins, H. (2009). *Emergency managers and first responders use twitter and facebook to update communities.* Retrieved August, 7, 2009, from http://www.emergencymgmt.com/safety/Emergency-Managers-and-First.html

Currion, P. (2005). *An ill wind? The role of accessible ITC following hurricane Katrina.* Retrieved February, 10, 2009, from http://www.humanitarian.info/itc-and_katrina

Dandoulaki, M., & Andritsos, F. (2007). Autonomous sensors for just in time information supporting search and rescue in the event of a building collapse. *International Journal of Emergency Management, 4*(4), 704–725. doi:10.1504/IJEM.2007.015737

Donath, J. (2007). Signals in social supernets. *Journal of Computer-Mediated Communication, 13*(1), article 12. Retrieved August 3, 2009, from http://jcmc.indiana.edu/vol13/issue1/donath.html

Fischhof, B. (1995). Risk Perception and Communication Unplugged: Twenty Years of Process. *Risk Analysis, 15*(2), 137–145. doi:10.1111/j.1539-6924.1995.tb00308.x

Fischhoff, B. (2006). Bevaviorally realistic risk management. In Daniels, R. J., Kettl, D. F., & Kunreuther, H. (Eds.), *On risk and disaster: Lessons from hurricane Katrina* (pp. 78–88). Philadelphia: University Pennsylvania Press.

Gavriilidis, A. (2009). Greek riots 2008: A mobile Tiananmen. In Economides, S., & Monastiriotis, V. (Eds.), *The return of street politics? Essays on the December riots in Greece* (pp. 15–19). London: LSE Reprographics Department.

Gundel, S. (2005). Towards a new typology of crises. *Journal of Contingencies and Crisis Management, 13*(3), 106–115. doi:10.1111/j.1468-5973.2005.00465.x

Harrald, J. R. (2002). Web enabled disaster and crisis response: What have we learned from the September 11th. In *15th Bled eCommerce Conference Proceedings "eReality: Constructing the eEconomy* (17-19th June 2002). Retrieved October 4, 2008, from http://domino.fov.uni-mb.si/proceedings.nsf/Proceedings/D3A6817C6CC6C4B5C1256E9F003BB2BD/$File/Harrald.pdf

Harrald, J. R. (2009). Achieving agility in disaster management. *International Journal of Information Systems for Crisis Response Management, 1*(1), 1–11.

Hewitt, K. (1998). Excluded perspectives in the social construction of disaster. In Quarantelli, E. L. (Ed.), *What is a disaster? Perspectives on the question* (pp. 75–91). London: Routledge.

Hughes, A. L., Palen, L., Sutton, J., & Vieweg, S. (2008). "Site-Seeing" in Disaster: An examination of on-line social convergence. In F. Fiedrich & B. Van de Walle, (Eds.), *Proceedings of the 5th International ISCRAM Conference.* Washington, DC, USA, May 2008.

International Risk Governance Council. (2005). *Risk governance: Towards and integrative approach.* White paper no.1. Retrieved August 5, 2009, from http://www.irgc.org/IMG/pdf/IRGC_WP_No_1_Risk_Governance__reprinted_version_.pdf

James, A. M., & Rashed, T. (2006). In their own words: Utilizing weblogs in quick response research. In Guibert, G. (Ed.), *Learning from Catastrophe: Quick Response Research in the Wake of Hurricane Katrina* (pp. 57–84). Boulder: Natural Hazards Center Press.

Joyce, M. (2008). *Campaign: Digital tools and the Greek riots.* Retrieved August 1, 2009, from http://www.digiactive.org/2008/12/22/digital-tools-and-the-greek_riots

Kittur, A., Suh, B., & Chi, E. H. (2008). Can you ever trust a wiki?: impacting perceived trustworthiness in wikipedia. In *Proceedings of the ACM 2008 Conference on Computer Supported Cooperative Work (San Diego, CA, USA, November 08 - 12, 2008). CSCW '08.* ACM, New York, NY, (pp.477-480). Retrieved August 7, 2009, from http://doi.acm.org/10.1145/1460563.1460639

Krimsky, S. (2007). Risk communication in the internet age: The rise of disorganized skepticism. *Environmental Hazards*, *7*, 157–164. doi:10.1016/j.envhaz.2007.05.006

Lagadec, P. (2005). *Crisis management in the 21st century: "Unthinkable" events in "inconceivable" contexts.* Ecole Polytechnique - Centre National de la Recherche Scientifique, Cahier No 2005-003. Retrieved August 5, 2009, from http://hal.archives-ouvertes.fr/docs/00/24/29/62/PDF/2005-03-14-219.pdf

Lem, A. (2008). *Letter from Athens: Greek riots and the news media in the age of twitter.* Retrieved August 3, 2009, from http://www.alternet.org/media/113389/letter_from_athens:_greek_riots_and_the_news_media_in_the_age_of_twitter

Linder, R. (2006). *Wikis, webs, and networks: Creating connections for conflict-prone settings.* Washington: Council for Strategic and International Studies. Retrieved August 5, 2009, from http://www.csis.org/component/option,com_csis_pubs/task,view/id,3542/type,1/

Liu, S. B., Palen, L., Sutton, J., Hughes, A. L., & Vieweg, S. (2008). In search of the bigger picture: The emergent role of on-line photo sharing in times of disaster. In F. Fiedrich & B. Van de Walle, (Eds.), *Proceedings of the 5th International ISCRAM Conference – Washington*, DC, USA, May 2008.

Moss, M. L., & Townsend, A. M. (2006). Disaster forensics: Leveraging crisis information systems for social science. In Van de Walle, B. and Turoff, M. (Eds.) *Proceedings of the 3rd International IS-CRAM Conference*, Newark, NJ (USA), May 2006.

Palen, L. (2008). On line social media in crisis events. *Educase Quartely*, *3*, 76–78.

Palen, L., Hiltz, S. R., & Liu, S. B. (2007). On line forums supporting grassroots participation in emergency preparedness and response. *Communications of the ACM*, *50*(3), 54–58. doi:10.1145/1226736.1226766

Palen, L., & Liu, S. B. (2007). Citizen communications in crisis: Anticipating the future of ITC-supported public participation. In CHI 2007 Proceedings (pp. 727-735), San Jose, CA, USA.

Palen, L., Vieweg, S., Sutton, J., Liu, S. B., & Hughes, A. (2007). *Crisis informatics: Studying crisis in a networked world.* Paper presented at the Third International Conference on e-Social Science (e-SS). Retrieved August 5, 2009, from http://www.cs.colorado.edu/~palen/Papers/iscram08/CollectiveIntelligenceISCRAM08.pdf

Palmer, J. (2008, May 3). Emergency 2.0 is coming to a website near you. *New Scientist*, 24–25. doi:10.1016/S0262-4079(08)61097-0

Perry, R. W., & Quarantelli, E. L. (Eds.). (2005). What is a disaster? New answers to old questions. Philadelphia: XLibris.

Quarantelli, E. L. (1998). *What is a disaster? Perspectives on the question.* New York: Routledge.

Quarantelli, E. L. (2007). Problematical aspects of the information/ communication revolution for disaster planning and research: ten non-technical issues and questions. *Disaster Prevention and Management*, *6*(2), 94–106. doi:10.1108/09653569710164053

Renn, O., & Walker, K. (Eds.). (2008). *Global Risk Governance: Concept and Practice Using the IRGC Framework*. Dordrecht, The Netherlands: Springer. doi:10.1007/978-1-4020-6799-0

Rettew, J. (2009). *Crisis communication and social media*. Paper presented at the AIM2009 Conference. Retrieved August 4, 2009, from http://www.slideshare.net/AIM_Conference/crisis-communications-and-social-media-jim-rettew-the-red-cross-2009-aim-conference

Shankar, K. (2008). Wind, Water, and Wi-Fi: New Trends in Community Informatics and Disaster Management. *The Information Society*, *24*(2), 116–120. doi:10.1080/01972240701883963

Shaw, M. (1996). *Civil society and media in global crisis: Representing distant violence*. London: Pinter.

Sorensen, J. H., Sorensen, B. V., Smith, A., & Williams, Z. (2009). *Results of An Investigation of the Effectiveness of Using Reverse Telephone Emergency Warning Systems in the October 2007 San Diego Wildfires*. Report prepared for the U.S. Department of Homeland Security. Retrieved August 4, 2009, from http://galainsolutions.com/resources/San$2520DiegoWildfires$2520Report.pdf

Stephenson, W. D., & Bonabeau, E. (2007). Expecting the unexpected: The need for a networked terrorism and disaster response strategy. *Homeland Security Affairs*, *III*(1). Retrieved August 5, 2009, from http://www.hsaj.org/?article=3.1.3

Sutton, J., Palen, L., & Shklovski, I. (2008). Backchannels on the front lines: Emergent uses of social media in the 2007 Southern California Wildfires. In F. Fiedrich & B. Van de Walle (Eds.), *Proceedings of the 5th International ISCRAM Conference*, Washington, DC, USA, May 2008.

The Risk Communicator. (2008). Social media and your emergency communication efforts. The *Risk Communicator*, *1*, 3-6. Retrieved June 15, 2009, from http://emergency.cdc.gov

Tinker, T., & Fouse, D. (Eds.). (2009). *Expert round table on social media and risk communication during times of crisis: Strategic challenges and opportunities*. Special report. Retrieved August 6, 2009, from http://www.boozallen.com/media/file/Risk_Communications_Times_of_Crisis.pdf

Tziros, T. (2008). *The riots in new media*. Article in Newspaper "Makedonia" on December 15, 2008. [In Greek]. Retrieved August 5, 2009, from http://www.makthes.gr/index.php?name=News&file=article&sid=30225

Vieweg, S. (2008). Social networking sites: Reinterpretation in crisis situations. *Workshop on Social Networking in Organizations CSCW08*, San Diego USA, November 9. Retrieved August 5, 2009, from http://research.ihost.com/cscw08-socialnetworkinginorgs/papers/vieweg_cscw08_workshop.pdf

Vieweg, S., Palen, L., Liu, S. B., Hughes, A. L., & Sutton, J. (2008). Collective intelligence in disaster: Examination of the phenomenon in the aftermath of the 2007 Virginia Tech shooting. In F. Fiedrich & B. Van de Walle (Eds.), *Proceedings of the 5th International ISCRAM Conference*, Washington, DC, USA, May 2008.

von Lubitz, D. K. J. E., Beakley, J. E., & Patricelli, F. (2008). All hazards approach' to disaster management: the role of information and knowledge management, Boyd's OODA Loop, and network-centricity. *Disasters*, *32*(4), 561–585.

von Lubitz, D. K. J. E., Beakley, J. E., & Patricelli, F. (2008). Disaster management: The structure, function, and significance of network-centric operations. *Journal of Homeland Security and Emergency Management, 5*(1). Retrieved August 5, 2009, from http://www.bepress.com/jhsem/vol5/iss1/42/

ENDNOTES

[1] Harrald (2009) advocates for agility in USA emergency management that has emphasized doctrine, process and structure after 9/11.

[2] Such tools serve mainly for alert and early warning, monitoring, decision support etc. To mention only a few, GDACS (Global Disaster Alert and Coordination System), the Indian Ocean Tsunami Warning System, various GIS based Decision Support Systems for disaster management at different scales and levels of administration.

[3] Tinker and Fouse (2009:17) provide a provisional list of social media.

[4] The issue "What is a disaster?" is said to be a "definitional minefield" (Alexander 2005:26).

[5] A google search in the internet (keywords: social media + crisis or disaster or emergency) returns mainly blogs and wikis.

[6] The connectivIT Laboratory of Natural hazards Center, University of Colorado investigated on-site and on-line citizen-site information generation and dissemination activities.

[7] See blog "The riots for the murder of Alexis Grigoropoulos in social media" [in Greek]. Retrieved August, 5, 2009, from http://oneiros.gr/blog/2008/12/07/griotscoverage/

[8] A similar tactic was followed during the Mumbai attacks.

[9] M.Tsimitakis was one of the few journalists who actually used new media to transmit information from the streets. There were claims that twittering, although extensive, contributed little to information as it either reproduced news from the mass media or was used for expressing thoughts and opinions (Tziros 2008).

[10] Blog "Twittering away". Retrieved August 6, 2009, from http://oneiros.gr/blog/2009/01/13/gazasocialmediane/

[11] Local newspaper "Il Centro". Retrieved August 3, 2009, from http://racconta.kataweb.it/terremotoabruzzo/index.php

[12] Announcement of EERI's briefing on the Abruzzo, Italy (L'Aquila) Earthquake. Retrieved August 3, 2009, from http://www.eeri.org/site/meetings/laquila-eq-briefing

[13] http://www.geosynthetica.net/news/article/2009/Interview_Rimoldi_080309.aspx

[14] See for example chat in Italian in the blog http://www.earthquake.it/blog/2009/terremoto-provincia-de-laquila-05-aprile-2009/#more-123 (retrieved 6th April 2009)

[15] See for example in English chat in http://www.italymag.co.uk/forums/general-chat-about-italy/12088-earthquake.html (retrieved on 6th April 2009) and

[16] http://it.wikipedia.org/wiki/Terremoto_dell'Aquila_del_2009

[17] E.Holdman (2009) presents his experiences on the advances of emergency management technology in his blog "Typewriters to Twitter: Emergency Management Technology Through the Decades". Retrieved August 5, 2009, from http://www.emergencymgmt.com/disaster/From-Typewriters-to-Twitter.html

[18] Governments use twitter for emergency alerts, traffic notices and more. Retrieved August 4, 2009, from http://www.govtech.

com/gt/579338?id=579338&full=1&story_pg=2

[19] Alexandra Rampy published a list of US government Twitter accounts on October 2008. Retrieved August 5, 2009, from http://fly4change.wordpress.com/2008/10/08/the-governments-a-twitter-take-2-its-official/

[20] Retrieved August 1, 2009, from http://www.fema.gov/emergency/ipaws

[21] http://www.redcross.org/en/connect

[22] The Institute of Commercial Management on 14th July 2009. Retrieved August 5, 2009, from http://news.icm.ac.uk/technology/australia-to-use-twitter-for-bush-fire-alerts/2516/

[23] D. Fleet lists Ontario government agencies trying Twitter. Retrieved August 5, 2009, from http://davefleet.com/2008/10/twitter-as-a-hyper-local-emergency-information-tool/

[24] M.Kujawski compiles a list of current initiatives involving social media and government in several countries. Retrieved August 5, 2009, from http://government20bestpractices.pbworks.com/

[25] Retrieved August 5, 2009, from http://www.guardian.co.uk/uk/2009/jul/27/twitter-socialnetworking/

[26] Al-Jazeera and Palestinian journalists were reporting; their broadcasts being, overall, popular in the Arab World; Al-Jazeera, broadcasting in English, does have potentially an international audience. See also http://www.haaretz.com/hasen/spages/1054282.html (retrieved August 4, 2009)

[27] Presuming the lack of an effective control of the territory by the Palestinian Authority.

[28] Facebook query on causes (keyword: Gaza) on 29/7/2009

[29] By donating one's status, a Facebook member becomes a personal advocate of a cause.

[30] The different roles and actors in the creation of the news message will be discussed below.

[31] http://www.pbs.org/mediashift/2009/02/how-social-media-war-was-waged-in-gaza-israel-conflict044.html (retrieved August 3, 2009).

[32] On the number of dead and wounded for example; or on the use of human shields.

[33] In January 2009, the Israeli Ministry of Immigration announced the international recruitment of multilingual volunteers to post comments in blogs and websites in defence of Israeli policies in Gaza. See also 19th January 2009 "Israel recruits 'army of bloggers' to combat anti-Zionist Web sites" article on Haaretz. Retrieved August 6, 2009, from http://haaretz.com/hasen/spages/1056648.html

[34] Israeli journalist Gideon Levy received death threats for instigating social media action over the Gaza crisis, as BBS reports on posts after Gideon Levy's article on Haaretz (retrieved August 5, 2009, from http://bbsnews.net/article.php/20090111162755263); so did Facebook members Joel Leyden and Hamzeh Abu-Abed (Article "Facebook Users Go to War" in Time. Retrieved August 5, 2009, from http://www.time.com/time/world/article/0,8599,1871302,00.html

[35] The presence of an internet site calling for the murder of Italian blogger Vittorio Arrigoni instigated a European Parliament member to formally address the Italian Minister of Foreign Affairs on the protection of Italian citizens. The offensive site was eventually closed down.

[36] As early as 2006, state control was exercised in limiting access to social media. State control of Facebook or Twitter in the 2009 upheavals in Iran and China is an unequivocal indication of the power of social media, perceived as antagonistic to crisis mitigation and management. The US Military recently (August 2009) considered a ban of Web 2.0 technologies in the armed forces.

[37]   EU High Representative for Common Foreign and Security Policy on the Gaza Crisis press remarks of 21st January 2009

[38]   See for example the MIT Center for International Studies' CIS Starr Forum video (13th January) "Chomsky on Gaza". Retrieved March 10, 2009, from http://web.mit.edu/cis/starr.html

[39]   http://www.nytimes.com/2009/01/16/world/middleeast/16mideast.html, http://www.guardian.co.uk/world/2009/jan/07/gaza-israel-obama, http://news.bbc.co.uk/2/hi/middle_east/7833919.stm. All three retrieved August 5, 2009.

[40]   http://it.blogbabel.com/metrics/

[41]   Arrigoni's personal communication to the authors (4.8.2009 -- original in Italian).

Chapter 11
Utilizing Web 2.0 for Decision Support in Disaster Mitigation

Kumaresh Rajan
The State University of New York at Buffalo, USA

Rui Chen
Ball State University, USA

Hejamadi Raghav Rao
The State University of New York at Buffalo, USA

JinKyu Lee
Oklahoma State University, USA

ABSTRACT

The principles of Web 2.0 such as transparency, security, community, usability, and availability are well suited to help effectively manage the effects of a disaster. Many Web 2.0 technologies rely on social collaboration, and as a result these technologies are built with robust communication channels. Utilizing this existing framework will help to create software systems that can efficiently manage disasters. This chapter will examine differing Web 2.0 innovations through the use of Activity Theory, and the benefits and drawbacks of each technology will be analyzed. From this analysis, recommendations and conclusions will be presented to the reader.

INTRODUCTION

The recent advances in computer systems have increased the ubiquity of Information Technology (IT), but it has not been fully utilized to solve some of the issues that affect our lives. Namely,

the umbrella of Web 2.0 technologies has not been effectively employed to manage disasters. Web 2.0 can support disaster management because it efficiently manages large amounts of data and connects individuals through multiple networks for informed decision making. This chapter reviews a variety of Web 2.0 technologies and discusses their attributes and implications through the use

DOI: 10.4018/978-1-61520-987-3.ch011

of Activity Theory. The chapter, also, identifies the remaining challenges facing these technologies and proposes future improvements that may overcome these remaining challenges.

Natural disasters can have devastatingly broad reaching affects on a population and behave unpredictably. For example, in 2008 Hurricane Ike formed off the Gulf Coast, and it was predicted at first to make landfall off the Florida Keys. Instead, Ike moved on a different path after it passed over Haiti and Cuba, and as a result the weather models related to Hurricane Ike needed to be quickly revised (Berger, 2008). This scenario illustrates the need of having robust information systems in place during an emergency response for a natural disaster. Incorporating the features of Information Communication Technology to disaster mitigation would allow for a more constructive and productive response to a disaster. The basic principles of Information Technology such as transparency, reliability, accuracy, and timeliness are well suited to manage the affects of an emergency situation.

In this chapter, we examine Web 2.0 technologies and their potential roles in disaster mitigation. The chapter reviews several different Web 2.0 systems and discusses their attributes through the use of Activity Theory. The chapter is organized as follows. A description of the individual Web 2.0 technologies is first presented to the reader. This is followed by a discussion section. Background information is presented about Activity Theory in the discussion section, so that the reader is familiar with the context in which the differing Web 2.0 technologies will be evaluated. This section continues by discussing the specific Activity Theory principles used to evaluate the differing Web 2.0 technologies, and this is followed by an evaluation of these technologies though the use of the previously defined principles. Next, the remaining challenges are identified and discussed, and the chapter concludes with a recommendation section that outlines possible solutions to these challenges.

WEB 2.0 APPLICATIONS AND DISASTER MANAGEMENT

Social Tagging

Social tagging is a typical technology that is part of the Web 2.0 umbrella. This technology allows others individuals to associate keywords with any given topic. These keywords are used to categorize the different topics, and this user created index is used during keyword searches to aggregate similar topics as search results. All users have access to this information which in turn emphasizes the collaborative aspect of this Web 2.0 technology by shifting the responsibility of tagging new material from the content creator to content consumers. Allowing users to tag and sort different content can create more robust search results because multiple people are managing how the content is categorized.

Essentially users are creating a sizeable repository of highly specific metadata. For example, one user called for members of Flickr, an online photo sharing site, to tag all pictures related to Hurricane Katrina with a unique identifying keyword. Essentially, this user instructed other individuals to use the keywords "katrinamissing", "katrinafound", and "katrinaokay" to indicate the status of the individuals portrayed in the uploaded images. Similarly, after the collapse of the Minneapolis bridge in 2007, people were instructed by Flickr members to use the specific identifier of "mpls35W". The selection of the keyword is only effective if non-generic terms are used. Generic keywords cause images to appear in multiple searches, and these retrieved images maybe irrelevant to the search criteria. For example, during the California wildfires and admin asked the users in his group to use the tags of "SanDiego" and "fire" to categorize the uploaded images, but the selection of these keywords caused the images to be retrieved in unrelated searches. The key to

Figure 1. Photo Sharing (©2008, Flickr)

these initiatives is to create a system that enables community social interaction (Palen, 2008).

Photo Sharing

Documenting the effects of a disaster can be a time consuming process, but the use of images to convey the impact of a natural disaster can allow first responders and victims to quickly assess the scope of the situation (see Figure 1). The use of photographs can describe a disaster scenario more vividly and accurately than most other forms of communication, and due to this a photograph can be considered an unbiased and impartial source of information. The information presented can also convey the overall devastation and size of the disaster to the public. These images can be used to motivate individuals to donate to disaster relief, to notify emergency response teams, to advise disaster management organizations, and to help estimate the total number of resources and labor needed to rebuild.

More importantly images help to connect people to a disaster, so that the public can connect to the victims of the tragedy by associating a face to the disaster. Images are a vital tool that can manage the response to a disaster, but this type of information needs to be quickly disseminated

to all agencies and individuals affected by the disaster. Online photo sharing sites already exist, and these sites index and track numerous images that span multiple areas. Utilizing this existing technology into a disaster management site would help to ensure that the personal side of the tragedy reaches a wide audience. Images would also help in the effort of reuniting families that were separated. Images could be uploaded into a site by family members, and multiple agencies could be given access to this information in order to see if a match could be found to the uploaded image.

Users are using images to create an amalgamation of multiple sources. This mashup brings together different image sources and creates a new graphical representation of the disaster. For example, location tagging in Google Maps can be cross-referenced with images uploaded by different users to show the spatial location of each of the uploaded images (see Figure 2). Using a mashup in this way will allow first responders and the media to quickly visualize the scope of the disaster. Overlaying pictures onto a map can also show which major streets are blocked or

Figure 2. Google Map Tagging (©2008, Google Maps)

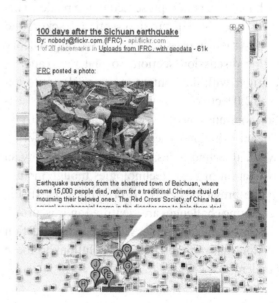

Table 1. Content management system categories

Type	Access	User Restriction
I	Fully Open	No restriction
II	Lockable	All pages public, but editing restricted in various ways
III	Gate	Some pages public, others restricted to registered users
IV	Restricted	All users must be registered; may involve further group restrictions
V	Firewalled	All users must be on a specific network
VI	Personal	Notebook usage on one system or private website directory
VII	Guarded	Access to viewing and editing pages is determined by assigned user roles

Adapted from "Leveraging a Wiki to Enhance Virtual Collaboration in the Emergency Domain"

damaged. Figure 2 illustrates how users of Google Maps worked together to create a mashup for the 2008 Chinese Earthquake. The pictures, the tagged locations, and the cross referenced news stories are all combined to create a clear and cohesive view of the situations.

Any system deployed to mitigate the effects of a natural disaster needs to meet certain minimum requirements. The system should be straightforward, readily available (e.g. platform independent), customizable, standardized, and used in the evaluation of a disaster (Murali, 2006). Content management systems such as wikis are platform independent web applications that are maintained by community groups and support customizability through the use of modules. Also, content management systems can allow for the creation of user roles to oversee the accuracy of the information entered into the system. An online content management system has the additional benefit of freely allowing for the collaboration of multiple parties that may be physically separated by large distances (Meier, 2008). Such a system can be used to archive past revisions, and these systems can easily search all the information available on the system. Content management systems fall into multiple categories depending

on how these systems handle access rights and privileges. Table 1 illustrates these categories.

Blogs

Blogs have become a more popular means of communication (see Figure 3). "A blog is a website where entries are written in chronological order and displayed in reverse chronological order, much like a message board or chat room" (Prentice, 2008). Blogs may combine text, images, graphics, video, and links to related web sites or blogs. For example, the recent trend has been the

Figure 3. Blogs (©2009, Red Cross)

inclusion of video blogging and audio blogging (e.g. podcasts). Blogs emphasize the social and community element by creating a dialog between the author and the readers. The author can post his or her views on a specific topic, and a large number of readers can post their comments to the author's original post. Corporations have started utilizing blogs as a fast way to release information to the public and respond to the concerns of the public. The Red Cross has begun using this technology to aide its ability to communicate with volunteers, the media, and the victims in an emergency situation. The Red Cross developed a blog as a way to organize information related to a specific emergency situation. "Beyond just providing important information to the public and media, the Red Cross has gained the ability to correct information that the media may have gotten wrong and increase fundraising and awareness efforts" (Prentice, 2008). This use of technology has greatly aided the organization's ability to react to multiple different scenarios, and due to its success the Red Cross has started training volunteers to use blogs and post information concerning emergency situations.

XML

Many of the broad ideas related to a disaster management system can help develop the role of the application, but any system developed for such a task must be coded with the strict enforcement of standards because multiple parties will interact with the Web 2.0 application during a disaster scenario. Encoding transmitted data in the form of standardized XML will allow for the application to integrate with other third party software and make standard services available to other web applications (Zisman, 2000). Coding standards will help a disaster management application enforce the use of a standard set of disaster related terminology (Herold, 2005). For example, the severity of each disaster could be codified to

show the strength of the disaster as well as the impact it has had to individuals located around the epicenter (Herold, 2005). This standardized terminology could convey the exact status of the victims, monetary cost of the disaster, weather information, logistical information, and scope of the damage (Careem, 2006). Communication among disaster management computer systems could be opened with the use of an XML standard. Adopting a standard would allow information to freely disseminate to the proper channels where it would eventually be handled. Also, collaboration among web applications would greatly increase through the use of a common standardized XML structure.

Ajax

Asynchronous JavaScript and XML (Ajax) in and of itself is not a new social networking web application. Instead Ajax is the framework on which other social web applications and Web 2.0 applications are based. This new programming model utilizes XML and client based JavaScript to emulate a rich desktop experience on the web. Ajax can be described as an amalgamation of many different component technologies such as JavaScript, Document Object Model (DOM), XML, CSS, and DHTML. Ajax based applications use DOM and JavaScript to make a more responsive user interface, and any meaningful changes that the user makes can then be communicated to the server by using a request/response structure that relies on the XMLHttpRequest. The XML-HttpRequest object is a standards based object that encapsulates the application data as it is being transmitted between the client machine and the server. Web applications can then be updated with the new information without having to refresh the entire page. This technology is utilized by social tagging sites like del.ic.ious and Geographical Information Systems (GIS) like Google Maps. Ajax has been utilized in many different varia-

Figure 4. Original activity theory model

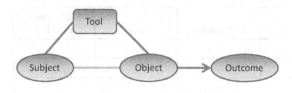

Figure 5. Revised activity theory model

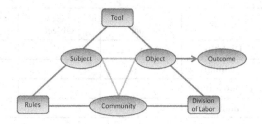

tions for many years, but only recently has all the different technologies been placed together in a standardized way (MindLeaders, 2007). A benefit of Ajax is that smaller packets of information are transmitted between the server and the client because Ajax based applications only refreshes a small portion of a page instead of rendering the whole page like a traditional web application. However, this benefit adds extra complexity to an application. This additional complexity could introduce additional defects into the system because web developers may not know how to properly design Ajax applications (Garrett, 2005).

DISCUSSION

Activity Theory

Vogotsky, Rubinshtein, Leont'ev, and other Soviet psychologists originally developed Activity Theory (Kuuti, 1996). Activity Theory was designed as a framework for understanding how humans achieve a desired goal by manipulating their environment. Recently, Activity Theory has been applied to human computer interaction to understand and evaluate the way in which humans use computer systems (Kuuti, 1996).

Activity Theory is comprised of three main principles. One of the principles of Activity Theory is that the activity is treated as the basic unit of analysis. Other psychological theories use human actions instead of activities as the basic unit of analysis. The use of human actions as the basic unit of analysis can create confusion because the context in which the action occurs is missing. The

analysis of an action without the proper context to frame that action introduces ambiguity into the analysis and makes the results of the analysis less meaningful. Another principle of Activity Theory is that activities develop over time. Activities cannot be thought of as static. Instead activities have a history, and the past variations of an activity stay embedded in the current incarnation of an activity. Another key principle of Activity Theory is that activities have artifacts that have a mediating effect on activities. Artifacts include rules, tools, procedures, laws, methods, and other human constructs that affect the way in which the activity occurs. The existence of these artifacts mediates the way in which the actor of the activity interacts with the object of the activity. This mediation does not allow the actor to have direct contact with the object of an activity (Kuuti, 1996).

Originally, the model of Activity Theory showed how the actor manipulated an entity through the use of a tool. The result of this manipulation was the outcome. In this model the actor was called the subject, and the entity was called the object. The object can either be a tangible or intangible entity. The only requirement for the object is that it must be able to be manipulated. The tool is what both limits and enables this manipulation, and the subject can either be an individual or a group. This relationship is illustrated in Figure 4.

This existing model was later revised by adding a new component called community. The community component was added to include the affects of the environment on the activity. Due to the inclusion of community, two new relationships

Table 2. Activity theory mapping

	Tool	Object	Subject	Community	Rules	Division of Labor
Usability	√					
Credibility			√			
Collaboration				√		
Standards					√	
Hierarchy						√

had to be added. These new relationships included rules and division of labor. This new model is illustrated in Figure 5.

The two new mediating relationships affect the way the community interacts with either the subject or the object. For example, rules (e.g. conventions and norms) show the acceptable way for the subject to interact with members of the community. Division of labor shows how the community is organized in order to transform the object to a given outcome (Kuuti, 1996).

Activity Theory Principles

Applying Activity Theory to how humans use computers can offer insights into the requirements for a computer system used in disaster mitigation. Activity Theory has been used by others in the past to better understand new developments in Information Technology. This approach can be employed with disaster management applications. The table below illustrates how the requirements of a disaster related application closely match the main principles of Activity Theory. The essence of Activity Theory states that the subject (user) uses a tool (disaster mitigation software) on an object (disaster situation) to achieve a specified output (managed disaster). Throughout this process the community (public), rules (mores and laws), and division of labor (user roles/organizational jurisdiction) help shape the way in which the subject achieves the desired outcome (Martins, 1999). Each of the components of Activity Theory maps

to a corresponding principle of software design. This mapping can be seen by looking at Table 2.

Disaster mitigation software needs to take into account the different constructs of Activity Theory. This entails that the software should be designed so that each element of Activity Theory exhibits the specific property listed in Table 2. For example, the tool must be usable. Usability entails that the tool easily and effectively allows the subject to manipulate the object into the desired outcome. The tool should have an intuitive interface, so that it can quickly be used by the intended user base during a disaster scenario. Also, a Web 2.0 disaster application must be scalable in order for this application to be considered usable. A potentially large number of users will want to use this tool during a disaster. As a result, the tool should have enough resources to handle a large volume of user requests. Another requirement of a Web 2.0 disaster application is that the subjects using the tool should be credible. Everyone must be able to view the information on the web application, but users that change the content on the system must be either affected by or connected to the disaster. For example, Disaster Management Organizations (DMOs), victims, volunteers, and fund raisers could all be potential users that legitimately have the right to change the content on the web application. Enforcing credibility will ensure that the information contained in the web application is correct, and this will allow users to be more comfortable in relying on the data within the application. The application must also allow

Table 3. Comparison of Web 2.0 technologies

	Example Instrument	Example Object	Example Outcome	Example Subject	Example Community	Example Rules	Example Division of Labor
Social Tagging	del.icio.us	Disaster scenario	Categorization of an emergency scenario	Individual/Group	Victims/Public	Meta data must relate to content	Users modify metadata in order to classify added content
Photo Sharing	Flickr	Disaster scenario	Disaster scenario visualization	Individual/Group	Victims/Public	Images must be accurately tagged and relate to disaster scenario	Users add images related to disaster and modify meta data
GIS	Google Maps	Disaster scenario	Visualization of a disaster through the use of a geographic information	Individual/Group	Victims/Public	Geographic information must be correct and relevant to disaster	Users modify system to reflect current situation
Blogs	Wordpress	Disaster scenario	Creating two way communication between content creator and content consumers	Individual/Group	Victims/Public	Owner of blog is moderator and content creator	Content creator introduces new topics, and content consumers respond to these topics
Content Management	Wikipedia	Disaster scenario	Creating a viewable and usable warehouse of disaster related information	Individual/Group	Victims/Public	All users must be given a role that outlines their duties and privileges	Content consumers and creators interact directly to create all content on system

for collaboration to occur among multiple users (Bessis, 2009). Pertinent information should be allowed to be freely exchanged among the affected parties, and this type of exchange can be made more meaningful through the enforcement of standardization. Standardization helps govern the way information is encapsulated and sent among different users. Standards simplify the building of common interfaces by allowing different applications to communicate among each other. The disaster management application must also enforce the principle of hierarchy. Users of the system must be given different privileges depending on their role in the system and their level of authority. The principle of least privilege should be enforced to ensure that users have the minimum level of access needed to perform their duties. Hierarchy lets the system hide information

from users who do not have or need access to the information, and it helps reinforce the principle of credibility (Nielsen, 1995).

Comparison of Web 2.0 Applications

Activity Theory is a theory that offers a framework in which the critical issues of context can be taken into account for system design (Bertelsen et al. 2003; Vygotsky 1978). It suggests that human activity such as disaster management is directed toward a material or ideal object, mediated by artifacts or instruments, and socially constituted within the surrounding environment. Using this framework we compare the Web 2.0 applications mentioned above.

Communication technology has been used as a means to reduce the effects of a disaster. For

Table 4. Summary of Web 2.0 in disaster management

Summary	Technology	Reference
Discusses the use of social media namely blogs in emergency situations.	Blogs	Huffman and Prentice (2008)
Article focuses on use of wiki for collaboration and group support in a disaster.	Wiki	White et al. (2008)
Discusses the use of open source software in the implementation of an internet based GIS system.	Geographical Information System	Herold et al. (2005)
Discusses Sahana a free open source application that manages information related to relief, recovery, and rehabilitation operations.	Content Management	Careem et al. (2006)

example, amateur radio operators were used in disaster situations when normal telecommunication networks failed. Work has been done to incorporate the advances in Information Technology to disaster mitigation (Shankar, 2008). This field of study has been garnering more attention from the research community after the events of the Asian tsunami, hurricane Katrina, and the Chinese earthquake (Murphy, 2006).

Currently there is a push to blend contemporary web applications and disaster management. Generally, user centered applications that emphasize collective intelligence, offer network-enabled services, and empower users with control over their own data are referred to as Web 2.0 technologies (Madden and Fox, 2006). The user experience for the Web 2.0 applications has become richer, and these gains in usability can significantly aid the implementation of websites dedicated to decision-making in disaster management. Wikis, social networks, and content sharing sites are examples of Web 2.0 applications; each of these web technologies provide unique features to facilitate decision making through the offering of data and collaboration. Despite these merits, the existing Web 2.0 applications still need to effectively incorporate the individual systems into an integrated disaster management platform that can rapidly and efficiently respond to different disaster situations. Moreover, disasters can behave unpredictably, strike without much warning, and affect large numbers of people. As a result, the

next generation of Web 2.0 applications needs to be able to quickly react to real time changes and handle a potentially large user base. They need to enforce standards and incorporate the principles of responsibility (user roles), reliability, collaboration, and accuracy in order to gain acceptance and eventually widespread usage. Table 4 presents a summarization of some of the technologies that are currently being pursued.

In the article entitled "Social Media's New Role in Emergency Management", blogs are discussed as a way for an organization to directly manage the transfer of information to the audience (Huffman and Prentice, 2008). This direct transfer removes any possible media bias or misinformation that could arise when an intermediary is used. Blogs can aggregate all the content that organizations need to convey during a disaster scenario. For example, pictures, press releases, and announcements can all be quickly distributed through blogs. When creating a blog, the scope and intent of the blog should be well defined. This ensures that the nature of the information on the blog relate to these defined guidelines. Also, the blog should be monitored to ensure the validity of the content, and the blog should be updated regularly when new incidents occur. Other forms of media such as videos and photos should be added to the blog in order to enhance the existing content (Huffman and Prentice, 2008). Doing this helps to enforce the principles of credibility and usability.

Connie White looks at using wikis for disaster management in her paper entitled, "Leveraging a Wiki to Enhance Virtual Collaboration in the Emergency Domain". The article elaborates on the different types of wikis that exist (see Table 1), and it overviews some of the characteristics and benefits of wikis. Different wikis in the emergency domain are also discussed. One specific wiki that the authors elaborate on is emergenciwiki.org. This wiki was designed as proof of concept research project that can be used by a community during a disaster scenario. The authors created this wiki by utilizing open source technology and outsourcing. The wiki was designed around a basic structure. The wiki was setup to handle a disaster situation by allowing members to add and modify content on the wiki as new incidents occurred. Initially, community involvement has added more elements to the original wiki framework, but currently the wiki no longer is available to the public (White, 2008). A wiki (e.g. content management site) can only be an effective tool when there is a high volume of community interaction. Collaboration can help to organize, classify, and moderate the information contained in the wiki. Also, strongly enforcing user roles will help to increase the credibility of a content management site. The use of standardization is another critical element of content management sites. Standards can help content creators judge the validity of information in the system by providing guidelines on acceptable content. Inclusion of articles, revision of information, and resolving disputes can all be more easily managed by creating rules and standards that govern the operation of the site.

Herold, Swada, and Wellar discuss their Internet based Geographical Information System (GIS) in their paper entitled, "Integrating geographic information systems, spatial databases and the internet: a framework for disaster management". The web application was developed using open source software including MapServer and Chameleon. MapServer is a development environment that can construct spatially enabled web applica-

tions. Chameleon is built on top of MapServer. Chameleon was designed to help create web mapping applications. Both of the applications follow the specifications of the Open Geospatial Constorium (OGC), and as a result these applications can work with any data format compatible with the OGC specifications. Users can use the GIS system to create areas or points on a map that specify disaster areas, relief centers, or distribution points. Users can manipulate areas of the map to help facilitate relief efforts by adding labels and other forms of textual information (Herold, 2005).

Sahana is discussed in detail in the article entitled "Sahana: Overview of a disaster management system". Sahana was created as a way to manage information associated with the relief, recovery, and rehabilitation operations after a disaster has struck. This application is web based and runs on Linux, Apache, Mysql, and PHP. Sahana was initially designed after the Asian Tsunami hit Sri Lanka. Volunteers from the Sri Lankan IT industry created Sahana. The core modules of this web application include an organizational registry, request management system, shelter registry, and a missing person registry. The optional modules in the system include a volunteer coordination system, child protection system, inventory control and catalog system, and mobile messaging system. These optional modules can be used to customize the functionality of the Sahana application. Sahana has been deployed to numerous disasters including the Sri Lanka Tsunami, Pakistan/Kashmir Earthquake, Guinsaugon landslides in the Philippines, and the Jogjakarta Earthquake in Indonesia (Careem, 2006).

Each of these technologies offers some unique advantage in relation to the others, but there is a tradeoff between the perceived benefits and the potential drawbacks. A benefit of all the systems is that the user base can be limited to only individuals who have proper access privileges. This restriction ensures that the data on the system is reliable, and that there is no misuse the system. This benefit also increases the credibility of the system. Social tagging's major benefit is that it can aggregate the diverse backgrounds of the user base in order to create a system based on majority opinion. Though, the problem with this is that it requires an active user base that can continually refine old information while also tagging new data entering into the system. Also, the content is only as good as the users of the system make it. These same benefits and flaws can also be applied to photo sharing applications due to the emphasis on community management. Enforcing user roles and closely adhering to a standard set of rules will allow help alleviate this problem. GIS systems are good at showing the scope and magnitude of a disaster by overlaying the disaster with geographical information. The epicenter, the distance, and potential damage can all be readily seen in a GIS application. The problem with displaying information in this manner is that everything becomes disjointed, and the other aspects of a disaster cannot be properly identified. Also, the system does not offer any channel of communication from a DMO to the victims. This lack of communication is also present in social tagging and photo sharing. Content management and blog applications tackle the communication problems by providing a way to organize and view information among multiple parties. These types of applications are ideal ways to quickly describe the entire situation. The only problem is that both these technologies are very general, and they cannot meet all the specific needs of a disaster. Blogs have the additional drawback of not giving equal emphasize on two way communication because the ideas from the author will be visible to everyone, but the user comments to the original post will be less easily accessible.

Taking this into consideration it seems that a hybrid approach that incorporates the functionality of each technology would be an adequate way to deal with a disaster scenario. In such a system, the content management would form the backbone, and additional technologies could be added into the content management system through the use of modules. These modules could be interchangeable depending on the user requirements. Such a design is already being used in the business world. Content management systems such as Joomla create the framework for which other modules can be added. These modules can make the initial system into a variety of other applications. For example, modules in Joomla can turn the initial base system into an online store application, a social network, or a message board system. With this variety of applications, user requirements can be better met (Joomla, 2009).

Each different technology that was discussed has its own negative and positive attributes. Currently computer systems exist that try to overcome the negative attributes of different Web 2.0 technologies by integrating the information provided by different technologies into a single system. The Humboldt-Universität zu Berlin has created such a system called HUODINI (HUmbOldt DIsaster MaNagement Interface). HUODINI integrates and visualizes the information from heterogeneous sources by scouring free data sources on the web. These sources include news feeds, personal blogs, user tagged images, and seismographic images. The compiled information is brought into the system as a schema less Resource Description Framework, and then the information is tagged with time and location metadata. Users run queries on the system, and the metadata tags allow the system to retrieve the relevant information. All selected information is then displayed to the user using Google Maps.

HUODINI queries data in multiple different formats. For this reason, HUODINI has to use wrapper objects that can take the disparate data objects and create a common interface that the system can use. The system's mediation layer then handles the wrapper objects. The mediator layer controls all the wrappers, stores all the information from wrappers in a central repository, and executes triggers that periodically query selected data sources for updated information. The mediation layer also contains an information extraction module that parses and categorizes the retrieved data into events, relationships, and objects. The data is then handled by an information integration module that filters duplicate information and houses the information for potential SQL queries that can be run by a user. This information is accessed by the user through a Web 2.0 application and Google Maps. The web application uses Ajax and XML to communicate with the server backend. Google Maps is used to visualize the retrieved information by marking the location of tagged images, news events, and disaster specific information (Fahland, 2007).

Remaining Challenges

Using Web 2.0 technologies for a disaster management system can seem like a very promising proposal, but there are also some issues that arise with adopting this design choice. Applying Activity Theory, we identify a number of issues that not been fully addressed by the existing Web 2.0 applications.

A strong user base is a fundamental requirement of an online disaster management system. Past research has shown that people naturally join both volunteer and community based groups during the wake of a natural disaster (Laituri, 2008). These local groups can consist of both first responders and victims of a disaster. Their proximity to the center of the disaster and their knowledge concerning the surrounding area are both invaluable pieces of information that could aide other organizations. Disseminating this information from these individuals could be achieved through the use of an online application. Ordinary people can effectively become sensors for a disaster. Connecting these sensors to each other and to the greater response community can provide for a quicker and more targeted response to the disaster. The specific information that can be provided by these human sensors will ensure that resources and funding will be fully utilized, and this information can allow for a more directed and

measured response to the disaster without needlessly wasting tangible and intangible resources. Strong community involvement can also aid in the raising of awareness and funds to begin the recovery and reconstruction phase after the initial affects of a disaster. "The notion of "people as sensors" - people collecting information to aid in the recovery process and posting this information for broad dissemination outside of the established traditional channels of emergency response is an aspect of disaster response that needs to be examined" (Laituri, 2008).

Effective organizational communication among different agencies in a disaster can be difficult to achieve given that some organizations do not share a common vocabulary. Classification schemes indicating the severity of a disaster may be different, and the use of these different classifications may either elicit the wrong response or may be entirely meaningless to another organization (Manoj, 2007). Developing standards to describe a disaster would be a useful way to further the meaningful exchange of information among Disaster Management Organizations (DMO), and it could possibly lead to a quicker and more unified response.

Organizations often have to deal with the problems of changing their chain of command when they work closely with other agencies at the local, state, and national level. In these situations sometimes there is no clear indication of the boundaries among the jurisdictions of different DMOs. Also, organizations may have to deal with the formation of ad-hoc or temporary agencies during a disaster. Lacking a clear hierarchy can reduce the ability to prepare and respond to a disaster because there may be redundant or wasted resources (Manoj, 2007). It is critical that during a disaster the chain of command and relationships among DMOs is clearly stated to everyone. Otherwise, the effectiveness of a disaster management application would be reduced.

Victims also have to be actively engaged to ensure that they clearly understand the ramifications of a disaster, and that they know all the available services and relief programs that are available to them. This can be difficult because during a disaster people are less likely to trust others. This lack of trust can be further compounded due to either the lack of information or the availability of numerous amounts of erroneous information (Manoj, 2007). A government run disaster website may ease these problems because it could house all the information available for the disaster, and more importantly it could add the credibility of a government for all the information available on the website. This website could also serve as a portal to other similar sites, and this could potentially allow citizens to quickly access both disaster funds and information concerning the natural disaster.

CONCLUSION

Disaster management systems are expensive to program, and their development is less profitable in comparison with business applications.. Yet, there is still a need to have a system that can help to increase an organization's ability to respond and recover from a disaster. Designing an open source software package would be the ideal way of creating a disaster management system. An open source project would allow organizations that lack proper funding to acquire a vital system that they might not have been able to afford. Also, organizations can enhance the original design of the system to suit their specific needs because the original source code is freely available and fully modifiable. Another advantage would be that different organizations could collaborate on enhancements that would make the application interoperable among multiple disaster response groups. If the development base for the application was active then regular iterations of the code could be made available, and the application could evolve to incorporate the use of currently available technologies as well as reflect the current needs of many disaster relief groups (Careem, 2006).

Sahana is an example of a disaster management system that illustrates the effectiveness of open source software. The scope and features of Sahana can be customized to meet the needs of a wide variety of disaster scenarios such as tsunamis, earthquakes, and landslides (Careem, 2006). Yet, there are some problems with open source software. In order for open source software to be effective certain requirements need to be met. The developers that work on open source programs need to be actively involved through the life of the project and need to have the time and resources needed to effectively contribute to open source programs. Testing, enhancing, and developing code can take a significant investment in time and resources, and open source projects are more likely to fail without this strong level of commitment. A large number of open source projects have less than a handful of developers that are actively working on the project. Some projects only have one developer. This low number illustrates the difficulty of recruiting qualified developers to a project. Another issue that can arise with open source software is picking an appropriate license for the project. There is a wide variety of licenses that can be applicable to open source software, and choosing the appropriate license can be difficult given the amount legal terminology that is present in most software licenses. Open source software has certain drawbacks, but overall this development philosophy allows projects to be created both cheaply and collaboratively (Stahl, 2005).

Using Web 2.0 to design an application to support disaster mitigation is relatively new. Such a software system would greatly help our ability to cope with the severity and unpredictability of disasters. Many other systems have been created specifically for disaster management, and each of these different systems was created to solve different problems. Adherence to a XML standard would enable these systems to connect to each other in a meaningful way. Freely sharing information among applications would allow for improved decision making.

Natural disasters can affect many people, and multiple agencies usually respond to a disaster as separate entities. There is little coordination or sharing of resources or responsibilities, and as a result multiple management and recovery initiatives are undertaken. The lack of a centralized and coordinated network can hinder the response to a disaster. Different agencies may work on the same problem, or they may undertake programs that conflict with the plans of other organizations. Organizations must collaborate with each other at various levels in order to fully form a unified and effective response (Meissner, 2002). Disaster management websites can aide in organizational communication and can ensure that organizational efforts are not duplicated. Online systems ensure that important information concerning the disaster is available to all of the involved organizations, and these organizations can collaboratively update such a system in order to ensure that real time data is available (Bessis, 2009). Users see which areas are affected, which actions need to be taken, and how the current situation is being handled. Connecting the responders to the victims allows rescuers to instruct the effected population to shelter, record vital information that could be later used for family reunification, or provide disaster relief supplies. The use of Information Technology in disaster management can help alleviate the problems of managing a disaster by bridging the divide between affected parties and allowing for informed decision making.

REFERENCES

Berger, E. (2008, Sept 13). Hurricane Ike: Targeting Florida? Houston Chronicle. Retrieved Sept 16, 2008, from http://blogs.chron.com/sciguy/archives/2008/09/hurricane_ike_t_1.html.

Bertelsen, O. W., & Bodker, S. (2003). Activity Theory. In Caroll, J. M. (Ed.), *HCI Models Theories, and Frameworks: Toward A Multidisciplinary Science* (pp. 291–324). San Francisco: Morgan Kaufmann. doi:10.1016/B978-155860808-5/50011-3

Bessis, N. (2009). *Grid Technology for Maximizing Collaborative Decision Management and Support: Advancing Effective Virtual Organizations*. Hershey, PA: IGI Publishing.

Careem, M., Silva, C. D., Silva, R. D., Raschid, L., & Weerawarana, S. (2006). Sahana: Overview of a disaster management system. In *IEEE International Conference on Information and Automation*.

Fahland, D., Glaber, T., Quilitz, B., Weibleder, S., & Leser, U. (2007). HUODINI – Flexible Information Integration for Disaster Management. In *Proceedings of the 4 International ISCRAM Conference, Delft, the Netherlands*.

Garrett, J. J. (2005). *Ajax: A New Approach to Web Applications*. Retrieved Oct 14, 2008, from http://www.adaptivepath.com/publications/essays/archives/000385.php.

Herold, S., Sawada, M., & Wellar, B. (2005, June). Integrating geographic information systems, spatial databases and the internet: a framework for disaster management. In *Proceedings of the 98th Annual Canadian Institute of Geomatics Conference, Ottawa, Canada*.

Huffman, E., & Prentice, S. (2008, March). Social Media's New Role in Emergency Management. *Emergency Management and Robotics for Hazardous Environments, Albuquerque, New Mexico*.

Joomla. (2009). Retrieved May 1, 2009, from http://www.joomla.org/.

Kuutti, K. (1996). Activity Theory as a Potential Framework for Human-Computer Interaction Research. In Nardi, B. (Ed.), *Context and Consciousness: Activity Theory and Human-Computer Interaction*. Cambridge, MA: MIT Press.

Laituri, M., & Kodrich, K. (2008). On line disaster response community: People as sensors of high magnitude disasters using internet GIS. *Sensors (Basel, Switzerland)*, *8*(5), 3037–3055. doi:10.3390/s8053037

Liu, S., Palen, L., Sutton, J., Hughes, A., & Vieweg, S. (2008) In Search of the Bigger Picture: The Emergent Role of On-Line Photo Sharing in Times of Disaster. In *Proceedings of the 2008 ISCRAM Conference, Washington, DC*.

Madden, M., & Fox, S. (2006). Riding the Waves of "Web 2.0." *Pew Internet Project*, October 5, 2006.

Manoj, B. S., & Baker, A. H. (2007). Communication Challenges in Emergency Response. *Communications of the ACM*, *50*(3), 51–53. doi:10.1145/1226736.1226765

Martins, L., & Daltrini, B. M. (1999). *Activity Theory: a Framework to Software a JAIIO – Requirements Elicitation. WER'99 - Workshop en Requerimentos, 28 Jornadas Argentinas de Informática e Investigación Operativa*. SADIO – IFIP.

Meier, P. (2008, Mar 26). *Upgrading the Role of ICT in Conflict Early Warning/Response*. Paper presented at the annual meeting of the ISA's 49th Annual Convention, Bridging Multiple Divides, Hilton San Francisco, San Francisco, CA, USA Online from http://www.allacademic.com/meta/p254277_index.html

Meissner, A., Luckenbach, T., Risse, T., Kirste, T., & Kirchner, H. (2002). Design challenges for an integrated disaster management communication and information system. In *DIREN 2002. The First IEEE Workshop on Disaster Recovery Networks*. New York, June 24, 2002, IEEE Computer Society Press, Los Alamitos.

Murphy, T. (2006). Knowledge Management, Emergency Response, and Hurricane Katrina. *International Journal of Intelligent Control and Systems*, *11*(4), 199–208.

Nielsen, P. J. (1995). *Ten Usability Heuristics*. Retrieved Oct 14, 2008, from http://www.useit.com/papers/heuristicsheuristic_list.html.

Raman, M., Terry, R., & Lorne, O. (2006). Knowledge Management System for Emergency Preparedness: An Action Research Study. In *Proceedings of the 39th Hawaii International Conference on System Sciences* (pp. 1-10).

Shankar, K. (2008). Wind, Water, and Wi-Fi: New Trends in Community Informatics and Disaster Management. *The Information Society*, *24*(2), 116–120. doi:10.1080/01972240701883963

Stahl, M. T. (2005). Open-source software: not quite endsville. *Drug Discovery Today*, *10*(3), 219–222. doi:10.1016/S1359-6446(04)03364-1

Vygotsky, L. S. (1978). *Mind and Society*. Cambridge, MA: Harvard University Press.

White, C., Plotnick, L., Aadams-Moring, R., Turoff, M., & Hiltz, S. R. (2008). Leveraging a Wiki to Enhance Virtual Collaboration in the Emergency Domain. In *Proceedings of the 41st HICSS*.

Zisman, A. (2000). An Overview of XML. *Computing & Control Engineering Journal*, *11*(4), 165–167. doi:10.1049/cce:20000405

Chapter 12
Incident and Disaster Management Training:
Collaborative Learning Opportunities Using Virtual World Scenarios

Anne M. Hewitt
Seton Hall University, USA

Susan S. Spencer
SetonWorldWide, USA

Danielle Mirliss
Seton Hall University, USA

Riad Twal
Seton Hall University, USA

ABSTRACT

The maturation of incident and disaster management training has led to opportunities for the inclusion of multi-modal learning frameworks. Virtual reality technology, specifically multi-user virtual environments (MUVEs) such as virtual worlds (VW), offers the potential, through carefully crafted applications, for increasing collaboration, leadership, and decision making skills of diverse adult learners. This chapter presents a review of ICT appropriate learning theories and a synopsis of the educational benefits and practices. A case study, offered as part of a Master of Healthcare Administration (MHA) course for health care managers, demonstrates the application of a virtual world training scenario hosted in Second Life® and using a Play2Train simulation. Students report a strong positive reaction to virtual learning and demonstrate improved crisis communication skills and decision making competencies. Additional research is recommended to demonstrate the utility of virtual world learning as compared to standard training options such as tabletop exercises.

DOI: 10.4018/978-1-61520-987-3.ch012

INTRODUCTION

As the importance of emergency preparedness escalates during the 21st century, due to both natural and man-made disasters, the U.S. government has responded with several initiatives highlighting the need for incident and disaster management training. First, the National Response Plan (NRP) details the crucial role of incident management across multiple government sectors (Jain & McLean, 2006). Second, to support these NRP multi-sector goals, an Eight Step Training Model for improving public agency disaster management leadership has been recently published (Slattery, Syverston, & Krill, 2009). Third, the National Emergency Training Center and the Emergency Management Institute together offer professional development through formal courses, grants, and training opportunities for first responders and allied professionals (USDHS, 2003; FEMA, 2009). Fourth, the National Exercise Simulation Center (NESC) now provides facilities and systems that are designed to offer live, virtual and constructive simulations for disaster management training (FEMA, 2009a). These primary federal initiatives, part of the National Response Framework, form the incident and disaster management learning network for government entities.

The majority of non-government organizations (NGOs), including commercial industry, the health and nonprofit sectors, and private volunteer organizations, appear to be tailoring their training efforts to follow the federal government's guidance. This is especially true for integration of the National Incident Management System-Incident Command System (NIMS-ICS) into emergency preparedness plans (Jarventaus, 2007). One example is the Hospital Emergency Incident Command System (HEICS) which serves as the framework for incident command within a healthcare setting (McLaughlin, 2003).

The health care system sector has become a key player in providing incident management training to all levels of healthcare employees throughout the nation as well as serving as a local community disaster planning and coordinating resource. Community hospitals still represent the largest number and type of health care institutions in the country (AHA, 2009), and they function effectively in disseminating emergency preparedness knowledge through internal learning and development programs within their institutions. The Joint Commission on Accreditation of Health Organizations (JCAHO) recently published eight new regulations that provide a clear requirement for hospital incident management training. These mandates include the integration of community partners for purposes of drills and practices (Joint Commission International, 2008). Other community agencies, including local public health departments, have begun to view the medical care system as a primary leader for emergency planning and response (Guidotti, 2004). The public also now sees the hospital and its spokesperson as a trusted resource for crisis communication during a health incident or disaster (CDC, 2006). All of the diverse and varied health institutions, as well as public health agencies, consistently try to emulate emergency preparedness and disaster management training protocols as established by the U.S. government.

Emergency preparedness training (EPT) has been considered essential for all health care employees for several years, although incident management training represents a newer subset of administrative skills within the formal health management curriculum. Hospitals routinely provide internal, in-service training that is on-going, dynamic and fully integrated within the healthcare system. This contrasts significantly with the majority of EPT that is typically government or regulatory mandated, tends to be top-down driven, and requires collaboration with community partners (Schafer, Carroll, Haynes & Abrams, 2008). A continuing challenge for health organizations involves balancing the need for real-life disaster preparation and training for handling emergency incidents without significantly disrupting the day-to-day health service delivery. Although tabletop exercises, where participants enact a disaster scenario in a face-to-face activity, have been routinely used for disaster training (Jarventaus, 2007), they are limited in their ability to facilitate community interaction and participation beyond the role of first responders. Technology enabled learning (TEL) activities, such as immersive digital scenarios and simulations, can be used as a training device to enhance crisis management competency development. Specifically, virtual world (VW) simulations have the potential to effectively integrate and highlight the role of community partners and collaboration activities. They also offer the added benefit of increasing opportunities for collaborative practice without disrupting everyday medical care. As academia begins to integrate incident and disaster management training into core health management curriculums, opportunities exist through these new learning technologies to facilitate collaborative training.

This chapter begins with a brief review of relevant learning theories and their integration with emerging information and communication technologies (ICT) to support the development of incident and disaster management training simulations. Next, an overview of current virtual learning practices identifies key benefits for EPT and health management education. To further illustrate the potential role of virtual world (VW) technology, as both a learning and collaboration tool, a case study using a disaster management scenario hosted in Second Life® is presented. This case study demonstrates how graduate Masters in Healthcare Administration (MHA) students participated in a digital simulation that was designed to enhance leadership, decision-making and teamwork skills. This paper presents an analysis of learning outcomes, including pre- and post- skill level ratings from this VW simulation experience, identifies critical characteristics for incident and disaster management training, and suggests opportunities for collaborative network implementations. This

case-study builds on two previous publications that reported on crisis communication training (Hewitt, Spencer, Ramloll & Trotta, 2008) and preparing graduate students for virtual world experiences (Hewitt, Spencer, Mirliss & Twal, 2009). Finally, recommendations are presented for educators and trainers who are interested in facilitating similar educational exercises.

BACKGROUND

The challenges inherent in designing a learning environment appropriate for incident and disaster management training encompass decisions related to learning theory, scenario design and the type of technology enabled learning. A brief review of each of these instructional components provides a rationale for the inclusion of virtual worlds as a preferred instructional option. Each component is described briefly to provide context for an incident and disaster management training pilot case study designed for graduate students in a MHA program.

Integration of Learning Theories with Information and Communication Technology

The increased adoption of learning theories appropriate for ICT training is a reflection of demands from today's sophisticated learners and can be attributed to the 21st century's reliance on digital technologies (Angerou, Ciborra & Land, 2004). For example, the Pew Internet & American Life project reports that 31% of American adults are now classified as elite tech users (Horrigan, 2007), and that 97% of American teens report playing computer, console or cell phone games (Lenhart et al., 2008). This important paradigm shift, the addition of technology literacy to simple information literacy, has resulted in a major change in learning theory with significant implications for adult learners and management training. Education is moving away from authoritative learning

Figure 1. Application of learning theories to emergency preparedness training scenarios

Learning Theory	General Application	Case-Study Application
Constructivism: Students construct their own meanings/knowledge through experiences and reflection.	Training experiences should favor real-life participation versus theoretical concepts and allow for individual discovery and reflection. Structure and pace of learning activity designed to permit individual information processing.	Students were active participants in the EP scenario; they took on various roles and interacted with one another in order to provide appropriate information to the public (CERC) regarding a crisis situation.
Situativity: Students learn best when immersed in scenarios and simulations that emulate real life and allows for social interaction and collaboration.	Training activities need to genuinely reflect (visual, audio, sensory, etc.) the EP incident or crisis environment in order to allow students to fully encompass all dimensions. Compatibility with student expectations of an environment is key.	Using a virtual world (MUVE), students learned skills in an applied setting that more closely resembled the culture of the actual setting in which CERC skills would be used.
Visualization: "Seeing" complete representations of a situation allow for big picture learning	3D or multi-user simulation environments that offer vivid depictions that show complete situational representations. The big picture allows students to make critical decisions.	Immersion in a virtual world scenario enabled students to visualize an actual crisis scenario by including the town, restaurant, street scene and complete hospital. Visual information about the crisis was provided allowing students to make critical decisions.
Interactivity: Communication opportunities are essential: two-way, command (decision making) and participatory (collaborative) engagement.	Scenarios and simulations provide consistent, integrated and scalable communication opportunities for all participants.	Students were able to communicate with each other in both broadcast and private messages. This was essential to their ability to collect important information regarding the crisis and their ability to collaborate on their communication to the public.

approaches founded on simple information sharing to a more social framework that is based on networks and connective technologies (Schaller & Allison-Bunnell, 2003). This shift also suggests a distancing from traditional command and authority-based learning, also known as "broadcast learning" or "banking style of education" to one integrating both discovery and experiential learning (King, 2008; Friere, 2007). Sontag (2009) and Gee (2003) report that students exposed to the omnipresent digital technology clearly favor teamwork, experiential activities and the use of technology in their learning. Additionally, Oblinger's (2004) research has shown that these students have specific strengths in skill areas of

multi-tasking and collaborative style. Clearly, the trend has moved from a teacher centered perspective to a learner centered environment.

Incident and disaster management training, while extremely important and content specific, also rely on universal learning theories that are applicable to all levels of training. Figure 1 presents four learning concepts that are especially relevant to the collaborative nature of EPT: constructivism, situativity, visualization, and interactivity. Two learning theories, constructivism and situativity, have been identified by educational experts as better suited to the needs of adults in the information age (Sontag, 2009). Constructivism suggests that a hands-on or experiential experience is the

Figure 2. A Relationship continuum for aligning teaching approaches, student engagement and multi-dimension learning.

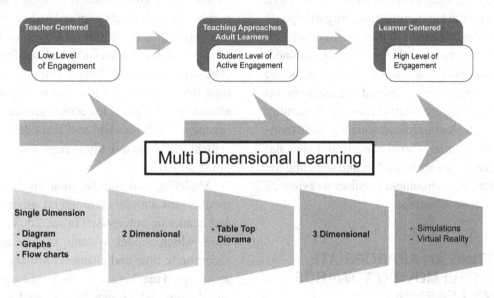

best method of learning, especially when students make their own discoveries and inferences (Duffy & Cunningham, 1996). Situativity asserts that students learn when knowledge is embedded within activity, context and culture (Lave & Wenger, 1990). Visualization plays an integral role in active learning and enables the student to improve decoding and encoding skills which contribute to more meaningful insights (NDT Resource Center, 2009). Interactivity is a learning theory that suggests communication is essential for any participation and subsequently increases participation in learning and problem solving. All of these learning theories are particularly applicable to gaming activities, scenarios and simulations as they involve the active creation and immersion of the student into a learning environment.

Figure 1 illustrates these learning theories and their application to a recently developed EPT scenario. In this virtual world scenario, a terrorist bombing has occurred at a local restaurant and has sent several victims to the community hospital. The educational goal of the scenario is to enhance the development of crisis communication skills for healthcare management professionals.

This learning approach, using the key elements of constructivism, situativity, interactivity, and visualization, forms a suitable framework for the development of an EPT scenario.

Applications to Scenario Design in Training and Development

With an instructional framework established, the next training development goal involves increasing the level of student engagement. Moving from a teacher centered approach to a learner centered environment facilitates student interaction as they now become the drivers for learning. To support this learner centered environment, a technology enabled activity that integrates the appropriate learning theories and aligns them with key concepts such as interactivity and collaboration is needed. A multi-dimensional learning platform offers the optimal choice of immersive virtual reality. Figure 2 presents scenario design options along a learning and engagement continuum.

This diagram schematically displays a relationship continuum aligning the various teaching approaches, student levels of engagement, and multi-dimensional learning strategies. Learning progresses from teacher centered instruction to a learner centered environment, where participants become fully immersed in their training activity through collaboration and interactivity. Combined, these concepts form a unified foundation for the development of student centered learning environments that integrate multi-dimensional learning and enrich incident and disaster management training. However, many forms of multi-dimensional learning activities are available, and selecting the right learning environment requires a review of potential student benefits.

SELECTING AN APPROPRIATE TEL MODEL: MOVING TOWARDS VIRTUAL LEARNING

From Tabletop Exercises to Simulations to Virtual Worlds

Effective incident management training requires leaders to successfully engage and immerse themselves in an emergency or crisis. Exercises serve as key elements in emergency preparedness training and are appropriate for first responders as well as senior officials (US DHS, nd.). Face-to-face tabletop exercises remain a standard training tool as they encourage participants to describe how they would respond to different scenarios, promote group learning (Mayer, 2003), and involve multiple disciplines. Tabletop exercises are an interactive group and skill-building format that require participants to collaborate in developing solutions to real-life problems (FEMA, 2007). Usually, participants in a tabletop exercise describe the evolving scenario and their responses at specific, planned points in time during the activity. For example, a tabletop exercise for pandemic influenza emergency preparedness would be led by a trained facilitator who presents participants with chronological segments of a scenario detailing

the spread of the disease. The scenario develops as additional information is provided about the number of individuals exposed and the mortality rate (Hewitt et al., 2009). This type of learning relies on a forced decision-making framework (Dausey, Aledort, & Lurie, 2005) as the real-time development of the scenario is essential to allowing participants to become immersed in the urgency of the scenario and practice delivering critical decisions quickly in response to a developing situation.

Modeling and simulation software development is now a major commercial, governmental, and academic enterprise (Taber, 2008). Simulations which model systems or processes that incorporate time and changes have also become an accepted teaching tool. For training purposes, simulations, which can be designed to present past or projected events based on algorithms and statistical projections, are exemplary at illustrating the big picture or macro views and perspectives of an incident. Another important benefit of simulations for EPT and incident management learning is their appeal and usefulness for first responders. The rapid acceptance of simulation training in multiple disciplines has led to the emergence of virtual world simulations for EPT.

Developed in the late 1980s by the Department of Defense, distributed simulations were recognized by educators as a potentially powerful learning tool over face-to-face learning activities (Dede, 1995). Virtual world simulations were initially piloted and accepted in the professional fields of aviation and applied clinical medicine (Lee & Wong, 2008). A recent and emerging body of research literature is supporting the use of Multiuser Virtual Environments (MUVE), such as virtual worlds, delivered via the internet and accessed on personal computers for emergency preparedness and incident management exercises (Ramloll, 2008). During the past few years, virtual simulations have entered the mainstream of emergency response training (Jarventaus, 2007) as they emphasize and integrate interaction and collaboration within a specially designed learning environment. The two key learning characteristics (interaction and collaboration) exist in virtual worlds through the use of avatars, graphical representations of individuals within the virtual space, and immersion, in which the student participates in a world that is realistic enough to suspend disbelief (Dede, 1995). Now, multi-user simulation environments (MUSE) exemplify the concepts of shared learning and learner centered environments.

Virtual Worlds as a Training Tool

Virtual world software, as experienced on one's personal computer through various 3D web applications such as Active Worlds™ and Second Life®, provides an immersive, multimedia experience which allows the learner to be in the center of a simulated environment (Lee & Wong, 2008). The benefits of computer-based virtual simulations outweigh the costs of a live drill when considering manpower, time, and overall expense (Jarventaus, 2007). Another frequently reported benefit is the opportunity for participants to experience dangerous situations in a controlled environment where mistakes lead to learning and not to possible injury or death (Bos, 2005). Students can practice with a realism that is not provided with tabletop exercises (Erich, 2008). Even more importantly, benefits of virtual worlds include: specificity of environment, availability and variety of simulations, multi-locations for users, and repeated opportunities for practice over time. From a pedagogical standpoint, VW offers multiple options for reflective learning and can easily include embedded pre, post and follow-up assessments. Studies are now confirming the impact of VW learning for teaching emergency preparedness and disaster management skills and competencies. Lessons learned and shared by researchers can be placed in the following outcome categories:

Collaboration Outcomes:

- VW technology allows for a heightened sense of co-presence (Schoeder, 2002), immersion in the learning experience (Barab et al. 2005; Dede, 1995), and opportunities for collaboration (Gardner, Scott, &Horan, 2008).
- VW emergency response training clearly integrates incident management training by the National Guard and US Defense Dept and for Law Enforcement officers (Erwin, 2001, Bos, 2005).
- The formation of virtual teams across departments results in better emergency responses and continued improvements (Dove, 2007).

Skill Improvement Outcomes:

- New emergency preparedness virtual simulations, such as SIMergency, integrate appropriate learning and software design theories in activities that engage learners in critical decision-making processes (Taber, 2008).
- VW environments that integrate fully immersive scenarios help novice learners improve in mass casualty triage skills (Vincent, Sherstyul, Burgess, & Connolly, 2008).
- Strong documentation continues to support the advantages of virtual world environments for clinical training in improving self-efficacy and decision making and improving patient safety (Hohenhaus et al., 2008).
- VW environments for team training provide repeated practice opportunities resulting in better learning outcomes (Heinrichs, Youngblood, Harter, & Deve, 2008).

VW Training Benefits and Outcomes:

- Second Life® continues to emerge as a platform for next generation employee training and functions as both an educational and social networking tool (US Fed News Service, 2008).
- Virtual learning simulations offer a cost-effective method for training (Jarventaus, 2007).
- Virtual training systems have been described as a successful bridge between classroom and live field experience (Erich, 2008).

With credible and robust evidence in support of using VW as an EPT educational tool, academic institutions can now begin to develop, pilot, and integrate VW scenarios to enhance their incident and disaster management courses.

CASE STUDY APPLICATION

Incident and Disaster Management Training in a Graduate Program

Realizing the need to investigate the educational potential of virtual environments, Seton Hall University's Teaching, Learning and Technology Center (TLT Center) set out to partner with faculty to pilot the use of virtual worlds to meet course learning objectives. The increased interest in the use of virtual worlds for communication and training by healthcare organizations has been well documented. National health agencies already have a presence in the virtual world Second Life®, including the US Centers for Disease Control and Prevention and the National Institutes of Health (NIH), a division of the US Department of Health and Human Services. Both examples utilize the virtual world space to disseminate information to the public in the form of web links, images and videos.

Virtual worlds are proving to be powerful spaces for both formal and informal learning to occur (Barab et al., 2005; Clarke & Dede, 2005). These virtual environments also support a wide-range of interactivity between avatars, as well as the ability for avatars to interact with the environment. Play2Train, a sophisticated Second Life® build spanning multiple islands (virtual "real-estate"), is one such environment that seeks to use the interactive potential of this technology to provide experiential learning opportunities to users. Play2Train was constructed under the purview of the Idaho Bioterrorism Awareness and Preparedness Program (IBAPP) and the Institute of Rural Health at Idaho State University to support Strategic National Stockpile (SNS), Simple Triage Rapid Transportation (START), Risk Communication and Incident Command System (ICS) training. The virtual facility includes a hospital, town, emergency vehicles, uniforms, triage stations and decontamination units. In addition, Play2Train has developed a number of EP training scenarios that can be used or customized to fit the training requirements of various organizations and academic programs.

The MHA program at Seton Hall University, which serves both on-campus and online students, had been searching for ways to replicate standard tabletop exercises commonly used to deliver EP training in order to offer this type of collaborative experience to both on-campus and online students. Previously published reports have documented the design, implementation and outcomes of the VW pilot study (Hewitt et al., 2008; Hewitt et al., 2009). (Figure 3)

The students enrolled in the online MHA program are leaders in healthcare and their professional titles are as diverse as their geographical locations. This diversity allows for rich collaboration between learners, enabling students to gain multiple perspectives from interacting with their peers. (Figure 4)

The TLT Center worked with the MHA faculty and administration as well as Play2Train repre-

Figure 3. Health management positions of MHA students

Category	Position	State/Country
Administration/ Management	Medical Group Administrator (2)	California, New Jersey
	Office Manager/ Psychiatric Center	New Jersey
	Administrative Manager & Consultant	New York
	Medical Billing Manager	New Jersey
	Office Manager/Veterans Affairs/Counseling Center	New Jersey
	Vice President of Operations/ Large Physician Practice	New Jersey
	Senior Director Cardio-Pulmonary Program	Massachusetts
	Blood Bank Manager	Texas
	Director of Case Management	Pennsylvania
	Clinical Practice Manager/ Neonatology Associates	New Jersey
	COO Private Medical Practice	Maryland
	Manager, Clinical Coordination Medical Assessments	Ontario, Canada
	Administrative Resident/ Veterans Affairs	New Jersey
	Manager of Decision Support	Illinois
	Director of Operations and Food Service Director	New York
	Director, Acute Medical Services	California
	Administrator/ Continuing Care Center	New Jersey
	Director of Employee Relations/Healthcare System	Tennessee-Arkansas
Clinical	Clinical Research Associate	New Jersey
	Aviation Optometry/Head Ancillary Serv. & Sr. Med. Off.	Maine
	Director of Nursing Behavior Health Facility (2)	Indiana, Michigan
	Assistant Director Nursing / Nursing Home/Operations	Georgia, New Jersey, Texas
	Clinical Manager/Health Family Clinic	Arkansas
	Radiation Oncology Group Leader/ Major Cancer Ctr.	New Jersey
	Cell Culture Scientist	Pennsylvania
	Urological and Surgical Technician	Florida
	Cardiopulmonary Director	Georgia
	Molecular Diagnostic Services Coordinator/ Cancer Ctr.	New Jersey
	Registered Nurse/ Cancer Center	New Jersey
Finance	CFO Healthcare System	Pennsylvania
	Budget Analyst	New Jersey
	CPA/ Financial Management (2)	North Carolina, Tennessee
Services	Systems Director of Food/Nutrition	Washington, D.C.
	Materials Management	New Jersey
	Health Information Specialist/Government Contract	Virginia
	Senior Data and Reporting Analyst/ Visiting Nurse Serv.	New York
	Firefighter/Paramedic	Texas
	Provider Outreach Coordinator/ Insurance Company	Pennsylvania
	Operations Manager/ University	Tennessee
	Externship Director/ Volunteer Coordinator	Kansas, Virginia
Business/ Marketing	Business Development	Virginia
	Vice President Marketing and Public Relations	Florida, Arkansas
	Sales Representative/ Large Medical Facility	Michigan
	Adoption & Change Mgt. Lead/ Chronic Disease Mgt. Project	Vancouver, Canada
Consulting	Senior Analyst	New York
	United States Air Force Recruiter	Maine

sentatives to deliver a customized, collaborative EP training activity that can be used by both on-campus and online students. An additional benefit of understanding the relationship between learner-centered approaches and immersive virtual worlds is the opportunity to portray almost any training scenario with relative ease. The scenario is the story line that simulates an emergency event, and the purpose of the scenario is to act as a background for the learning lesson exercise (Decker

Figure 4. Location map of Play2Train participants (Adapted from nationalatlas.gov™)

& Holterman, 2009). A virtual world hospital can be the scene for a staged triage scenario involving mass casualties, serve as a facility and grounds for practicing surge capacity protocols, or offer basic step-by-step implementation of decontamination units. The only scenario development criterion was the mandate to provide an environment that is so realistic that it mimics visual, audio and sensory elements that reflect real life.

This activity has been piloted to three cohorts of online students during their mid-residency on-campus visit. The virtual scenario begins with students viewing avatars, operated by TLT Center staff and Play2Train representatives, dining at a restaurant located on Play2Train. One of the avatars identifies himself as the "mayor" and he is accompanied by an unknown woman, assumed to be his wife or girlfriend. The restaurant suddenly explodes and chaos erupts. The diners are dragged from the flames to a sidewalk that quickly becomes a triage unit as the emergency vehicles arrive and begin to take the wounded to the hospital (Hewitt et al., 2008). After viewing

the initial crisis scene, students, assuming roles taken from the hospital incident command system plan (HEICS), convene their avatars in the hospital command center.

Without knowing the cause of the explosion, students are expected to make critical decisions and manage this current crisis. Using the Crisis and Emergency Risk Communication (CERC) model, students generate appropriate communication from the hospital to the community. As the crisis unfolds, participants must quickly comprehend and synthesize the incoming reports and unsubstantiated rumors as well as manage the hospital's immediate response to the crisis. Collaboration and teamwork is required as each individual, who is assuming a hospital leadership role, seeks to validate and verify situation specifics, such as the number of mortalities and exposure to a potential dirty bomb situation. The scenario concludes with students completing the required group assignment using the CERC initial response worksheet.

Scenario Course Implementation and Learning Methodology

Two previously published results present detailed information on the actual preparation of students for the VW scenario implementation and the required course outcomes (Hewitt et al., 2008, Hewitt et al., 2009). Each of the three pilot study applications had both similarities and differences, despite a planned standard protocol. The most important similarity included the adherence to the planned learning format (pre-information provided online, face-to-face lecture, guided Virtual World practice, completion of scenario activity using the Play2Train platform, de-briefing, evaluation, and follow-up online-discussion). Another similarity was that the primary faculty and facilitators for the didactic content material, Second Life® training, and Play2Train Scenario remained the same. Differences included: student background and familiarity of with technology, length of time spent on guided practice, availability of facilities and the addition of practitioner perspectives.

Scenario Scripting

The scripting of the scenario allowed for many constructivist principles to be incorporated into the teaching and learning experience. From the teaching perspective, the faculty member was able to act as a facilitator, guiding the students' learning throughout the entire process. For example, several times during the scenario the faculty member found opportunities to stop the activity and discuss important issues or redirect the class. In addition, immediate feedback from the executive in residence, a practicing healthcare administrator who acts as a mentor to students, stimulated additional critical thinking and helped direct the flow of the scenario toward the activity objectives.

Facilitating Scenario Communication Opportunities

The scenario also allowed learning to become an active and social experience. Duffy and Jonassen (1991) concluded that diverse learners should collaborate in order to come to a shared understanding of an experience. The utilization of a virtual world allowed students to share the same space, via avatars, and collaborate with one another in order to develop a shared understanding of the crisis and the development of the necessary communication messages to the public. The virtual world also allowed users from remote locations to synchronously participate in the scenario. For example, Play2Train representatives were able to assist in the execution of the initial crisis scene and the potential for additional community partners joining the activity would add to the authenticity of this experience. The context in which learning occurs can be critical to the transfer of knowledge to other domains such as the workplace (McMahon, 1997). The scenario allowed students to apply important concepts in a complex learning environment that makes explicit the interrelationships between key players (e.g. first-responders, hospital management, media and the public) during a crisis situation. Even though the scenario took place in a virtual world, the authenticity of the task situated learners in a culture similar to a real-life crisis setting.

The goal of creating an engaging and challenging learning environment for a diverse group of healthcare leaders involved facilitating active engagement throughout the learning activity. Students were pre-assigned HEICS roles with their accompanying responsibility sheets and during the guided VW practice become familiar with their virtual world avatars. Once the scenario began, a chat log records all typed dialogue between each student in the Play2Train activity. The chat log facilitates communication between all participants and at the same time provides a real-time report that documents peer-to-peer interaction and student-to-faculty interaction. Provided below is a short example of the collaboration occurring during the scenario activity:

11:00 **EMS Command Weimes (Faculty):** I'm declaring this an MCI, contact hospital incident commander.

11:52 **Incident Commander Capelo:** There has been an explosion at a local restaurant and the fire dept has been dispatched.

11:53 **Public Information Officer Umia:** Joe, this is Dave the public info officer. Do you have the incident Mgt. team?

11:53 **Medical Specialist Jinx:** Respiratory and skill irritations have been reported.

11:55 **Risk Manager Yoontz:** Capelo, make sure we get the decontamination tents up before the patients arrive. We do not want to risk exposure to staff to whatever agent was deployed in the explosion.

11:56 **Public Information Officer Umia:** All media relations need to occur at the hotel. That will be the staging area for media comments.

11:57 **EMS Command Weimes (Faculty):** ETA 30 minutes for casualties.

11:57 **Emergency Dept Director Hannu:** I am in the library with the incident commander.

11:58 **Safety Officer Daines:** Personnel area assigned to victims for additional protection. All visitors need to go through security entrance at ED

12:00 **Staging Manager Jupiter:** Patient staging and triage will be in main parking area outside.

12:02 **Risk Manager Yoontz:** I need immediate updates in the event of employee exposure.

12:02 **Personnel Director Brucato:** Yes, I will have all available staff from labor pool and volunteers meet in the conference room for credentialing and dispatch to the areas where they are needed.

The chaos, stress and pressure of a real time crisis is reflected in the rapid communication collated in the chat log as well as the interaction between all the students in their various emergency preparedness roles. After approximately 10 to 15 minutes, the primary faculty halts the scenario to ask for feedback and immediate perceptions. The scenario is allowed to continue for another 10 minutes and then students report to their work groups and complete the CERC assignment (CDC, 2006). Each group then presents their CERC communication to the entire class and peer-to-peer suggestions and criticisms as well as faculty and practitioner recommendations are shared. The entire learning experience requires approximately three hours for completion.

Scenario Evaluation

After completion of the virtual learning scenario, students participated in a debriefing session similar to the type held after community drills. This step provided immediate feedback and allowed reflections to be shared among all participants. The open comment time also allowed students to identify missing resources in their preparation for virtual learning and for the scenario. Overall, participants felt that the Second Life® resources were sufficient and that they had received the initial information early enough to allow for adequate preparation. General reactions were very favorable from each group and students verbally expressed their satisfaction with the experience (Hewitt et al, 2008). Students also completed a paper/pencil survey addressing their satisfaction with the experience, as well as assessing changes in emergency preparedness knowledge and perceptions. Figure 5 presents the survey results from each of the three pilot groups.

Across all three pilot groups, students indicated that participation in the virtual world learning experience increased their EP awareness (X = 4.22), knowledge (X =4.20), understanding of team collaboration (X =4.39), crisis decision

Figure 5. Survey means for pilot groups: March 2008, July 2008 and March 2009.

Survey Item	Group 1*	Group 2**	Group 3***	Global Mean
1. I enjoyed the VL Experience.	4.64	4.53	3.72	4.30
2. The VL experience increased my awareness of crisis communication.	n/a	n/a	4.22	4.22
3. The VL helped me to gain further knowledge about managing emergency preparedness.	4.64	4.07	3.89	4.20
4. The VL scenario helped me to gain additional insight into team collaboration.	4.64	4.40	4.12	4.39
5. The VL helped me to understand the complexity of making real time decisions in a crisis situation.	4.64	4.67	3.72	4.34
6. The VL scenario could be used in place of a class discussion as an applied lesson or table top exercise.	4.55	4.00	3.44	4.00
7. Overall, I felt the virtual learning experience enhanced this particular unit.	4.25	4.25	3.89	4.13
8. I would recommend that this scenario continue to be part of the course.	4.50	4.50	4.00	4.33
9. I would like to see additional virtual learning experiences integrated into appropriate units and courses as part of my MHA experience.	4.75	4.75	3.83	4.44

Students used a modified Likert scale (Agree, Neutral, Disagree) with rating numbers assigned (Agree = 5, 4, Neutral = 3, Disagree = 2, 1) to indicate their agreement.

*Group 1: N =8, Response Rate = 73%, ** Group 2: N = 15, Response Rate = 88%, ***Group 3: N = 18, Response Rate = 90%

making (X =4.34) and communication (X =4.22). They also strongly supported the inclusion of the virtual world experience and recommended its continuation.

Faculty sought to verify that perceived knowledge change and skill development had occurred and that it was stable across time. To assess the stability of responses, an identical follow-up survey at eight weeks was administered by email to Group 1 following completion of the entire course. For Items #7 and #9 (See Figure 5), the mean increased from 4.25 to 5.00 (+15%) and from 4.75 to 4.80 (+2%) respectively. Item #8, which focused on recommending that this scenario continue to be part of the course, decreased slightly from 4.50 to 4.40 (-3%). These results suggested that students still supported the use of the Play2Train activity after experiencing the entire course.

Faculty and training staff were also interested in ascertaining whether the reported knowledge and skill changes were linked to course competen-

cies. Group 2 was asked at pre and post activity to rank their level of experience on ten EP course competencies using the rating categories of no experience, minimal, moderate and extensive. The percentage of students indicating none or minimal EP competencies at pre-test was 80% and at post-test was 10%. The rankings of moderate and extensive EP competency levels increased from 17% to 75% for an overall self-report competency increase of +58%. The largest increases occurred in competencies that focused on basic EP protocols (86%), risk communication protocols (74%), disaster recovery steps (74%), and community preparedness (74%).

Participants in Group 3 were asked to rate their crisis communication skill abilities using the Dreyfus Model of Skill Acquisition scale of novice, advanced beginner, competent, proficient and expert categories (Dreyfus & Dreyfus, 1986). These students who participated in the virtual world scenario reported dramatic improvement

Figure 6. Student Pre- and Post- activity ratings of crisis communication skills (%).

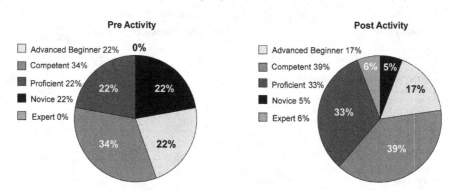

in their skill levels. Figure 6 presents the percent change across all skill levels. The largest positive percent changes was between the novice and advanced beginner level (pre = 44%, post = 23%, (+21%) and the competent and proficient levels (pre = 55% and post = 72%) with a percent change of +17%. These results demonstrate the students' self-reported progression on the skill continuum for emergency crisis communication. Overall, the training evaluation for the MHA virtual world scenario found a strong positive reaction to the learning environment, increased EP awareness and a demonstration of skill improvement that should be transferable to the worksite situation.

Scenario Limitations

Using Kirkpatrick's four levels of training evaluation as a model (Kirkpatrick, 1975), our efforts only focused on gauging satisfaction with the virtual world experience and learning transfer through EP skill self-assessment. Although results from the self-report surveys demonstrate increases in collaboration, teamwork skills and emergency preparedness competencies, individual six month follow-ups were not available in order to assess distal outcomes. In addition, a comparison with students completing a similar table-top exercise was not possible given course scheduling constraints. Clearly, our intention was not to demonstrate a learning preference for virtual

world training sessions over any other type of EPT. Although results from this non-experimental evaluation are not appropriate for generalizing to any other population, findings were beneficial in guiding the MHA's program efforts in refining future applications for this and other courses.

Lessons Learned: Solutions to the Training Challenges

Introducing the virtual world scenario into a graduate academic course was a novel experience and resulted in three obvious training challenges: (a) creating a personalized learning environment, (b) increasing student interaction within the virtual world and (c) facilitating and capturing collaboration activities. All training activities face the challenge of meeting individual learning needs based on a wide range of student pre-experiences. This dilemma was especially apparent in two areas for the virtual world training. First, a broad range of past EPT experience and competency levels was apparent between Group 2, which had highly experienced group members including some with military background, and Group 3 whose members had significantly less EPT exposure. This situation altered the balance of time spent on the actual scenario versus background lecture preparation and didactic instruction. We were able to alleviate some of the student differences in past EPT experience by providing the pre-information for their HEICS role and a copy of a sample job action sheet. This step helped to instill student confidence in their ability to participate fully.

The second area of concern was the diversity of virtual technology skill levels across all groups, with a smaller percentage of students exhibiting basic virtual world navigation skills and the majority of students displaying novice or elementary level skills. Again, the level of instruction was modified to accommodate these differences and to increase students' comfortability with the virtual world.

- **Recommendation #1:** *Personalized learning environments designed for virtual world training need flexibility to accommodate baseline levels of EP experience as well as student technology familiarity.*

Throughout all three pilot administrations, students responded positively to the virtual learning experience, and the use of avatars clearly contributed to their interactivity within the constructed scenario. However, students suggested that they would have liked additional opportunity for their avatars to interact with the environment rather than just maintain a presence in a single location in the hospital command center. This request for additional activity and "active actor" participation resulted in refinement of the scenario outline and situation pacing.

- **Recommendation #2:** *Scenario scripting should include pause points to allow for avatar exploration and discovery rather than relying only on periodic narrative updates as the scenario unfolds.*

The final challenge emerged as an unexpected result of the complete virtual world experience. As students were participating in the scenario and completing their role assignments, they shared pertinent perspectives, individual comments and past experiences with other classmates – all outside of the original structure of the planned EP

scenario. This ad hoc and complementary learning contributed enormously to the teamwork of the group and helped to produce higher quality collaborative decision-making because of the peer interaction.

- **Recommendation #3:** *Enhance collaboration and allow peer teamwork and other non-structured interactions by using multiple communication channels (i.e. broadcast chat, group chat, personal chat (IM) or voice chat). Multiple communication channels will support both the formal scenario communication as well as peripheral learning that might occur between participants.*

Each of these recommendations can be directly linked to the previously described information and communication learning (ICT) theories. In recommendation #1, the need to create a personalized learning environment is directly related to the idea of constructivism where discovery and reflection should be aligned with the student's capabilities. Both recommendations #2 and #3 encompass the theories of situativity, visualization and interactivity, as they encourage students to become active actors using their avatars and increasing scenario participation, communication and immersion. Adherence to the principles of these ICT learning theories will increase positive EP training outcomes.

IDENTIFICATION OF CHALLENGES AND FUTURE RESEARCH

Challenges

One of the major faculty challenges for EPT planning is the gap in published research specific to health management curriculums. The majority of rigorous research in this area highlights scenarios that involve first responder/civilian communities

(Jones & Hicks, 2004; Chen et al., 2001; Louka & Balducelli, 2001), medical trauma (Hospital Business Week, 2008), or preparing for mass casualties such as an influenza pandemic (Decker & Holtermann, 2009). However, abundant literature is available from the public administration perspective and related articles in publications such as the Journal of Homeland Security and Emergency Management and trade publications from software and scenario developers.

A second challenge is identifying appropriate evaluation methodologies that establish the efficiency and effectiveness of virtual world learning. Clearly, distinctions need to be established between educational uses of virtual learning and those focused on organizational training such as EPT (Learning Light, 2007). A recent publication on the transfer of learning in virtual environments suggests this concept can be a key to measuring the success of the virtual environment in facilitating both internal and external contexts that promote learning (Bossard, Kermarrec, Buche & Tisseau, 2008).

While these issues represent serious concerns for selecting a virtual learning technology, recent evidence is emerging on the widespread adoption of virtual world platforms such as Second Life®. A significant number of web reports are now available highlighting the increased use of Second Life® for EPT and reporting training successes (Loschiavo, 2009; Yellowlees, 2009; Schlauch, B, 2008; Wagner, M., 2009). As a greater number of academic institutions and community health agencies such as hospitals and public health departments integrate virtual learning activities into their training repertoire, opportunities for sharing best practices through informal conversations and formal presentations should become more evident.

FUTURE RESEARCH

The purpose of incident and disaster management training is to prepare a diverse group of individuals to successfully meet unknown challenges and crises. Opportunities to assess this type of training are readily available, but appropriate and standard methodologies and instruments do not appear to be uniform or universally adopted. Evaluations of emergency preparedness trainings and traditional tabletop exercises usually rely on informal "lessons learned" discussions and/or checklist type debriefings. Sponsored sessions by academic or credentialing agencies include knowledge post-tests and assess general comprehension. Additional research, involving controlled trials and follow-up surveys, is recommended to demonstrate the effectiveness of virtual world training as compared to standard table-top exercises. Further research efforts should focus on examining which learning theories are most appropriate for ICT and identifying the various pedagogical characteristics of successful training exercises including criteria for scenario development and implementation. Even more important is the need to establish where virtual world learning is most appropriate in the exercise continuum: pre-table-top, post-table-top, pre-real life enactment or post-real-life enactment. As virtual world learning opportunities expand and simulations become more commonplace, special attention should be placed on assessing students' familiarity with the technology, as well as categorizing changes in collaboration capabilities, teamwork skills and decision-making outcomes.

CONCLUSION

Incident and disaster management training efforts will continue to evolve as academia and non-government agencies begin to provide effective and efficient educational opportunities for a broader community audience. Virtual world scenario training offers an educational enhancement to standard table-top exercises. However, the appropriate use and place of this enhanced learning technology in the spectrum of emergency preparedness training methodologies has not been fully established.

In this chapter, a review of ICT learning theories and documentation of the emergence of virtual world scenario and simulation option supports the use of an instructional framework for designing learner-centered scenarios using immersive virtual worlds. A case study presents results from three groups of MHA students who successfully completed pilot administrations of a disaster management training using the Play2Train simulation in Second Life®. Overall, the training evaluation for the MHA Play2Train scenario found a strong positive reaction to the virtual learning, increased EP awareness and a demonstration of skill improvement that should be transferable to the worksite situation. Students with a diversity of work backgrounds and positions, representing several states across the United States, successfully collaborated to complete an incident and disaster management scenario that focused on crisis communication and sought to increase teamwork and decision making skills. Aligning disaster management training goals with the students' level of experience and optimizing virtual world technology options led to increases in leadership, decision-making and teamwork skills. Additional research, involving controlled trials and follow-up surveys, is recommended to demonstrate the effectiveness of virtual world training as compared to standard table-top exercises. The delineation of academic role in incident and disaster training and development of standardized incident and disaster management competencies will further aid in the development of virtual world learning.

REFERENCES

American Hospital Association. (2009). Fast Facts on US Hospitals. Retrieved on May 11, 2009 from http://www.aha.org/aha/resource-center/Statistics-and-Studies/fast-facts.html

Angerou, C., Ciborra, C., & Land, F. (Eds.). (2004). *The Social Study of Information and Communication Technology: Innovation, Actors and Contexts*. New York: Oxford University Press.

Barab, S. A., Thomas, M., Dodge, T., Carteaux, R., & Tuzun, H. (2005). Making learning fun: Quest Atlantis, a game without guns. *Educational Technology Research and Development, 53* (1), 86-108. Retrieved January 28, 2009 from http://inkido.indiana.edu/research/onlinemanu/papers/QA_ETRD.pdf.

Bos, T. (2005). The impact of using virtual reality technology to train for law enforcement critical incidents. *Journal of California Law Enforcement, 39*(2), 5–14.

Bossard, C., Kermarrec, G., Buche, C., & Tisseau, J. (2008). Transfer of learning in virtual environments: A new challenge. *Virtual Reality (Waltham Cross), 12*(3), 151–161. doi:10.1007/s10055-008-0093-y

Center for Disease Control and Prevention (CDC). (2006). Crisis and emergency risk communication (CERC): Pandemic influenza. "Checklist: Scientific risk communication or the public". Retrieved October 10, 2007 from http://www.bt.cdc.gov/erc/panflu/

Chen, Y. F., Rebolledo-Mendez, G., Liarokapis, F., Freitas, S., & Parker, E. (2008). The use of virtual world platforms for supporting an emergency response training exercise. In *Proceedings of the 13th International Conference on Computer Games: AI, Animation, Mobile, Interactive Multimedia, Educational & Serious Games*.

Clark, J., & Dede, C. (2005, April). *Making Learning Meaningful: An Exploratory Study of Using Multi-User Environments (MUVEs) in Middle School Science*. Paper presented at the meeting of the American Educational Research Association, Montreal, Canada.

Dausey, D. Aledort, J., & Lurie, M. (2005). *Table-top exercises for pandemic influenza preparedness in local public health agencies*. U.S. Department of Health and Human Services Office of the Assistant Secretary for Public Health Emergency Preparedness. Retrieved April 15, 2009 from http://www. pandemicflu.gov/plan/states/tr319.html.

Decker, K. C., & Holtermann, K. (2009). The role of exercises in senior policy pandemic influenza preparedness. *Journal of Homeland Security and Emergency Management, 6*(1), 32. doi:10.2202/1547-7355.1521

Dede, C. (1995). The evolution of constructivist learning environments: Immersion in distributed virtual worlds. *Educational Technology, 35*, 46-52. Retrieved January 28, 2009 from http://www. virtual.gmu. edu/ss_pdf/constr.pdf.

Dove, K. (2007). *Emergency management information systems: Application of an intranet portal for disaster training and response*. An examination of emerging technologies in a local emergency operations center. (Unpublished dissertation). Pepperdine University. 256 pages. AAT3293112.

Dreyfus, H., & Dreyfus, S. (1986). *Mind over Machine: The power of human intuition and expertise in the era of the computer*. Oxford: Basil Blackwell Pub.

Duffy, T. M., & Cunningham, D. J. (1996). Constructivism: Implications for the design and delivery of instruction. In Jonassen, D. H. (Ed.), *Educational Communications and Technology* (pp. 170–199). New York: Simon & Schuster Macmillan.

Duffy, T. M., & Jonassen, D. H. (1991). Constructivism: New implications for instructional technology? *Educational Technology, 31*(5), 7–11.

Erich, J. (2008). Virtually REAL. Fort Atkinson. *EMS Product News, 16*(5), 24.

Erwin, S. (2001). Virtual metropolis underpins emergency response trainer. *National Defense, 86*(576), 42–32.

Federal Emergency Management Agency (FEMA). 2007. *NIMS Compliance Metrics Terms of Reference*. Retrieved January 29, 2008 from http://www. fema.gov/pdf/emergency/nims/comp_met_terms. pdf

Federal Emergency Management Agency (FEMA). (2009). *Emergency Management Institute (EMI). Home Page*. Retrieved May 11, 2009 from http:// training.fema.gov.

Federal Emergency Management Agency (FEMA). (2009a). *The National Exercise Simulation Center (NESC)* (2009). Retrieved February 6, 2009 from http://www.fema.gov/news/newsrelease.fema? id=47280.

Freire, P. (2007). *Pedagogy of the Oppressed*. New York: Continuum.

Gardner, M., Scott, J., & Horan, B. (2008). MiRTLE. *EDUCAUSE Review, 43*(5). Retrieved January 28, 2009 from http://connect. educause.edu/Library/EDUCAUSE+Review/ MiRTLE/47238?time=122340170.

Gee, J. P. (2003). *What video games have to teach us about learning and literacy*. New York: Palgrave Macmillan. Retrieved May 20, 2009 from http://www.amazon.com/Video-Games-Learning-Literacy Second/dp/1403984530/ref=sr_1_1?ie=U TF8&s=books&qid=1245466968&sr=8-1.

Guidotti, T. L. (2004). Why do public health practitioners hesitate? *Journal of Homeland Security and Emergency Management, 1*(4), 403. doi:10.2202/1547-7355.1069

Heinrichs, W., Youngblood, P., Harter, P., & Deve, P. (2008). Simulation for team training and assessment: Case studies of online training in virtual worlds. *World Journal of Surgery, 32*, 161–170. doi:10.1007/s00268-007-9354-2

Hewitt, A., Spencer, S., Mirliss, D., & Twal, R. (2009). Preparing the graduate student for virtual world simulations: Exploring the potential of an emerging technology. *Innovate Journal of Online Education, 5*(6). Retrieved September 20, 2009 from http://innovateonline.info/index.php?view=article&id=690.

Hewitt, A., Spencer, S., Ramloll, R., & Trotta, H. (2008). Expanding CERC beyond public health: Sharing best practices with healthcare managers via virtual learning. *Health Promotion Practice, 9*(4), (Special Supplement, Pandemic), 83-87.

Hohenhasu, S., Hohenhas, J., Saunders, M., Vadergrift, J., Kohler, T., & Manikowskie, M. (2008). Emergency response: lessons learned during a community hospital's in situ fire simulation. *Journal of Emergency Nursing: JEN, 34*(4), 352–354. doi:10.1016/j.jen.2008.04.025

Horrigan, J. B. (2007). *A Typology of Information and Communication Technology Users. PEW/INTERNET & AMERICAN LIFE PROJECT.* Washington, DC: PEW/INTERNET.

Hospital Business Week. (2008). UFS Burn Center tests virtual reality disaster training: Gaming culture offers endless options for training. Retrieved September 20, 2009 from http://www.homeland1.com/ homeland-security-products/cpr-training-simulation-equipment/articles/429022-ufs-burn-center-tests-virtual-reality-disaster-training/.

Jain, S. & McLean, C.R. (2006). An integrating framework for modeling and simulation for incident management. *Journal of Homeland Security and Emergency Management, 3*(1), Article 9.

Jarventaus, J. (2007). Virtual threat, real sweat. *Training & Development, 61*(5), 72–78.

Joint Commission International. (2008). *Emergency Management in Healthcare: An All Hazards Approach.* Chicago, IL: JCAHO.

Jones, G., & Hicks, J. (2004). 3D online learning environments for emergency preparedness and homeland security training. Retrieved September 20, 2009 from http://courseweb.unt.edu/gjones/pdf/Jones_ elearn04.pdf.

King, B. J. (2008). Web 2.0 and education: Literature review of tools & technologies to enhance education. Retrieved from http://www.scribd.com/doc/9010043/Web-20-Tools-for-Higher-Education-Literature-Review-by-Brian-J-King.

Kirkpatrick, D. (1975). "Techniques for evaluating programs." Parts 1, 2, 3 and 4. *Evaluating Training Programs.* ASTD.

Lave, J., & Wenger, E. (1990). *Situated Learning: Legitimate Peripheral Participation.* Cambridge, UK: Cambridge University Press.

Learning Light. (2007). Second Life® *and Virtual worlds.*

Lee, E., & Wong, K. W. (2008). A review of using virtual reality for learning. In (Z. Pan et al. Eds.) Transactions on Edutainment 1, LINCS 5080, 231-241.

Lenhart, A., Kahne, J., Middaugh, E., Macqill, A., Evans, C., & Vitak, J. (2008). Teens, video games, and civics: Teens gaming experiences are diverse and include significant social interaction and civic engagement. *Pew Internet & American Life Project.* Retrieved May 12, 2009 from http://www.pewinternet.org/Presentations/2009/17-Teens-and-Social-Media-An-Overview.aspx

Loschiavo (2009). The future of emergency preparedness training. BAM INTEL: Leadership through thinking. Retrieved on September 20, 2009 from http://bamintel.blogspot.com/2009/06/future-of-emergency-preparedness.html

Louka, M., & Balducelli, C. (2001). Virtual reality tools for emergency operations support and training. In *Proceedings of TIEMS 2001* (The International Emergency Management Society), Oslo, June 2001. Retrieved on September 20, 2009 from http://www2.hrp.no/vr/publications/tiems2001.pdf

Mayer, H. (2003). Strengthening our homeland security: A collaborative planning approach to readiness enhancement. *Journal of City and State Public Affairs, 10*, 18–23.

McLaughlin, S. (2003). *Emergency Management Handbook. American Society for Healthcare Engineering of the American Hospital Association.* Chicago: ASHE.

McMahon, M. (1997, December). Social Constructivism and the World Wide Web- A Paradigm for Learning. Paper presented at the meeting of the ASCILITE, Perth Australia.

NDT Resource Center. (2009). Teaching with the Constructionivist Learning Theory. Retrieved May 12, 2009 from http://www.ndt-ed.org/TeachingResources/ClassroomTips/Constructivist%20_Learning.htm.

Oblinger, D. 2004. The next generation of educational engagement. *Journal of Interactive Media in Education, 8*. Retrieved May 6, 2009 from http://jime.open.ac.uk/2004/8/oblinger-2004080duspaper.html.

Ramloll, R. (2008, July). Using Web 2.0 & Table Top exercises in Safety and Health Profession Topics. Webinar presented for American Society of Safety Engineers.

Schafer, W., Carroll, J. M., Haynes, S. R., & Abrams, S. (2008). Emergency management planning as collaborative work. *Journal of Homeland Security and Emergency Management, 5*(1), 10. doi:10.2202/1547-7355.1396

Schaller, D. R., & Allison-Bunnell, S. (2003). Practicing what we teach: How learning theory can guide development of online educational activities. Paper presented at Museums and the Web 2003, Charlotte, NC, March. Retrieved May 6, 209 from http://www.archimuse.com/msw2003papers/schaller/ schaller.html.

Schlauch, B. (2008). Game theory 2.0: Second life for disaster training. Retrieved September 20, 2009 from http://www.homeland1.com/homeland-security-products/technology/articles/427835-game-theory-2-0-second-life-for-disaster-training/

Schoeder, R. (2002). Social Interaction in Virtual Environments: Key Issues, Common Themes, and a Framework for Research. In Schoeder, R. (Ed.), *The Social Life of Avatars: Presence and Interaction in Shared Virtual Environments* (pp. 1–18). London: Springer-Verlag.

September 20, 2009 from http://www.submityourarticle.com/articles/Peter-Yellowlees-5256/virtual-reality-67144.php.

Slattery, C., Syvertson, R., & Krill, S. Jr. (2009). The eight step training model: Improving disaster management leadership. *Journal of Homeland Security and Emergency Management, 6*(1), Article 8.

Sontag, M. (2009). A learning theory for 21st – century students. *Innovate, 5*(4). Retrieved May 6, 2009 from http://www.innovateonline.info/index.php?view=article&id=524&action=article.

Taber, N. (2008). Emergency response: E-learning for paramedics and firefighters. *Simulation & Gaming, 39*(4), 515–527. doi:10.1177/1046878107306669

US Department of Homeland Security (USDHS). (2003). *Fact Sheet: National Emergency Training Center.* Retrieved May 11, 2009 from http://www.dhs.gov/xnews/releases/press_release_0192.shtm.

US Department of Homeland Security (USDHS). (n.d.). *About HSEEP.* Retrieved March 15, 2009 from https://hssep.dhs.gov/pages/1001_About.aspx.

US Fed News Service. Including US State News. (2008, July 14). Using Web 2.0 & Table Top Exercises in Safety and Health Professional Topics of Upcoming American Society of Safety Engineers' Webinars. Washington, DC.

Vincent, D., Sherstyul, A., Burgess, L., & Connolly, K. (2008). Teaching mass casualty triage skills using immersive three dimensional virtual reality. *Academic Emergency Medicine, 15*(11), 1160–1171. doi:10.1111/j.1553-2712.2008.00191.x

Wagner, M. (2009). Second Life® helps save, improve lives. *InformationWeek Healthcare.* Retrieved September 20, 2009 from http://www.informationweek.com/news/healthcare/patient/showArticle. jhtml?articleID=22.

Yellowlees, P. (2009). *Virtual reality in medicine – many evolving uses and advantages.* Retrieved.

Section 3
Next Generation Approaches and Distributed Frameworks for Disaster Management

Chapter 13
Mathematical Models Generators of Decision Support Systems for Help in Case of Catastrophes:
An Experience from Venezuela

José G. Hernández R.
Universidad Metropolitana, Venezuela

María J. García G.
Minimax Consultores C.A., Venezuela

ABSTRACT

Immediately after the catastrophes that affected Venezuela at the end of 1999, especially the flood of the State of Vargas, a group of investigators of a consultancy company and of a private university of Caracas Venezuela, started working in decisions support systems (DSS) that could be useful in the moment of a catastrophe, helping to minimize the impact of its three principal stages: Pre-catastrophe, Impact and Post-catastrophe. Clearly, for the development of these DSS, it was indispensable to construct mathematical models to support them. The objective of this chapter is to disclose this experience by presenting some of these mathematical models and its conversion in DSS that supports decision making in the case of catastrophes.

INTRODUCTION

The contribution of this chapter is centred in two aspects: the mathematical models, especially emphasizing on the models, particularly those of minor publication and the Decision Support System (DSS) and their application for help in case

of catastrophes. The constructed DSS commented here briefly, only reflects a part of the situation in Venezuela, and it is intended for them to be a starting point for future reference in other places and societies, with the necessary adapting.

Among the models involved in the development of DSS can be mentioned: Problems of shorter routes, particularly its use of the Dijkstra algorithm; Problems of flows, especially

DOI: 10.4018/978-1-61520-987-3.ch013

maximum flow and minimal cost flows; Goal programming (GP); Multiattribute Models (MM) with multiplicative factors; Matrixes Of Weighing (MOW); Structures of decision trees; Inventory models; A, B, C Models, or 80/20 or Pareto model; Decreasing digits (Dd); Transportation and Transhipment problems and Assignment model and Fuzzy set. However, only some of them would be commented, especially those that for some reason appear slightly in the literature.

On the other hand there will be brief comments of at least twelve decision support systems, for aiding in cases of catastrophes, as performed by students of the Metropolitan University in Caracas, Venezuela.

Given that with this chapter it is intended to give a greater coverage to the mathematical models, as to the DSS developed by students of the Metropolitan University in Caracas, Venezuela, the objective of this chapter could be enunciated as: Disclose this experience when presenting some of these mathematical models and its conversion in DSS that supports decision making in the case of catastrophes.

This general objective implies two specific objectives: Construct the mathematical models and integrate them to a support system for decision making that could aide in case of catastrophe.

To achieve the general objective and the specific objectives that are generated, the followed methodology takes the scientific method as a base for the Operations Research or for decision making (Hernández & García, 2006; 2007; Hernández, García & Hernández, 2009), which tackles the problems of making decisions without pass for the exposition of hypothesis, but, does it across the following steps:

a. Defining the problem, as indicated in the objectives, present the mathematical models and DSS that can aide in case of catastrophes,

b. Searching for data, in particular mathematical models and support systems for decision making, that can aide in case of catastrophes,

c. Establishing the alternatives, that would be different mathematical models to use in case of catastrophes,

d. Evaluate alternatives, according to the raised objectives, deciding which of the proposed alternatives is feasible,

e. Selecting the best alternative, as product of previous evaluation process, and based on the secondary objectives, tacit or explicit, being considered,

f. Implementing the best alternative, meaning, choosing the better mathematical models and with them constructing the DSS that can aide in case of catastrophe and

g. Establishing controls, or mechanisms that allow recognizing if the developed systems are still valid over time.

The results will be given by the presentation of some models and brief commentaries of the rest, just like it will be done for the DSS, whom will be summarize (Figure 4) and only one of them will be discussed with a little more depth.

Catastrophes, Models and DSS

Since it is desired to present a previous experience with the use of DSS in an university of Caracas, Venezuela, the three main aspects of this chapter are the catastrophes, the mathematical models and the DSS, since in this book deep discussions are due to be presented on catastrophes, the commentaries about them will be brief and more emphasis will be given to the use of mathematical models and the DSS.

Catastrophes

Of catastrophe it is only interesting to point out some aspects, being the first that "Confidence in catastrophic situations is determined by means of getting some knowledge on catastrophes and the ways of reducing their impact" (Rohweder & Virtanen, 2008, 105), since the support systems to be presented in this chapter have as objective to facilitate knowledge of catastrophes to the population, with the belief that as more information is gained about them, it would be easier to face them.

The second aspect is that the terms catastrophe and disaster would be used as synonyms, and would be understood as those events that alter the population everyday life and to reestablish it, in general, external help would be needed.

On the other hand, about classification of catastrophes, the words of Noji (1997): "Disasters can be further divided into two broad categories – those caused by natural forces and those that are caused by people or generated by the humans…", would be taken into account, however, although some of the DSS are directed or inspired, in an specific kind of catastrophe, in this chapter, unless it is necessary, catastrophe will be used in general, without clarifying if they are caused by man or by nature.

It is important to point out that when searching references about catastrophe, some related to DSS are found, but in this chapter, the DSS and mathematical models who sustain them, are the main theme.

Mathematical Models

As commented one of the main objectives of this chapter is to present some mathematical models that had been or could be of great help to develop DSS in case of catastrophes. Following a brief presentation will be dome of some of the mentioned models in the introduction, those of smaller figuration in the literature, and then, in Figure 4, a summary of the used models in the different DSS will be presented.

Decreasing Digits

This concept, Decreasing digits (Dd), taken from accountable sciences and although previously used, in the field of Research Operations, it only has been referenced in restricted circulation works

(Hernández, García & Nieto, 2005), which is why it can be said that is one of the concepts or models of lesser use.

Surely, there is no administrator than during his professional life, has not performed depreciation from a good, and although because of its simplicity, it is possibly that he decides to do it linearly, but also, if desiring to perform an early or lately depreciation, he must have used Dd, whom Anthony (1987) points were legally allowed in the United States of America before 1954.

As can be intuited, Decreasing digits is no more that a succession of n numbers di (d1, d2,…,di, …, dn-1, dn), where, in the case of depreciation, n represents the years in which the value is wished to depreciate, and consist in assigning to the first in the established hierarchy order the weight $n / \sum i_{i=1, n}$, to the next one $(n-1) / \sum i_{i=1, n}$, and so, until $(n - (n-1)) / \sum i_{i=1, n}$, where n represents, in its general use, the number of aspects that are being weighed and $\sum i_{i=1, n}$, is the sum: $n + (n-1) + (n-2) + … + i + … + (n - (n-1))$, of the number of aspects.

In general the series will be: $n / \sum i_{i=1, n}$, $(n-1) / \sum i_{i=1, n}$,…, $i / \sum i_{i=1, n}$,…, $1/ \sum i_{i=1, n}$. As can be seen, with i varying from n to 1. In particular the Dd had been used with Multiattribute Models (MM) with multiplicative factors, in whose case it is had equation (1), which is useful for the calculus of the criteria weight, as the attributes.

$$P_i = (n + 1 - i) / \sum i_{i=1, n} \qquad (1)$$

Where Pi be the weight that is assigned to the criterion or to the attribute in the i position, into the established hierarchy, and n is the number of parameters that is being evaluated just then.

Goal Programming

The Goal programming (GP) or multiobjective programming or global programming figures with frequency in literature (Charnes & Cooper, 1961; Ignizio, 1976; Lin, Nagalingam & Lin, 2009;

Figure 1. A, B, C model or Pareto law or 80/ 20 relationship

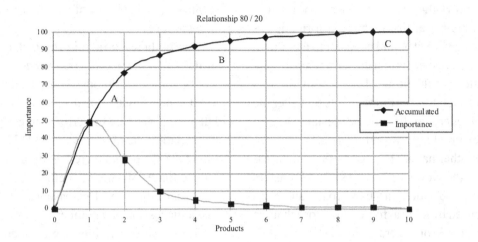

Scott, 2002; Stewart, Janssen & Herwijnen, 2004; Weber, Current & Benton, 1991)

Inventory Models and A, B, C Model

Other highly studied models in research operations are the models for inventory handling (Archibald, Black & Glazebrook, 2009; Beasley, 2002; Chang & Lo, 2009; Luo, 2007; Rabinovich, Dresner & Evers, 2003; Ravichandran, 2007; Skouri, Konstantaras, Papachristos & Ganas, 2009). For this work is relevant to point out the particular case of the model A, B, C, next commented as the Pareto Law, or relation 80/20.

A, B, C Model

It is recognized as Law of Pareto or the relation (80/20) that at heart is equivalent to the A, B, C model (Figure 1) and that is encountered, at least approximated, in many situations of the economic life of a nation, a company, and even of an individual, nevertheless the Law of Pareto following the expressed by Samuelson & Nordhaus (1986), is a law of the economy that can be enunciated: "the rent shows an inevitable tendency to distribute itself, in the same way, independent of the social

and political institutions and the tributary systems" (p. 685); only that the observations of Pareto can be deduced that the greater percentage of the wealth (it is spoken approximately of an eighty percent [80%]) is generated by a small group (esteemed at approximately the twenty percent [20%]) of the population and from which it is necessary to speak of the Law of Pareto, when referencing to the relation 80/20, that is encountered frequently in the daily life becomes. By the previous, in this chapter it will be spoken of the Law of Pareto, relation 80/20 and A, B, C model like synonymous.

Because A, B, C model (Cardós & García-Sabater, 2006; Guo, Fang & Whinston, 2006), it is discussed to establish models for the handling of inventories, and consists of differentiating products in three categories: A, those of greater importance, B, those of medium importance and C, those of smaller importance. In Figure 1, this distribution can be visualized, the first products, the twenty percent (20%) of the total, accumulate the eighty percent (80%) of the importance.

Matrixes of Weighing (MOW)

The Matrixes Of Weighing (MOW) are an adjustment of rows and columns, as can be seen in Figure

Figure 2. A general scheme of a Matrix of Weighing

		Criteria (C_j) and its weights				
		Criterion 1	...	Criterion n-1	Criterion n	Total
	Weight $\longrightarrow$	PC_{1i} a PC_{1f}		PC_{n-1i} a PC_{n-1f}	PC_{ni} a PC_{nf}	---
Alternatives (A_k)	A_1	$P_{1,1}$	...	$P_{1,n-1}$	$P_{1,n}$	Total A_1
	A_2	$P_{2,1}$	...	$P_{2,n-1}$	$P_{2,n}$	Total A_2
	..	...	$P_{k,j}$	...	...	...
	A_{m-1}	$P_{m-1,1}$	...	$P_{m-1,n-1}$	$P_{m-1,n}$	Total A_{m-1}
	A_m	$P_{m,1}$	...	$P_{m,n-1}$	$P_{m,n}$	Total A_m

2, and is a model cited few in literature (Hernández et al., 2009), in some cases only an idea of them is had by references to works with similar or equivalent matrices (Environmental Management Systems, 1997; Monteiro & Rodrigues, 2006; Prasad, 1995; Shen & O'Hare, 2004).

Characteristics of the Matrixes of Weighing

Usually in a Matrix Of Weighing (MOW), as it is seen in Figure 2, where the variables with subscript i mean initial value and with subscript f the final value, is that in the first column appears the alternatives (A_k) to be evaluated and in the following columns the criteria (C_j), leaving the first row to identify the respective criteria and the ranks of their weights and the remaining squares of the matrix to make the valuation itself, the P_{kj}, that would be the weight assigned to alternative k according to criterion j, and the last column is conserved to complete the evaluation of each alternative, adding the points of its respective row. However in some cases, when there is a higher number of criteria that alternatives the order is inverted and the alternatives are in the row.

In Figure 2, normally ranks are equal, meaning X_i, ..., W_i, Z_i are equal to zero for all criteria j and X_f,..., W_f, Z_f equal to all criteria j for a steady value, generally ten, or a multiple of ten, increasing according to how difficult can an alternative be differenced from the other.

A variant is to give different values to the ranks of each criterion, in this case the variables of subscript i, not necessarily zero or one, and the values of the variables of subscript f, usually are different, being understood that the greater of this last value, greater weight is desired to be given to that criterion and as greater it is the difference of the greater value minus the smaller value (Xf - Xi), more differentiating as far as this criterion is desired.

Another variant of the MOW leaves from some of the two structures presented before, with equal or differentiated ranks, but whose sum it is not done directly, instead points of each criterion are multiplied by their respective value of criteria (v_j), generally a whole number between one and ten, that is to say, each one of the totals would be calculated by (2), which would apply for all alternatives i

$$\text{Total } A_i = \sum_{j=1,n} v_j * P_{k,j} \qquad (2)$$

In (2), n represents the number of criteria.

A fifth case, is the one of standardized percentage, where all the ranks (X_f - X_i) are equal, and its sum of the factors of criteria v_j is forced to be equal to the unit. It is possible to see that (2), is applied for all the five cases, even the first two considered, only that for them v_j will be all equal to one.

Models Based in Fuzzy Set

The fuzzy logic, although transformed into models, more than a model is an approach and has its origins with the work Fuzzy Sets presented Lofti Zadeh in 1965, and in spite that it is a relative recent focus it figures highly in the literature (Chen, 2001; Güngör & Arikan, 2000; Kojadinovic, 2005; Peneva & Popchev, 2003; Rommelfanger, 2007).

Multiattribute Models with Multiplicative Factors

The Multiattribute Models, appear a lot in literature (Almeida, 2005; Greco, Matarazzo & Slowinski, 2001; Jacquet-Lagreze & Siskos, 1982; Jessop, 2004; Jiménez, Ríos-Insua & Mateos, 2006; Li & Liao, 2007; Roy & Vincke, 1981; Tavana, 2004), but are more uncommon when referencing MM with multiplicative factors, this is why, these brief commentaries were taken from Hernández & García, 2007) and based on the concepts of Ben-Mena (2000), Ehrgott & Gandibleux, (2002) and of Roy & Vincke (1981), but particularly following Baucells & Sarin (2003), will define the Multiattribute Models (MM), as those designed to obtain the utility of alternatives through the valuable attributes, that must be evaluated as components of the criteria. In any case, the final result, as expressed in (3), will be an additive model (Hernández & García, 2007):

$$Pts = \sum_i pc_i * (\sum_j pa_jc_i * va_jc_i) \tag{3}$$

Where subscript i represents the criterion and subscript j the attribute, therefore pc_i will be the weight assigned to criterion i, pa_jc_i will be the weight of attribute j of criterion i, va_jc_i will correspond to the value assigned to attribute j of criterion i, and Pts will be the total value reached about the variable in study.

Thus under this definition and by its way to operate, for a MM, it is only needed to define criteria, attributes and a valuation mechanism for

such, which implies that the Multiattribute Models is very useful when choosing between different alternatives, or when them must be valued or to hierarchized. Nevertheless, which is its greater strength, the additvity, that makes it simple to operate, becomes its main weakness.

The Multiplicative Factors

This weakness, consequence of additivity, that manifest when there are different scales from evaluation, or values in very distant ranks, can be corrected, as seen in (4) through the multiplicative factors (Hernández & García, 2007), which transform the model into:

$$Pts = \prod_k fg_k * (\sum_i f_i * pc_i * (\sum_j pa_jc_i * va_jc_i)) \tag{4}$$

That maintains the previous variables in addition to the use of the multiplicative factors f_{gk} and f_i, where k enters the number of correction factors, which operate for the entire model, which will be called general factors, f_{gk}, and f_i would represent the product of the correction factors which operate for criterion i.

Nevertheless to have an equation that reflects better the reality of the multiplicative factors of the criteria, when they are more than one, although, just as for the general factors, never is recommended that when they are more than three, instead of (4) is preferable to use (5):

$$Pts = \prod_k fg_k * (\sum_i \prod_h f_{ih} * pc_i * (\sum_j pa_jc_i * va_jc_i)) \tag{5}$$

That maintains the previous nomenclature, where h will be the accountant of the factors of each one of the criteria, this way (5) happens to be the general expression, being able to use (4) if for each one of the criteria there is only a multiplicative factor and (3) in case of not having multiplicative factors of the criteria nor general multiplicative factors.

These multiplicative factors, which give greater flexibility to MM, that with this correction stop being an only additive model, are standardized generally between zero and one, and can be continuous, between 0 and 1, or discreet, that is to say, 0 or 1, or even intermediate values, and in very special cases multiplicative factors that surpass the unit could be used.

Problems of Maximum Flow and Flow of Minimal Cost

The flow problems, it is also widely treated in specialized literature, in particular in Sakarovitch (1979), a detailed study of the most commonly found problems appears: Maximum flow, feasible flow and flow of minimum cost. Presently, although not always the problems here presented are studied, special interest is given to the flow fields as logistics (Lieckens & Vandaele, 2007; Persona, Regattieri, Pham & Battini, 2007; Sarimveis, Patrinos, Tarantilis & Kiranoudis, 2007; Singhal, Singhal & Starr, 2007; Wang, Rivera & Kempf, 2007), where not only material flows are studied, but information and monetary flows are attended among others.

Problems of Shorter Routes and Dijkstra Algorithm

In Hernández & García, (2007), it is commented that although other cases can be presented, when speaking on the problems of routes, in general, it can be affirmed:

If in a network R (V, E, d), where d is a magnitude function called distance, and E represents the directed set of arcs or sides and V the set of vertices or nodes, are desired to know some of the following aspects:

a. the shortest route between any pair of vertices V,
b. the shortest route between a pair of them (x, y) perfectly defined, or

c. the shortest route between one of them, that can be called as origin or root (s) and the rest; speaking of the shortest route problem.

Evidently, the objective in the three cases, is to find the more economical way (shortest route), according to the established parameter of measure (cost, time, distance or any other), of achieving one of the previously mentioned cases.

In order to solve these problems it is had different algorithms, among them, perhaps the most known is Dijkstra (Eglese, Maden & Slater, 2006; Loudni, Boizumault & David, 2006; Sakarovitch, 1979), which solves the problem c) previously enunciated, of a given root node of a connected graph, preferable a strongly connected graph, to obtain the smaller distance by this root to each one of the nodes of the network.

Structures of Decisions Trees

When studying decision making under uncertainty and risk, in special under risk, usually decision trees are utilized, that more than models they are simple structures that allows detaching the problem, mainly when it can be divided in stages, each a consequence of the other. With all this simplicity, nevertheless the structure of the decision trees is commonly used, especially in problems where there is the need to make diagnoses or to be discarding alternatives, from which it is frequently used in medical sciences (Geurts, et al., 2005; Kosuda, Ichihara, Watanabe, Kobayashi & Kusano, 2000; Razavi, Gill, Ahlfeldt & Shahsavar, 2008).

In the decision trees there are only two types of elements, the nodes and the branches that unite them, the branches are presented by sides and in general they unite a pair of nodes, which can be of two types, of decision, which represented by a square and those of probability, represented by a circle, just as the terminal nodes, that are not represented, or is represented as a circle, a little smaller than those of the probability nodes (Figure 3).

Figure 3. Decision tree scheme

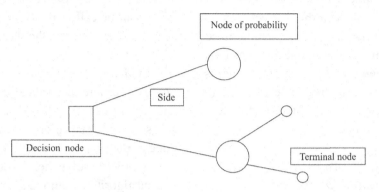

It is evident that in every node of decision a way to follow must be taken, whereas for each probability node, it is expected that different probabilities appear from existence of related events, thus is due to calculate its expected value.

Aside from its use in decision making problems under risk and those already indicated in other sciences, the tree decision structure is used in systems of support to the decisions when situations occur where several alternatives converge and there are different actions to follow, depending on the situation that appears.

Transportation Models, Transhipment Problems and Assignment Model

Three problems studied extensively in Operations Research and that can be solved starting from a common model are Transportation, Transhipment and Assignment, where the latter two can be solved applying the model and solution philosophy of the first. Although at the present time the initial models have been sophisticated, the problem of transport and its variations, specially when it is to diminish costs, are still being found with certain frequency in the literature(Araz, Ozfirat & Ozkarahan, 2007; Archibald et al., 2009; Chen, 2001; Dong, Carter & Dresner, 2001; Lee, 2001; Tsamboulas & Mikroudis, 2006; Vasant, Barsoum & Bhattacharya, 2008).

Culminated the revision of the different models, brief commentaries over the Decision Support Systems would be given.

Decision Support Systems (DSS)

It is not easy to enunciate a common definition of the DSS, perhaps two of the more accepted are:

A Decision Support System (DSS) can be defined as an interactive system based on computer, which helps the decision makers to use information and models to solve problems little or not structured (Turban, 2001).

The Decision Support System (DSS) they are systems based on computers that help to make decisions confronting the problems structured through the direct interaction with it dates and analytic models (Sprague & Watson, 1986).

Among the most common reasons to create and use a DSS are:

- Best prosecution speed.: Prosecution of information in a faster way and at low cost.
- Increase of the productivity: Reduction of costs and optimization of resources.
- Technician support: More efficient and in a quick and economic way.
- Better quality in the decisions: Bigger alternatives to carry out evaluations and bigger facilities for the analyses of risks.

On the other hand among the most outstanding characteristics in a Decision Support System (DSS) they highlight:

- To present the necessary information to support to the making of decisions and of this form to solve problems.
- To present to the decision maker different focuses and opportunities of a certain problem, with the purpose of forcing him to analyze this through different perspectives.
- To possess interfaces, graphic, charts and diagrams that facilitates the better understanding of the problem.
- To manage mixed models (qualitative-quantitative) or at least one of them.
- To allow the handling of scenarios (Analysis What it if?).

Evidently the DSS have been created for many areas of the knowledge, nevertheless in this chapter are only of interest the ones related to disasters, as presented by Tian, Wang, Li, Li & Wang (2007), or in a set of works, that one way or another will lead to DSS for support in case of disasters (Jotshi, Gong & Batta, 2009; Lu, Chou & Yuan, 2005; Mustajoki, Hämäläinen & Sinkko, 2007). And as it was already indicated in the objectives brief comments about the DSS developed in the Metropolitan University in Caracas, for aid in the case of catastrophes.

Application of the Decision Support Systems in Catastrophes

The three main aspects of this chapter are integrated in a single one, the created DSS to aide in case of catastrophes, which necessarily are constructed on one or several mathematical models.

In order to visualize how these models have been used, brief commentaries will be made of the use that had been given to each of the commented models and it will soon appear the Figure 4 to summarize them, where will be listed the works of degree, the profits of such and the used mathematical models. And in order to understand how this DSS Works, they will be illustrated using only one of them.

Decreasing Digits

The Decreasing digits (Dd) were used to obtain an estimate of the population that remains within a refuge or lodges, in each of the days, immediately later to a catastrophe (Belozercovsky & Sensel, 2002; Gamboa & Peña, 2004). The necessity to have a good estimate of this population is in order to being able to calculate the goods, mainly foods that must be provided, avoiding wastes, but mainly the deficiency, when they are necessary.

Goal Programming

The Goal programming (GP) has been used to generate DSS that aides to recover the conditions of a location affected by a catastrophe (Alessio & Vicuña, 2003), using for this case two sub-models, one in the benefits, that produced each alternative of improvement and other that took in consideration the costs associated to it.

Also the GP was used to decide the precedence of the different types of infrastructure that must be recovered after a catastrophe (Lecuna & Roa, 2005), for it, goals are settled related to the type of catastrophe, the size of the affected population, infrastructure to recover, age and number of the affected persons.

A pair of similar applications was done in the works of López & Sánchez (2005) and Dos Santos & Rojas (2006), the first centred in the recovery of logistics after an industrial accident and the latter in the more sensible aspects of an enterprise to achieve its recovery after a catastrophe.

Inventory Models and A, B, C Model

The inventory models were used in direct form in Gamboa & Peña (2004), for the handling of food inventories, in particular those that arrive by donation, in Belozercovsky & Sensel, (2002), taking care of the necessary calories by the population that remain within a shelter and with smaller depth in Gómez & Zapata (2001).

However when speaking of inventories, where it has been done more work with greater depth and obtained better contributions it is in the use of the A, B, C, Model (Belozercovsky & Sensel, 2002; Gamboa & Peña, 2004), in both cases, combined with Decreasing digits it has been estimated that the population in a given time could remain in a shelter, according to the population that would arrive at this shelter in the case of a catastrophe

Matrixes of Weighing (MOW)

The MOW, as commented already, has not been used in any DSS for aid in catastrophes, but they can be used in all the cases where MM with multiplicative factors was used, in particular in the selection of possible shelters, when it is wanted to make a selection quickly.

Models Based in Fuzzy Set

The models based in Fuzzy Set, used by Gamboa & Peña (2004), in joint form to the decreasing A, B, C Model and Dd, and the obtained estimates by these two models through the concepts of fuzzy logic are corrected and additionally this technique allowed them to fit to the results day to day.

Multiattribute Models with Multiplicative Factors

The applications of Multiattribute Models with multiplicative factors, are given for being multicriteria models, thus have been used in cases where evaluations based on multiple criteria and attributes are necessary, between the more important are: Rodriguez & Zabala (2002), for the location of shelters, although ended up using an own algorithm, based on the philosophy of

MM with multiplicative factors, also Clavaud & Navarro (2006), ended up using an own model, from MM, when constructing their model of cost of communication channels, when estimating the costs of the transfers considering multiple criteria as the distance, traffic, type of road and conditions in which they are, finally in their model for the evaluation of shelters, Lopez & Perez (2006), also use MM with multiplicative factors, where the used criteria was: Space, Services and Security and communication and were used the multiplicative factors: Capacity, Energy and light, Ventilation, Distance, Water and Sanitary facilities, being the four first factors of criteria and the remaining two general factors.

Flow Problems

The flow problems, were of the first in being applied, already in the graduate special works of Gómez & Zapata (2001), were used to establish the minimum routes between immediate refuges and shelters, to support the transfer of the affected people, from the immediate refuges to the shelters and establishing for them a network of flows, where the parameter of costs contemplated the unitary value by kilometre to cross.

Also applied flows (Gómez & Zapata, 2001) when assigning transportations, where it was desired the best possible distribution of the transports arranged to evacuate the affected population, in which case an algorithm that consisted in a mixture of flows and routes was created.

Lopez & Perez (2006), also handled a model for the flow of affected persons, where they use a modification of the algorithm of Ford and Fulkerson (Sakarovitch, 1979) to solve the problem of maximum flow, that way the displacement of the affected persons was carried out, with two versions of the algorithm, one for the affected persons that do not need special transportation, since their conditions are healthy or slightly wounded and another, for those, severely wounded, if they require a special transportation.

On the other hand Moleiro & Rojas (2006), constructed a DSS that is centred in helping in the displacement of the affected persons by a catastrophe from the immediate refuges to the shelters, in principle this problem is a transportation problem, nevertheless the used algorithm to solve it, although it is inspired by it, is an own algorithm, with the structure of a flows problem.

A separate case is the one of Goldsztajn, (2005), that took care of the problem of establishing early alerts, to deviate the vehicles, when overflowing of gorges occur, avoiding so they were catch by the flood. Although it was inspired by the flow algorithms, the used model was of own character.

Problems of Shorter Routes

Just as the flow models, which sometimes have been used in joint form, the models of shorter routes have been used since the first DSS, thus Gomez & Zapata (2001), make use of an inverted Dijkstra when they present their algorithm for the primary evacuation, that is to say, from the homes or centers of work, to the immediate refuges. As it was already indicated, they also used (Gomez & Zapata, 2001) a combination of routes algorithms and distribution flows for the used transportation to evacuate the affected persons.

Of the same form, Rodriguez & Zabala (2002), in the allocation of transportation, parted from a graph where the transportation centers, the refuges, the attention centers and the shelters are represented. In the evacuation from the homes to the refuges Belozercovsky & Sensel (2002) consider the refuges of finite capacity and apply Dijkstra to make the allocation of the affected persons according to the shortest distances, but they are saturating the refuges, when its capacity is worn, which would be equivalent to close them, thus the remaining persons must be assigned to the refuges that still are left with capacity surplus, until assigning them all. López & Pérez (2006), to establish, the access routes to immediate refuges in each zone and the evacuation routes, use Dijkstra.

Structures of Decisions Trees

The structure of decision trees was used by Rodriguez & Zabala (2002) to handle the access to inaccessible zones, and among others they created relative branches: (1) The product of the Triage, where they were possible to had green, that was unharmed or slightly wounded, red, the severely wounded, or black, who were the deceased, (2) Zone, about if they were safe or unsafe, (3) Possibility or not of constructing access routes, (4) Possibility or not of landing.

For Alessio & Vicuña (2003) the structure of decision tree was the body of the general model, on it the structure settles down and all the desired fields are verified: Type of catastrophe, Type of affected area, Topography of the affected area, Dimension of the affected area, Size of the affected population, Sector to recover, Elements for the reconstruction and Activities for the recovery.

On case of Hernández & Rodriguez (2005), their DSS is based on structure of decision tree, they had created concepts of family nuclei, which is desired to remain united during the any transfer consequence of the catastrophe, and where there are members which need to be protected or to be taken care of by others and of layers, that were members of a familiar group, which also was desired that remained united, but conformed by members that enjoyed greater autonomy.

In the DSS of Clavaud & Navarro (2006), for the model of assigning the priority of attention of hurt people, which is translated like a triage, they follow a structure of decision tree, so that each person, survivor or not, must culminate with a card that identifies it as colour green, yellow, red or black.

Transportation Models and Their Varieties

Although the problem of transportation and its two varieties the Transhipment problems and Assignment model, have been used in several DSS, in general the algorithms had not been maintained as such, but they have been modified or combined with other algorithms. It is the case of Gomez & Zapata (2001), in its model of second stage of evacuation, when they make the transfer of the affected persons, it appears as a transportation problem, but is solved as a minimum cost flow problem. Similarly Rodriguez & Zabala (2002) centred their DSS in the problem of the transportation of the affected persons, but only after obtaining results with Dijkstra they resorted to the allocation of transportation algorithm, and even in the modulate of transfer of wounded and affected, use a transportation graph, on which the algorithm of minimum cost flow is run.

Also in its model of distribution of support personnel Belozercovsky & Sensel (2002), distribute the personnel of the different organizations that are prepared to take part in the case of catastrophes, using a small variation of the transportation problem. Moleiro & Rojas (2006), center their DSS in helping in the displacement of the affected persons by a catastrophe from the refuges immediate to the shelters.

Finally Clark & Elvira (2003), in the modulate system SOPC (System for the Organization and Preparation of the Community in cases of catastrophes of natural origin), from which the allocation of the people to the tasks is done (Food administration, Medicine administration, Administration of human resources, Administration of available material resources, Administration of clothes, Analysis of necessities, Attention to missing, Medical attention of first order, Psychological attention, Combat against fire, Evacuation, Evaluation of damages, Information to the community, Civil works, First aid, Removal of remains, Rescue, Security and Transport) that must be taken care of in case of a catastrophe, they used the transportation model, in particular for the assignment problem.

In order summarized all the DSS and mathematical models that support them in Figure 4 it

is presented a summary of all the graduate special works which were mentioned in the previous paragraphs.

Next, in order to illustrate, the work by Lopez & Perez (2006) will be discussed to greater detail.

The Decision Support System of López & Pérez (2006)

With this brief description of López & Pérez (2006) DSS, it is intended to offer a general vision of how the DSS created in the Universidad Metropolitana in Caracas, Venezuela works, and whose presentation is the objective of this work.

For starters it is due to clarify that these systems are developed completely in Spanish and by their academic origin, more than being used at the moment of a catastrophe to make decisions involving the affected population, they must be used in the pre-catastrophe stage to train the civil employees of the competent organisms. In their majority this systems cover general situations, that is to say, they enjoy great flexibility to be adapted to a certain population, thus required.

In the particular case, of the commented DSS, it is used on a determined population, Baruta Municipality, in the Miranda State, who with the municipalities Sucre, Chacao, El Hatillo and Libertador, conform the Metropolitan zone of Caracas.

Before developing the system one of the first activities was to digitize the maps of the Baruta Municipality and to identify in them all the possible routes between the different points, it was also determined the risk zones in the municipality, as well as an inventory of its resources, with the purpose of knowing possible refuges, immediate shelters and transportation centers, as well as recognizing the welfare centers, and their capacities.

Also as a previous activity, the different models to be used in the DSS are developed, in this particular case brief modifications to the algorithms of Dijkstra were made, for the handling of escape routes, the one of Ford & Foulkerson, to handle the

Figure 4. Graduate special works, its models and characteristics of their DSS

Authors	Characteristics of DSS	Mathematical model used
Alessio & Vicuña(2003)	Handling the post-catastrophe, meaning the logistics aspects to follow after the occurrence of the catastrophe.	Structures of Decision trees; Goal programming & PERT-CPM.
Belozercovsky & Sensel (2002)	Handling the logistics aspects of catastrophes.	Shortest route problems; Transportation models; A, B, C model; Decreasing digits & Inventory models.
Clark & Elvira (2003)	Organizing communities and prepare them in case of catastrophes.	Transportation models.
Clavaud & Navarro (2006)	Distributing the population to the attention centers after a catastrophe in the State Miranda, Venezuela.	Structures of Decision trees & Multiattribute Models.
Dos Santos & Rojas (2006)	To reestablish the operation of the companies, analyzing the more sensible aspects of its infrastructure to be affected by a catastrophe.	Goal programming.
Gamboa & Peña (2004)	To deepen the study of handling the inventory in a shelter, mainly of foods and taking into account the donations.	A, B, C model; Decreasing digits; Inventory models & Models based in Fuzzy set.
Goldsztajn (2005)	To redistribute vehicles in case of overflow of gorges, when establishing early alert.	Problems of flow.
Gomez & Zapata (2001)	To establish plan that allow to program and to facilitate the attendance of the affected persons by a catastrophe, when evacuation is necessary.	Shortest route problems (Dijkstra); Problems of flow; Transportation models & Inventory models.
Hernández & Rodríguez (2005)	To help diminishing the disintegration of the familiar groups, independent to the displacements that can undergo the different members, at the moment of a catastrophe.	Structures of Decision trees.
Lecuna & Roa (2005)	To help making decision about infrastructure type to recover with high-priority character, considering a lot of factors.	Goal programming.
López & Pérez (2006)	To work in particular with the geography of a municipality and to study handling of affected persons.	Shortest route problems (Dijkstra); Problems of flow & Multiattribute Models.
López & Sánchez (2005)	To recover the logistics of an enterprise, once an industrial accident occurs.	Goal programming.
Moleiro & Rojas (2006)	To help in the displacement of the persons affected by a catastrophe from immediate refuge to the shelters.	Transportation models.
Rodríguez & Zabala (2002)	To transport the affected persons not only to the immediate refuges, but also to definitive shelters.	Multiattribute Models; Shortest route problems (Dijkstra); Transportation models; Problems of flow & Structures of Decision trees.

flow of affected and developing a Multiattribute Model with multiplicative factors, for the choosing of immediate shelters and refuges.

In Figures 5, 6 and 7, in Spanish, some general aspects of the system will be displayed.

In the first screen (Figure 5), it is possible to distinguish the access to three great areas: Emergency plan by zone, Shelter evaluation and Alarm generator, in this brief summary, by their greater relevance just the first two will be commented.

Also in Figure 5, it is possible to observe that it can: (1) in Risk zone: To add zone, To consult and to update zone and To consult emergency plan, which can be maintained loaded in the system; (2) in Organizations: To add organization and To consult and to update organization, which allows the system to be adapted to any other population different from the Baruta Municipality; (3) in Refuges: To add refuge, To consult and to update refuge, in the case that the conditions of a place that could serve as possible refuge have changed and (4) in Shelters: To add shelter, To consult and to update shelter and To consult and to update criteria, which gives a greater flexibility to the system, allowing to change the criteria, attributes and their weights, in case of it being necessary.

Also in Figure 5, it is observed that the Available Resources can be activated, allowing the data entrance of the organizations that comprise the available resources of the municipality and the routes related to them. The administrator of the system, besides loading all the previous information, is responsible to enter the different types of transportation needed.

In Figure 6, is a screen corresponding to the implementation of an emergency plan, where the map of the zone in study and its possible routes can be observed.

And in the Figure 7, it is displayed how the immediate evaluation of shelters or refuges would be carried out, which have been preloaded in the system. In any case the operator, for the situation in study will activate the refuges to be evaluated, loading the multiplicative valuation of each one of the attributes and factors and obtaining from the system the respective evaluation.

In summary the user, besides updating the Alarms generator, which was not commented, can access the system and enter: (1) Emergency plan by zone, from where besides loading zones and update them, including its routes, could see the affected zones and thence see the routes follow, excluding affected routes, to evacuate the affected population, towards immediate shelters or refuges, or welfare centers; (2) Shelters evaluation, where it would load the immediate shelters or refuges or would evaluate and activate those immediate shelters and refuges needed to handle a certain catastrophe.

In any case, this information can be stored, and changed in order to see the effects of its decisions in the moment of a catastrophe, which would serve as training to face them.

Figure 5. Starting screen of the system S.E.M.B.a, by López & Pérez (2006)

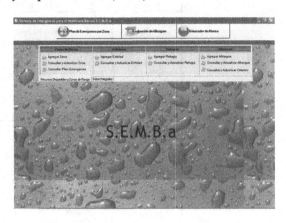

Figure 6. Emergency plan by zone in the system S.E.M.B.a, by López & Pérez (2006)

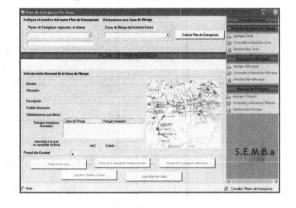

Figure 7. Evaluation of refuges in the system S.E.M.B.a, by López & Pérez (2006)

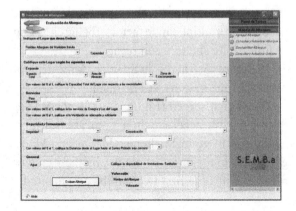

Concluded the presentation of the DSS of Lopez & Perez (2006) a general concept of how the DSS developed in the Metropolitan University in Caracas operate, to aid in case of catastrophes, thus some commentaries can be done on the future lines of investigation to follow and to present some conclusions.

FUTURE RESEARCH DIRECTIONS

Happened in Venezuela the tragedies of natural origin at the end of 1999, in the development of the DSS of the Metropolitan University in Caracas, Venezuela, for the aid in case of catastrophes, that appeared in this chapter, it was intended not to follow existing DSS in the national or international scope, but a reaction before a clear necessity of the country, which was not other than to produce DSS that served, at least of training to the organisms in charge of handling the catastrophes. Since one worked with students of a Systems Engineering School one had to take advantage of two great strengths that they had: handling of mathematical models and knowledge on the construction and development of DSS.

For the immediate future, one aspect that is due to review are these DSS which can exist in other latitudes, that have been created with the objective to confront catastrophes, to incorporate improvements to the future developments that can be undertaken.

Another aspect in this same line is to try to establish networks of national or international character with groups dedicated to the subject of handling of catastrophes, even if their investigations are including different aspects from the mathematical models, in this sense some interchanges had been established with group COMIR of the Central University of Venezuela, that had been handling the problem of population risk from a different perspective, including aspects relative to the architecture.

About lines of investigation to follow a first suggestion must be oriented towards the Matrixes Of Weighing (MOW), since they are a multi-criterion tool of easy application and that they would be easy to implement as DSS.

Additionally it is recommended to deepen all the here used models and doing combined use of such, it is also recommendable the application of other techniques of Operations Research not covered here, as they could be all those related to the Metaheuristic.

In any case, either with worked models or, new models, or combination of models, the work must be oriented to create DSS that can be used by the population or the competent organisms in immediate form.

CONCLUSION

In accord to the general objective, which was to present the mathematical models that sustained it and the DSS developed in the System Engineer School in the Universidad Metropolitana in Caracas, Venezuela, to aide in case of catastrophe, it is important to point out that a great amount of models of the OR that can be helpful at the moment of facing catastrophes, and more importantly being able to integrate them to DSS, that could help the specialized organisms in handling catastrophes or the population in general to be prepared to confront them.

Another aspect of importance is that most of these models and the DSS constructed of them are easy to implement and to use, as it was seen in the DSS model shown as an example, thus, the population or the competent organisms, with little previous training, could take benefit from them.

On the other hand it is important to emphasize the integration of techniques of the OR that are not of frequent appearance in specialized literature, as could be the Decreasing digits, MM with multiplicative factors and the MOW.

ACKNOWLEDGMENT

This investigation would not have been possible without the support given by: the Universidad Metropolitana, specially its deanship of research and development, the deanship of engineering, through the department of technology management and Professor Carlos Nieto of the course risk management and Minimax Consultores, C.A., through its research management.

REFERENCES

Alessio, M., & Vicuña, J. (2003). *Reestablecimiento de la normalidad a través de la gerencia de la logística de catástrofes de origen natural. Trabajo especial de grado*. Venezuela: Universidad Metropolitana.

Almeida de, A. T. (2005). Modelagem multicritério para seleção de intervalos de manutenção preventiva baseada na teoria da utilidade multiatributo. *Pesquisa Operacional, 25*(1), 69–81.

Anthony, R. N. (1987). Contabilidad Gerencial. Barcelona: Ediciones Orbis Colección Biblioteca de la Empresa Vol. I. N° 21. (English, 1974, Programmed learning aid for management accounting. Homewood, Illinois: Learning Systems).

Araz, C., Ozfirat, P. M., & Ozkarahan, I. (2007). An integrated multicriteria decision-making methodology for outsourcing management. *Computers & Operations Research, 34*(12), 3738–3756. doi:10.1016/j.cor.2006.01.014

Archibald, T. W., Black, D., & Glazebrook, K. D. (2009). An index heuristic for transshipment decisions in multi-location inventory systems based on a pairwise decomposition. *European Journal of Operational Research, 192*(1), 69–78. doi:10.1016/j.ejor.2007.09.019

Baucells, M., & Sarin, R. (2003). Group decisions with multiple criteria. *Management Science, 49*(8), 1105–1118. doi:10.1287/mnsc.49.8.1105.16400

Beasley, J. E. (2002). *O.R.-Notes. Inventory Control*. Retrieved October 2008, from http://people.brunel.ac.uk/~mastjjb/jeb/or/invent.html

Belozercovsky, C., & Sensel, D. (2002). *Sistema de apoyo a la toma de decisiones en caso de una catástrofe natural. Trabajo especial de grado*. Venezuela: Universidad Metropolitana.

Ben-Mena, S. (2000). Introduction aux méthodes multicritères d'aide à la decision. *Biotechnol. Agron. Soc. Environ, 4*(2), 83–93.

Cardós, M., & García-Sabater, J. P. (2006). Designing a consumer products retail chain inventory replenishment policy with the consideration of transportation costs. *International Journal of Production Economics, 104*(2), 525–535. doi:10.1016/j.ijpe.2004.12.022

Chang, Ch., & Lo, T. Y. (2009). On the inventory model with continuous and discrete lead time, backorders and lost sales. *Applied Mathematical Modelling, 33*(5), 2196–2206. doi:10.1016/j.apm.2008.05.028

Charnes, A., & Cooper, W. W. (1961). *Management models and industrial applications of linear programming*. New York: John Wiley and Sons.

Chen, Ch. (2001). A fuzzy approach to select the location of the distribution center. *Fuzzy Sets and Systems, 118*(1), 65–73. doi:10.1016/S0165-0114(98)00459-X

Clark, R., & Elvira, C. (2003). *Organización de la comunidad en caso de desastres de origen natural: Un sistema de apoyo. Trabajo especial de grado*. Venezuela: Universidad Metropolitana.

Clavaud, M., & Navarro, R. (2006). *Distribución de la población a los centros asistenciales después de una catástrofe, dirigida al estado Miranda. Trabajo especial de grado.* Venezuela: Universidad Metropolitana.

Dong, Y., Carter, C. R., & Dresner, M. E. (2001). JIT purchasing and performance: an exploratory analysis of buyer and supplier perspectives. *Journal of Operations Management, 19*(4), 471–483. doi:10.1016/S0272-6963(00)00066-8

Dos Santos, A., & Rojas, J. (2006). *Logística para reestablecer el ambiente de infraestructura en una empresa después de una catástrofe. Trabajo especial de grado.* Venezuela: Universidad Metropolitana.

Eglese, R., Maden, W., & Slater, A. (2006). A road Timetable ™ to aid vehicle routing and scheduling. *Computers & Operations Research, 33*(12), 3508–3519. doi:10.1016/j.cor.2005.03.029

Ehrgott, M., & Gandibleux, X. (Eds.). (2002). *Multiple criteria Optimization: State of the art annotated bibliographic surveys.* Kluwer Academic Publishers.

Gamboa, D., & Peña, D. (2004). *Sistema de apoyo para manejo de inventarios en presencia de donaciones. Trabajo especial de grado.* Venezuela: Universidad Metropolitana.

Geurts, P., Fillet, M., Seny, D., Meuwis, M., Malaise, M., Merville, M., & Wehenkel, L. (2005). Proteomic mass spectra classification using decision tree based ensemble methods. *Bioinformatics (Oxford, England), 21*(15), 3138–3145. doi:10.1093/bioinformatics/bti494

Goldsztajn, M. (2005). *Sistema de apoyo para la redistribución de vehículos en caso de desbordamiento de quebradas. Trabajo especial de grado.* Venezuela: Universidad Metropolitana.

Gómez, M., & Zapata, E. (2001). *Catástrofes naturales: Plan de acción para evaluación y asistencia. Trabajo especial de grado.* Venezuela: Universidad Metropolitana.

Greco, S., Matarazzo, B., & Slowinski, R. (2001). Rough sets theory for multicriteria decision analysis. *European Journal of Operational Research, 129*(1), 1–47. doi:10.1016/S0377-2217(00)00167-3

Güngör, Z., & Arikan, F. (2000). Application of fuzzy decision making in part-machine grouping. *International Journal of Production Economics, 63*(2), 181–193. doi:10.1016/S0925-5273(99)00010-9

Guo, Z., Fang, F., & Whinston, A. B. (2006). Supply chain information sharing in a macro prediction market. *Decision Support Systems, 42*(3), 1944–1958. doi:10.1016/j.dss.2006.05.003

Hernández, J. G., & García, M. J. (2006). *The Importance of the Procurement Function in Logistics.* Proceedings ICIL'2006, Kaunas University of Technology, Lithuania 149-157.

Hernández, J. G., & García, M. J. (2007). Investigación de operaciones y turismo en *Revista de Matemática. Teoría y Aplicaciones, 14*(2), 221–238.

Hernández, J. G., García, M. J., & Nieto, C. (2005). Modelos matemáticos y su ayuda en caso de catástrofes. Universidades y Riesgo. Una vitrina desde la UCV; Facultad de Arquitectura Universidad Central de Venezuela; Caracas, Venezuela; Publicado en CD del evento Hábitat y Riesgo el rol de las Universidades.

Hernández, L., & Rodríguez, V. (2005). *Herramienta de apoyo para la ayuda del control poblacional y preservación del núcleo familiar ante una catástrofe. Trabajo especial de grado.* Venezuela: Universidad Metropolitana.

Hernández, J. G., García, M. J., & Hernández, G. J. (2009). *Influence of Location Management in supply chain Management*. Global Business And Technology Association (GBATA 2009). Reading Book; Prague, Czech Republic (pp. 500-507).

Ignizio, J. (1976). *Goal Programming and extensions*. Massachusetts: Lexington Books.

Jacquet-Lagreze, E., & Siskos, J. (1982). Assessing a set of additive utility functions for multicriteria decision-making, the UTA method. *European Journal of Operational Research, 10*(2), 151–164. doi:10.1016/0377-2217(82)90155-2

Jessop, A. (2004). Sensitivity and robustness in selection problems. *Computers & Operations Research, 31*(4), 607–622. doi:10.1016/S0305-0548(03)00017-0

Jiménez, A., Ríos-Insua, S., & Mateos, A. (2006). A generic multi-attribute analysis system. *Computers & Operations Research, 33*(4), 1081–1101. doi:10.1016/j.cor.2004.09.003

Jotshi, A., Gong, Q., & Batta, R. (2009). Dispatching and routing of emergency vehicles in disaster mitigation using data fusion. *Socio-Economic Planning Sciences, 43*(1), 1–24. doi:10.1016/j.seps.2008.02.005

Kojadinovic, I. (2005). An axiomatic approach to the measurement of the amounts of interaction among criteria or players. *Fuzzy Sets and Systems, 152*(3), 417–435. doi:10.1016/j.fss.2004.11.006

Kosuda, S., Ichihara, K., Watanabe, M., Kobayashi, H., & Kusano, S. (2000). Decision-Tree sensitivity analysis for cost-effectiveness of Chest 2-Fluoro-2-D-[18 F] Fluorodeoxyglucose positron emission tomography in patients with pulmonary nodules (Non-small cell lung carcinoma) in Japan. *Chest, 117*, 346–353. doi:10.1378/chest.117.2.346

Lecuna, M., & Roa, L. (2005). *Logística para reestablecer el ambiente de infraestructura después de una catástrofe de origen natural. Trabajo especial de grado*. Venezuela: Universidad Metropolitana.

Lee, S. (2001). On solving unreliable planar location problems. *Computers & Operations Research, 28*(4), 329–344. doi:10.1016/S0305-0548(99)00120-3

Li, Y., & Liao, X. (2007). Decision support for risk analysis on dynamic alliance. *Decision Support Systems, 42*(4), 2043–2059. doi:10.1016/j.dss.2004.11.008

Lieckens, K., & Vandaele, N. (2007). Reverse logistics network design with stochastic lead times. *Computers & Operations Research, 34*(2), 395–416. doi:10.1016/j.cor.2005.03.006

Lin, H. W., Nagalingam, S. V., & Lin, G. C. I. (2009). An interactive meta-goal programming-based decision analysis methodology to support collaborative manufacturing. *Robotics and Computer-integrated Manufacturing, 25*(1), 135–154. doi:10.1016/j.rcim.2007.10.005

López, A., & Sánchez, C. (2005). *Empresas de riesgo mayor: Recuperación de su logística. Trabajo especial de grado*. Venezuela: Universidad Metropolitana.

López, L., & Pérez, M. (2006). *Deslaves e inundaciones: Sistema de emergencia para el municipio Baruta. Trabajo especial de grado*. Venezuela: Universidad Metropolitana.

Loudni, S., Boizumault, P., & David, P. (2006). On-line resources allocation for ATM networks with rerouting. *Computers & Operations Research, 33*(10), 2891–2917. doi:10.1016/j.cor.2005.01.016

Lu, D., Chou, Y., & Yuan, H. (2005). Paradigm in the institutional arrangement of protected areas management in Taiwan-a case study of Wu-Wei-Kang Waterfowl Wildlife Refuge in Ilan, Taiwan. *Environmental Science & Policy, 8*(4), 418–430. doi:10.1016/j.envsci.2005.03.013

Luo, J. (2007). Buyer-vendor inventory coordination with credit period incentives. *International Journal of Production Economics, 108*(1-2), 143–152. doi:10.1016/j.ijpe.2006.12.007

Moleiro, A., & Rojas, K. (2006). *Sistema de apoyo para el manejo de afectados en una catástrofe. Trabajo especial de grado.* Venezuela: Universidad Metropolitana.

Monteiro, R., & Rodrigues, G. (2006). A system of integrated indicators for socio-environmental assessment and eco-certification in agriculture-ambitec-agro. *Journal of Technology Management & Innovation, 1*(3), 47–59.

Mustajoki, J., Hämäläinen, R. P., & Sinkko, K. (2007). Interactive computer support in decision conferencing: Two cases on off-site nuclear emergency management. *Decision Support Systems, 42*(4), 2247–2260. doi:10.1016/j.dss.2006.07.003

Noji, E. K. (Ed.). (1997). *The public health consequences of disasters.* New York: Oxford University Press, Inc.

Peneva, V., & Popchev, I. (2003). Properties of the aggregation operators related with fuzzy relations. *Fuzzy Sets and Systems, 139*(3), 615–633. doi:10.1016/S0165-0114(03)00141-6

Persona, A., Regattieri, A., Pham, H., & Battini, D. (2007). Remote control and maintenance networks and its applications in supply chain management. *Journal of Operations Management, 25*(6), 1275–1291. doi:10.1016/j.jom.2007.01.018

Prasad, B. (1995). JIT quality matrices for strategic planning and implementation. *International Journal of Operations & Production Management, 15*(9), 116–142. doi:10.1108/01443579510099706

Rabinovich, E., Dresner, M. E., & Evers, P. T. (2003). Assessing the effects of operational processes and information systems on inventory performance. *Journal of Operations Management, 21*(1), 63–80. doi:10.1016/S0272-6963(02)00041-4

Ravichandran, N. (2007). A finite horizon inventory model: An operational framework. *International Journal of Production Economics, 108*(1-2), 406–415. doi:10.1016/j.ijpe.2006.12.057

Razavi, A. R., Gill, H., Ahlfeldt, H., & Shahsavar, N. (2008). Non-compliance with a postmastectomy radiotherapy guideline: Decision tree and cause analysis. *BMC Medical Informatics and Decision Making, 8*(41). doi:.doi:10.1186/1472-6947-8-41

Rodríguez, D., & Zabala, J. (2002). *Trasporte para el traslado de heridos y afectados desde refugios a centros de asistencia y albergues. Trabajo especial de grado.* Venezuela: Universidad Metropolitana.

Rohweder, L., & Virtanen, A. (2008). Learning for a sustainable future. Innovative solutions from the Baltic Sea region. Uppsala, Swedish: The Baltic University Press.

Rommelfanger, H. (2007). A general concept for solving linear multicriteria programming problems with crisp, fuzzy or stochastic values. *Fuzzy Sets and Systems, 158*(17), 1892–1904. doi:10.1016/j.fss.2007.04.005

Roy, B., & Vincke, P. (1981). Multicriteria analysis: survey and new directions. *European Journal of Operational Research, 8*(3), 207–218. doi:10.1016/0377-2217(81)90168-5

Sakarovitch, M. (1979). *Techniques mathématiques de la recherche opérationnelle III – Optimisation dans les réseaux*. France: Université Scientifique et Médicale Institut National Polytechnique de Grenoble.

Samuelson, P. A., & Nordhaus, W. D. (1986). Economía Duodécima edición. España: McGraw-Hill. (English, 1985, Economics, USA: McGraw-Hill).

Sarimveis, H., Patrinos, P., Tarantilis, Ch. D., & Kiranoudis, Ch. T. (2007). Dynamic modelling and control of supply chain systems: A review. *Computers & Operations Research*, *35*(11), 3530–3561. doi:10.1016/j.cor.2007.01.017

Scott, M. A. (2002). *Modeling and analysis of multicommodity network flows via goal programming*. Master of Science in Operations Research dissertation, Air Force Institute of Technology.

Shen, S., & O'Hare, G. M. P. (2004). Agent-Based resource selection for grid computing. *Workshop 4: International Workshop on Agents and Autonomic Computing and Grid Enabled Virtual Organizations* (pp. 658-672).

Singhal, K., Singhal, J., & Starr, M. K. (2007). The domain of production and operations management and the role of Elwood Buffa in its delineation. *Journal of Operations Management*, *25*(2), 310–327. doi:10.1016/j.jom.2006.06.004

Skouri, K., Konstantaras, I., Papachristos, S., & Ganas, I. (2009). Inventory models with ramp type demand rate, partial backlogging and Weibull deterioration rate. *European Journal of Operational Research*, *192*(1), 79–92. doi:10.1016/j.ejor.2007.09.003

Sprague, R., & Watson, H. (1986). *Decision Support Systems*. USA: Prentice Hall.

Stewart, T. J., Janssen, R., & Herwijnen, M. (2004). A genetic algorithm approach to multiobjective land use planning. *Computers & Operations Research*, *31*(14), 2293–2313. doi:10.1016/S0305-0548(03)00188-6

Systems, E. M. (1997). *EMS* (pp. 1–44). Project Evaluation Matrix.

Tavana, M. (2004). A subjective assessment of alternative mission architectures for the human exploration of Mars at NASA using multicriteria decision making. *Computers & Operations Research*, *31*(7), 1147–1164. doi:10.1016/S0305-0548(03)00074-1

Tian, J., Wang, Y., Li, H., Li, L., & Wang, K. (2007). DSS development and applications in China. *Decision Support Systems*, *42*(4), 2060–2077. doi:10.1016/j.dss.2004.11.009

Tsamboulas, D. A., & Mikroudis, G. K. (2006). TRANS-POL: A mediator between transportation model and decision makers' policies. *Decision Support Systems*, *42*(2), 879–897. doi:10.1016/j.dss.2005.07.010

Turban, E. (2001). *Decision Support Systems and Intelligent Systems*. New York: Prentice Hall.

Vasant, P. M., Barsoum, N. N., & Bhattacharya, A. (2008). Possibilistic optimization in planning decision of construction industry. *International Journal of Production Economics*, *111*(2), 664–675. doi:10.1016/j.ijpe.2007.03.006

Wang, W., Rivera, D. E., & Kempf, K. G. (2007). Model predictive control strategies for supply chain management in semiconductor manufacturing. *International Journal of Production Economics*, *107*(1), 56–77. doi:10.1016/j.ijpe.2006.05.013

Weber, Ch. A., Current, J. R., & Benton, W. C. (1991). Vendor selection criteria and methods. *European Journal of Operational Research*, *50*(1), 2–18. doi:10.1016/0377-2217(91)90033-R

Chapter 14
Integrating Scenario–Based Reasoning into a Multi–Criteria Decision Support System for Emergency Management

Tina Comes
Institute for Industrial Production (IIP), Karlsruhe Institute of Technology (KIT), Germany

Michael Hiete
Institute for Industrial Production (IIP), Karlsruhe Institute of Technology (KIT), Germany

Niek Wijngaards
Thales Research & Technology Netherlands/D-CIS Lab, The Netherlands

Frank Schultmann
Institute for Industrial Production (IIP), Karlsruhe Institute of Technology (KIT), Germany

ABSTRACT

Multi-criteria decision analysis (MCDA) is a technique for decision making among multiple alternatives for action providing transparent and coherent decision support for complex situations with conflicting objectives. Managing longer term decisions for environmental incidents is an application domain in which MCDA has proved useful. Yet a difficulty in applying MCDA is when uncertainties abound. Contrarily, scenario-based reasoning is a method allowing for the assessment of multiple possible future developments of the situation. In this way, the use of scenarios is a transparent and easily understandable way to integrate uncertainties into the reasoning process. We propose a mechanism to integrate scenarios. Our theoretical framework can be operationalised by decision support systems relying on both automated systems and human experts. These facilitate the assessment of consequences within a scenario, and may propose new scenarios. We illustrate this mechanism taking the decision making in emergency management after a train crash with potential release of chlorine as an example.

DOI: 10.4018/978-1-61520-987-3.ch014

INTRODUCTION AND BACKGROUND

While natural hazards like floods or earthquakes have disrupted society since ancient times, nowadays both, natural as well as man-made emergencies threaten society, economy and environment with devastating consequences. The spread of high-risk technologies (e.g., nuclear power plants, chemical refineries) exacerbates the situation (Perrow, 1984).

Decision making situations in emergency management are usually characterised by multiple objectives, participation of multiple actors with conflicting interests, competences and knowledge, and uncertain or lacking information (see e.g. French & Geldermann, 2005). As time plays often a critical role, the need for timely, coherent and effective decision support arises. Furthermore, emergency response is usually a collaborative effort of many actors. That is why the need for acceptance on all levels of hierarchy which play a role in the implementation of counter- or remediation measures – directly or indirectly – arises. This need encompasses for example the population affected who has to comply and respond to the measures as well as the heads of all emergency management agencies involved.

Furthermore, the experts, stakeholders and decision makers who have to find a consensus before implementing a decision are often geographically dispersed. In these cases, it can be necessary to use an ICT system – as bringing all actors involved together face to face would require a substantial amount of time and resources.

In situations with multiple objectives, multi-criteria decision analysis (MCDA) is often chosen as basis for decision support systems (see French, 1996), for its coherent support in complex situations. MCDA's popularity is mainly due to the intuitive and transparent evaluation of 'alternatives for action' (termed *'alternatives'* in this chapter) it offers.

Additionally, MCDA facilitates the communication of the results to stakeholders and other emergency management authorities. First, experts and decision makers may want to explain the justification of their decision to the public and thus to promote acceptance and trust. Secondly, stakeholders are more and more actively involved in the emergency management. With the rise of the new communication technologies like mobile phones and the internet, the public has not only to comply with the decisions made, but it is as well a source of information (Shim et. al., 2002). Therefore communication with the public becomes more and more important in emergency management.

Despite all the advantages described, the applicability of MCDA in emergency management is limited. Most importantly the use of MCDA can become cumbersome when uncertainties prevail. But uncertainties due to a lack of information, time pressure prohibiting the collection of further information and high uncertainty characterize most emergency management situations (Barker & Haimes, 2009; Bertsch, 2008; Cooke & Goossens, 2004). To facilitate reasoning under *severe uncertainty,* we integrate *scenarios* – possible, coherent and internally fully consistent descriptions of the situation and its future development – into the decision making framework.

This chapter starts with a presentation of deterministic state-of-the-art MCDA systems. Then, MCDA techniques for reasoning under uncertainty are discussed and their major drawbacks are described (e.g., the fact that they do not allow for handling situations which can develop in fundamentally different ways). To cope with highly varying situations, scenarios can be used (see e.g. Durbach & Stewart, 2003). Therefore, the next section of this chapter deals with scenario-based reasoning (SBR) approaches.

As MCDA is deficient in case of fundamentally changing situations and SBR does not evaluate alternatives, we then propose integrating scenarios into MCDA. A scenario selection technique based on the evaluation is developed in the next section. Then, the possibilities to analyse alternatives under a range of scenarios to arrive at robust decisions are discussed.

We illustrate our approach by means of an emergency management example, which has been developed together with experts from the Danish Emergency Management Agency (DEMA).

The situation is the following: A freight train crashed into an empty passenger train stationary at the central train station in Odense, Denmark. A *hazmat team* with particular training for coping with hazardous material and large incidents was called to handle the chlorine that is leaking from a ruptured tank wagon. The team has applied a plastic based cover and stabilized the situation. To resolve it definitively, the chlorine must be transferred to another tank. During the transfer, there is the risk that a new leak occurs or that one of the transfer buckets falls over. In this case, a large amount of chlorine is released quickly and floats over the downwind area. That is why a decision has to be made regarding evacuation or sheltering in the downwind area. On the basis of this example a set of scenarios is constructed and evaluated using MCDA techniques. Finally, we give a brief conclusion summarizing our main results and discussing future steps.

Information Systems for a Standard MCDA Approach: Multi-Attribute Value Theory and Tools

Numerous MC)DA methods have been developed. They all deal with the problem that a comparison between a set of decision alternatives must be made with respect to multiple objectives (for a general overview of MCDA techniques see e.g. Belton & Stewart, 2002). In emergency management, usually the number of alternatives is rather small; this makes multi-attribute decision making the technique of choice.

MCDA has proved useful in longer term strategic emergency management (French, 1996; Geldermann, Bertsch & Rentz, 2006; Linkov et. al., 2006). Using case-based reasoning, a default

decision support system can be developed *before* the actual accident happens. To this end, goals, preferences and alternatives are elicited from experts, stakeholders and users. This approach ensures that – unlike standard MCDA systems – the decision support system can be launched immediately when an emergency happens. Nevertheless, applying MCDA (and scenario analysis) takes time. First, the input has to be adapted to the current situation. It is important to keep the system adaptable, such that new ideas and criteria can be integrated or changing preferences can be taken into account. Secondly, the output has to be analysed in depth to benefit from the insights MCDA tools provide and to overcome cognitive biases which may arise – particularly when using computational decision support systems (Paprika & Kiss, 1985). Therefore, it is not recommended to use a standard MCDA approach on an operational level, but mainly for medium and longer term emergency management decisions (e.g. selection of remediation measures).

Multi-attribute value theory (MAVT) forms the basis of our approach as it allows evaluating alternatives with respect to different criteria. MAVT has been applied successfully in emergency management (see e.g. French, 1996; Mustajoki, Hämäläinen & Sinkko, 2007). A brief description of MAVT is provided below. Thereafter, typical approaches to capture uncertainties within multi-attribute decision frameworks are discussed namely *multi-attribute utility theory* (MAUT) and *sensitivity analysis*.

Problem Structuring

Each MCDA process starts by structuring the problem, taking it from an initially vague and intuitive understanding to a more formal description that can be analyzed mathematically. To this end, it is necessary to come to a consensual formulation of the problem, and to agree about performance measures, constraints, and the relations through which an action produces consequences (Mingers

Figure 1. Attribute tree

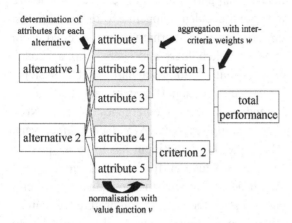

& Rosenhead, 2004). This phase is often described as one of the hardest, yet most crucial in MCDA (Belton & Stewart, 2002), as it gives a better understanding of both, the problem at hand and the value judgements affecting the decision.

In MAVT, he structure of the problem is represented in an attribute tree showing how the overall-objective ("total performance") is decomposed first into criteria and finally into measurable attributes (cf. Figure 1; reading from right to left). The attribute tree captures the understanding of all the issues involved, including the options available to the decision makers and the key factors that they felt to be important in the decision making process (Hodgkin, Belton & Koulouri, 2005).

To construct the attribute tree, (strategic) top-down or (tactical) bottom-up approaches can be used. While the first approaches start with the identification of the most general goals which are successively divided into sub-goals and attributes, the latter start with listing all differences between the decision alternatives (Belton & Stewart, 2002). Executing the process iteratively – and allowing all actors involved to reflect about the problem's structure and their own beliefs about it – has proved useful (von Winterfeldt & Fasolo, 2009). Furthermore, it is important to keep the tree adaptable to new insights and information avail-

able or to a changing perception of the problem and its structure.

To facilitate the MAVT's problem structuring process, software tools graphically supporting the construction of an attribute tree and its evaluation can be used (INFORMS, 2009; Weistroffer, Smith & Subhash, 2005). Additionally, these tools usually offer several different options to communicate the results and have easy-to-use user interfaces making it possible for the participants to use the software even by themselves (Mustajoki, Hämäläinen & Sinkko, 2007). Particularly, web-based decision support systems can be useful when the actors involved are geographically dispersed. These tools have reduced technological barriers and have made it easier and less costly to make decision-relevant information available to decision makers and experts (Shim et. al., 2002). Emergency management has been considered as one of the application areas where an application of web-based decision support is most needed and useful (Nunamaker, 1997). This, however, requires that experts still have web access, i.e. telecommunication and power supply must not be interrupted (Merz, Hiete & Bertsch 2009).

Preference Elicitation

In the next step of MAVT, preferential information is elicited (for preference elicitation techniques cf. e.g. von Winterfeldt & Edwards, 1986). This information usually consists of two components, namely *intra-* and *inter-criteria* preferences.

Intra-criteria preferences are used to make the attributes measured on different scales comparable. This is done by mapping the scores s_{ij} of an attribute i estimated for each alternative j to [0, 1] using value functions v_i such that the best score is mapped to 1, while the worst score is mapped to 0. Using linear value functions can be justified when the assumption is made that the difference between the outcomes of all alternatives are relatively small (see Bertsch, 2008). However, attention should be paid, whenever it is important

Figure 2. Linear and exponential value functions

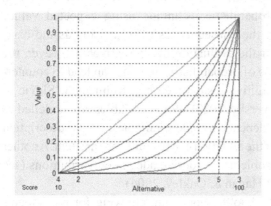

that the score actually attained is close to the best possible score or when outliers occur. In these cases exponential value functions can be used, expressing the importance of being sufficiently close to the optimal result by the steepness of curvature (Kirkwood, 1992). Figure 2 shows a screenshot of a user interface that has been developed in the Diadem[1] project and realised in Matlab®. The interface supports different types of value functions: The linear value function is depicted in red, whereas several types of exponential value functions (with varying steepness of curvature) are shown in blue. In Figure 2 the position of the alternatives are marked by the vertical dashed lines. The worst alternative is alternative 4 with a score of 10 and alternative 3 is the best alternative with a score of 100. As the scores of the worst and best alternatives are shown in the same figure, the user gets an idea about the scope of scores. Furthermore, it is possible to read off the value of each alternative for the attribute currently under scrutiny by looking at the intersections of the vertical lines (representing the score of the alternatives) and the value functions. Thus, it becomes possible to compare different alternatives and the selection of value functions is facilitated.

The preferences between attributes are captured in weights w_i for each attribute i. These weights are basically scaling factors, i.e. a weight w_i indicates the relative importance of attribute i

compared to all the other attributes. If an attribute i has a weight that is only half the weight of an attribute j, this means that for the decision maker a value of 1 in attribute i and *0* in attribute j is equally preferable to a value of 0 in i and 0.5 in j.

There is a variety of different weighting elicitation procedures. The most naïve approach is giving each attribute its weight *directly*. Other methods frequently used are SMART, where the least important attribute is used as the basis for assessing the weights (see e.g. Edwards, 1977), or SWING where the most important attribute is used as reference (von Winterfeldt & Edwards, 1986). To support the elicitation process visually, bar charts representing the currently chosen weights can be used (Mustajoki, Hämäläinen & Sinkko, 2007).

MCDA Results

The last step in MAVT) is for each alternative a the aggregation of all values $v_i(a)$ to the total performance p_a taking into account the weights w_i for each attribute i, i. e. $p_a = \sum_i w_i\, v_i(a)$. In this manner, a ranking of alternatives is achieved.

Finally, a number of different methods and tools to visualise and explain the results can be used (see e.g. Bertsch, Geldermann, Rentz & Raskob, 2006; French, 1996; Raskob, Gering & Bertsch, 2009). Figure 3 shows an exemplary plot of the most commonly used visualisation technique: a stacked bar chart. This plot has been created using the results of an MCDA problem realised with the Matlab®-tool developed in Diadem. The higher a bar is, the better the results are. This type of diagram allows not only for easily identifying the best possible results, but also for representing the contribution of each of the criterion to the overall performance.

MCDA and Uncertainty

Usually, MCDA models rely on the assumption that attributes and criteria are well-defined (i.e.

Figure 3. Stacked bar chart of MCDA results

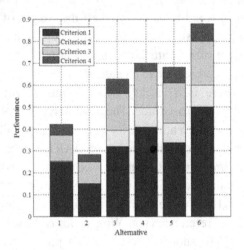

for each alternative the scores of all attributes can be determined) and known (i.e. the performance assigned can be treated as a *deterministic* variable) (cf. Fenton & Neil, 2001). This is justified as long as the complexity of a decision problem is dominated by the multiplicity of objectives (see Stewart, 2005). However, in emergency management the information underlying a decision is frequently limited and uncertain (cf. French & Niculae, 2005). Even if it was possible to reduce these uncertainties through further investigations, decisions in emergency management must often be made within short time forestalling any possibility to reduce uncertainty.

Multi-attribute decision theory techniques address the problem of uncertain information. However, they are based on the assumption of independence of criteria and attributes (see French, 1986). Furthermore, often only uncertainties concerning preferential information are covered (see e.g. Bertsch, Geldermann & Rentz, 2007). Nevertheless, problems of measurement error and lacking or noisy information (cf. Weber, 1987) as well as the uncertainties about the model itself should be considered. The latter arise, as a model is necessarily a simplified representation of the real word (cf. Ascough, Maier, Ravalico & Strudley, 2008).

Multi-attribute utility theory (MAUT) is a probabilistic technique using expected values of the consequences of a decision as the basis for the evaluation of alternatives. However, the use of probabilistic methods can lead to counter-intuitive results (see e.g. Kahneman, Slovic & Tversky, 1982). When variables are modelled as independent random variables, also the description of the situation can become inconsistent. Another problem is to derive adequate distributions (see O'Hagan & Oakley, 2004).

Another approach, allowing for the integration of uncertainties in MCDA without applying probabilistic techniques, is *sensitivity analysis* (e.g. Ríos-Insua & French, 1991). While standard sensitivity analysis focuses on variations of one parameter, approaches exist to analyse the simultaneous variation of multiple parameters using sampling techniques (e.g., Monte-Carlo methods, cf. Butler, Jia & Dyer, 1997). Although Monte-Carlo methods have linear complexity, simulations can become computationally expensive as the models themselves become increasingly complex (cf. Ascough et. al., 2008) so that its application to emergency management is limited.

Scenario-Based Reasoning (SBR)

Another possibility to take uncertainties into account is scenario-based reasoning (SBR). In strategic business planning, scenarios have been used successfully for many years (e.g. Bunn & Salo, 1993; van der Heijden, 2007). Unfortunately, the term 'scenario' is frequently encountered in everyday language and scientific literature (for an overview of different scenario definitions see Heugens & van Oosterhout, 2001). In our work, we focus on systematic aspects of scenario development and analysis, particularly on identifying impact factors and their interrelations. That is, our approach is based on *formative* SBR, where a scenario describes the state a system is currently in and how this state develops in the future.

SBR offers the possibility to consider and discuss several possible situation developments – regardless of their likelihood. SBR offers a way to take system changes (e.g. occurrence of new key drivers) into account. Hereby it is possible to overcome cognitive biases such as overconfidence and to integrate fundamental risks that may be of very little probability into the reasoning framework (Schoemaker, 1993). However, a major drawback of SBR is that it has no inherent technique for the evaluation of alternatives under multiple objectives. An integration of SBR into MCDA offers the possibility to systematically evaluate several alternatives taking into account uncertainties in a transparent and easily understandable way.

Constructing and Using Scenarios in Emergency Management

Ideally, the set of scenarios challenges the imagination of experts and decision makers. Therefore, scenarios framing the limits of possibility for a range of plausible futures are developed. In this way, scenarios can help adapting to a changing environment – even if unpredicted and sudden changes occur. However, when just a small number of people with related backgrounds is involved in the scenario construction process, cognitive biases can lead to the development of rather similar scenarios. These biases can e.g. emanate from the use of heuristics and intuitive logics like availability and representativeness (Kahneman, Slovic & Tversky, 1982). In some cases, they can be also the result of too optimistic or wishful thinking (Fildes, Goodwin, Lawrence & Nikolopoulos, 2008). To overcome these biases, a network of experts from various disciplines and key stakeholder groups should be involved in the scenario construction process (Roubelat, 2000).

New information and communication technologies facilitate the participation of multiple experts from various disciplines and stakeholders – even if they are dispersed geographically. These experts do not need to be involved in the process of developing the scenario as a whole (i.e. as a complete "what if …"-story), but they can also contribute their assessment of possible scores of key variables *given* all the possible scores of those key variables which already have been determined by other experts or the decision makers. In this way, the experts can also contribute to the scenario building process although they may have limited time and may not be available for longer discussions.

To ensure acceptance and trust, each scenario should fulfil three conditions: plausibility, coherence and consistency (cf. Heugens & van Oosterhout, 2001). A scenario is plausible, when it does not go beyond the realm of possibility. Coherence is ensured when the causal links explain how the system evolves are made explicit. This is particularly important when the scenario seems to be very unlikely. A scenario is internally fully consistent when the description of the situations' development is unambiguous.

To enhance the acceptance and trust further, visualisations and maps of the affected areas have proved useful (Tress & Tress, 2003). To this end, decision support software like the Accident Reporting and Guidance Operational System (ARGOS, see Baklanov, Sørensen, Hoe & Amstrup, 2006) which allows for the representation of areas affected and levels of concern can be used. Furthermore, critical infrastructure and areas of particular importance can be highlighted in the maps. By forwarding these maps to the incident commanders and emergency managers in the field, information about possible further developments of the situation can be transferred in an easily understandable way. This is very important, as the incident commanders in emergency are working under stress and time pressure.

To develop and construct scenarios,) several different techniques are available. Formative SBR reduces the performances of all impact variables identified to a crisp set of possible states, stored in a vector d_j (see e.g. Scholz & Tietje, 2002). Then, the scenarios are compiled by combining

Figure 4. Scenario selection. Left-hand side: Scenario selection for both alternatives, alt1 and alt2. Right-hand side: Combined representation of worst, medium and best scenario for both alternatives

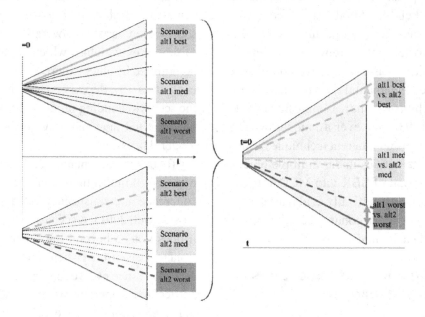

for all variables all possible states. To ensure the consistency of each scenario, filtering techniques *(consistency assessments)* assessing the consistency of all pairs of variables *(i, j)* states have to be applied ex post. This assessment answers the question whether the scores of *i* and *j* fit together or if a combination is unlikely or even impossible.

Scenario Selection and Evaluation of Decision Alternatives

The number of scenarios presented to the decision makers must be manageable. Ideally, the scenarios presented to the user cover the whole span of likely and relevant situation developments. Experience shows that users can cope with at most five to six scenarios (cf. Godet, 1990; Schnaars, 1987).

To reduce the number of scenarios, several techniques have been developed in scenario planning and formal scenario analysis. We propose an approach based on the *evaluation* of scenarios. Both, every-day language and scenario literature frequently refer to a "best" or "worst case" scenario

(cf. Mahmoud et. al., 2009; Schnaars & Ziamou, 2001). However it is often unclear *how* these scenarios are developed and how their performance is measured. To be able to base the evaluation of each scenario on the preferences, beliefs and value judgements of the actors involved in the decision making process, each scenario constructed is evaluated using the MAVT attribute tree. The aim of this approach is to select from the set S_a of all scenarios which assume that alternative *a* has been implemented, the subset S^*_a which contains the scenarios that are the most distinct with respect to their total performances. Thus, it is possible to select the "best" or "worst" scenarios according to the decision makers' evaluation.

Our proposed approach allows for a presentation of the broadest span of *total performances* (i.e. evaluations or impact assessments) of each decision alternative. Thus, the worst, best and median case for each alternative can be compared (cf. Figure 4, right side). It is furthermore possible to choose the decision alternative which has a sufficiently good total performance in *each* case.

Figure 5. Robust recommendation encompassing several scenarios

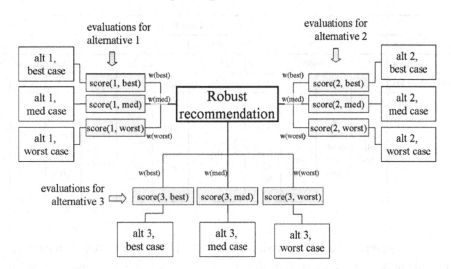

Additionally, the robustness of the alternatives under scrutiny becomes obvious: While certain decision alternatives may be more susceptible to changes in one of the key drivers, others will result in a more similar performance.

The selection of the scenarios presented above allows for comparing the performances of the selected scenarios for *all* alternatives a_l *(l=1,...,m)*, i.e. $S(a_l)^w$, $S(a_l)^m$ and $S(a_l)^b$. Figure 4 (right side) shows an example, where the performances of the scenarios selected for two decision alternatives are compared. While alternative 1 (continuous lines) has a better performance in the best and median case, alternative 2 (dashed lines) has a better worst case performance.

By presenting the performances of the scenarios selected for each alternative to the users, transparency and clarity is ensured. However, there is the risk that the thinking of the decision makers is heavily influenced by just one of the three scenarios (e.g. the most likely or impressing one, see French & Geldermann, 2005; O'Brien, 2004).

To facilitate decision making using the results under several scenarios MAVT techniques can be used (cf. Comes, Hiete, Wijngaards & Kempen, 2009). To this end preferences of the decision makers, which reflect the importance of the worst,

medium and best case, are elicited. These preferences are modelled as weights that are used in the aggregation procedure, cf. Figure 5. As for each alternative, three scenarios are considered, the total performance for each alternative a_i is:

$$p(a_i) = \sum_{j=1}^{3} w(j) \cdot score(a_i, j),$$ where a_i is the

alternative implemented in scenario j and $score(a_i, j)$ is its overall performance. This approach ensures the development of robust recommendations, i.e. the alternative recommended is not only optimal in one (deterministic) possible case, but sufficiently good or acceptable under each of the possible developments of the situation (Ben-Haim, 2000; Lempert, Groves, Popper & Bankes, 2006). As usual in MCDA it is important that this process can be used iteratively. Such, it becomes possible to use the results of the analysis for the definition of strategies and to stimulate discussions.

Illustrative Example

The approach presented is now illustrated by means of the example that was described in the introduction: Although the leak of chlorine is provisionally covered, a decision whether to

Table 1. Structure of attribute tree and weights

Overall goal	Total Performance											
Criteria weights	0.5				0.1	0.2		0.2				
Criteria	Health				Effort	Cost		Disruption of society				
Attribute Weights	0.5	0.3	0.1	0.1	1	0.6	0.4	0.2	0.2	0.2	0.2	0.2
Attributes	# Fatalities	# Seriously injured	# Reversibly injured	# People affected in long term	Manhours [h]	Material cost [€]	Direct economic losses [€]	Duration [h]	# Prisoner escapes	# Riots	Damages from looting [€]	Indirect economic losses [€]

evacuate or shelter the population in the downwind area needs to be made before the transfer of the chlorine to a new transportation device can start.

Chlorine Incident Scenario Construction

The scenarios presented have been developed together with experts from the Danish Emergency Management Agency. To start with the scenario construction, a number of assumptions needs to be made. Basically, there are two possibilities what could happen during the transfer of the chlorine:

A.   No chlorine release.

B.   The connection from the tank to the buckets gets loose, chlorine is released.

When the performances of all attributes have been estimated for each scenario, MAVT is applied. The value functions chosen here are linear. The structure of the attribute tree and the weights for the inter-criteria preferences used are summarized in Table 1. The value functions and weights can be adapted to the users' needs.

Scenario Selection and Representation of Results

After the evaluation of all scenarios, a representation of the worst, a medium (selected according to

Figure 6. Scores for evacuation (evac.) and sheltering (shelt.) in worst, medium and best scenario

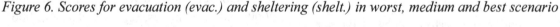
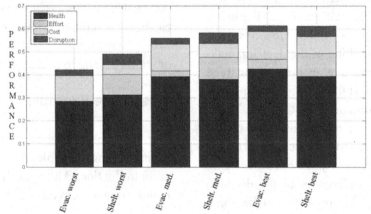

Figure 7. Sensitivity analysis for scenario weights

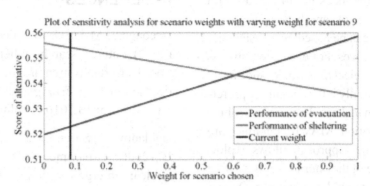

the median) and best scenarios for each decision alternative (according to the user's preferences) are shown to the decision makers. An example plot (created with MATLAB®) is shown in Figure 6. Comparing the bars' heights, the plot shows that in all three cases, sheltering has a better or very similar performance than evacuation (reflected in a higher or equally high bar). A detailed look at the plot shows, for example, that in the worst case the performance of sheltering is more than 0.1 points better than the performance of sheltering. Contrarily, in the best case both alternatives have highly similar performances. Figure 6 allows furthermore an analysis of the contribution of each criterion to the overall score. While in the worst case the performance of the criterion health is better in case of sheltering, evacuation has a better performance with respect to health in the median and best case.

Finally the results of the scenarios for each alternative are aggregated in an additional MAVT step. To this end, weights are elicited reflecting the importance of the respective scenarios to the decision makers. In the default version of the tool, equal weights are used (implying that all scenarios are equally important to the decision makers). As a comparison of the bar's heights in Figure 6 suggested, sheltering has a higher total performance than evacuation (0.56 vs. 0.52).

When the results are closely together and/or when the decision makers are unsure if the weights

chosen reflect their actual preferences correctly, a sensitivity analysis can be performed. This type of analysis allows for the ex-post analysis of parameter variation.

Varying e.g. the weight for one scenario, one deduces that a slight variation of weights does *not* change the hierarchy of alternatives. Figure 7 shows an exemplary application, where the red (blue) line represents the performance of sheltering (evacuation), while the black line marks the currently chosen position.

CONCLUSION AND DISCUSSION

We proposed a framework to integrate uncertainty into MCDA by allowing consideration of different possible future developments of the situation (i.e. scenarios). To arrive at a system for robust decision making, first consistent scenarios need to be developed. Scenarios, being coherent and consistent situation descriptions, are usually perceived as easily understandable (cf. Bunn & Salo, 1993). Their use circumvents the problem of counterintuitive results which can arise from probabilistic models (e.g., Kahneman et. al., 1982). This may overcome major drawbacks of standard MCDA techniques for reasoning under uncertainty. By integrating scenarios into MAVT, a transparent handling of uncertainties enhancing the results' acceptance is ensured.

Scenario selection needs not to be based on the users' intuition or on an abstract notion of 'distance'; rather it can be based on a systematic evaluation of alternatives. For each alternative is becomes possible to select the "worst" and "best" scenario taking explicitly into account the preferences of the decision makers. To come to a robust overall recommendation, MAVT-techniques are applied. The combined approach allows exploring the performance of the alternatives assuming several possible situation developments. Deeper insights into the decision situation can be gained than those provided by standard methods that base the evaluation on a best guess or the most likely development. An example has been used to illustrate our approach and highlight its advantages.

Our proposed approach enables further elicitation of reasoning under uncertainties during emergency situations within the Diadem project, e.g. with the Danish Emergency Management Agency (DEMA). The aim of the research is to continue tailoring our approach so that it fits the decision makers (currently perceived) needs.

The framework is not finished, a number of open questions remain, including the following. Care must be taken in order to avoid the double or multiple counting of parameters reflecting risk aversion. Another issue is how the dynamics can be taken into account and how multiple decision making processes can be coordinated. Finally, the results must be presented in a transparent and easily understandable way so that users and stakeholders will accept the recommendations found during the reasoning process.

ACKNOWLEDGMENT

The Diadem project is funded by the European Union under the Information and Communication Technologies (ICT) theme of the 7th Framework Programme for R&D, ref. no: 224318.

REFERENCES

Ascough, J., Maier, H., Ravalico, J., & Strudley, M. (2008). Future research challenges for incorporation of uncertainty in environmental and ecological decision-making. *Ecological Modelling, 219*(3-4), 383–399. doi:10.1016/j.ecolmodel.2008.07.015

Baklanov, A., Sørensen, J., Hoe, S., & Amstrup, B. (2006). Urban meteorological modelling for nuclear emergency preparedness. *Journal of Environmental Radioactivity, 85*(2-3), 154–170. doi:10.1016/j.jenvrad.2005.01.018

Barker, K., & Haimes, Y. Y. (2009). Assessing uncertainty in extreme events: Applications to risk-based decision making in interdependent infrastructure sectors. *Reliability Engineering & System Safety, 94*(4), 819–829. doi:10.1016/j.ress.2008.09.008

Bauer, H. (2001). Wahrscheinlichkeitstheorie. Berlin: de Gruyter.

Belton, V., & Stewart, T. (2002). *Multiple Criteria Decision Analysis. An Integrated Approach.* Boston: Kluwer Academic Publishers.

Ben-Haim, Y. (2000). Robust rationality and decisions under severe uncertainty. *Journal of the Franklin Institute, 337*(2-3), 171–199. doi:10.1016/S0016-0032(00)00016-8

Bertsch, V. (2008). Uncertainty Handling in Multi Attribute Decision Support for Industrial Risk Management. Karlsruhe.

Bertsch, V., Geldermann, J., & Rentz, O. (2007). Preference Sensitivity Analyses for Multi-Attribute Decision Support. In *Operations Research Proceedings 2006* (411-416).

Bertsch, V., Geldermann, J., Rentz, O., & Raskob, W. (2006). Multi-criteria decision support and stakeholder involvement in emergency management. *International Journal of Emergency Management, 3*(2/3), 114–130. doi:10.1504/IJEM.2006.011163

Bruckner, J., Keys, D., & Fisher, J. (2004). The Acute Exposure Guideline Level Program: Applications of Physiologically Based Pharmacokinetic Modeling. *Journal of Toxicology and Environmental Health. Part A.*, *67*(8-10), 621–634. doi:10.1080/15287390490428017

Bui, T., & Lee, J. (1999). An agent-based framework for building decision support systems. *Decision Support Systems*, *25*(3), 225–237. doi:10.1016/S0167-9236(99)00008-1

Bunn, D., & Salo, A. (1993). Forecasting with scenarios. *European Journal of Operational Research*, *68*(3), 291–303. doi:10.1016/0377-2217(93)90186-Q

Butler, J., Jia, J., & Dyer, J. (1997). Simulation techniques for the sensitivity analysis of multi-criteria decision models. *European Journal of Operational Research*, *103*(3), 531–546. doi:10.1016/S0377-2217(96)00307-4

Comes, T., Hiete, M., Wijngaards, N., & Kempen, M. (2009). Integrating Scenario-Based Reasoning into Multi-Criteria Decision Analysis. In *Proceedings of the 6th International Conference on Information Systems for Crisis Response and Management (ISCRAM 2009)*. Gothenburg.

Cooke, R. M., & Goossens, L. H. J. (2004). Expert judgement elicitation for risk assessments of critical infrastructures. *Journal of Risk Research*, *7*(6), 643–656. doi:10.1080/1366987042000192237

Durbach, I., & Stewart, T. (2003). Integrating scenario planning and goal programming. *Journal of Multi-Criteria Decision Analysis*, *12*(4-5), 261–271. .doi:10.1002/mcda.362

Edwards, W. (1977). How to Use Multiattribute Utility Measurement for Social Decisionmaking. *IEEE Transactions on Systems, Man, and Cybernetics*, *7*(5), 326–340. doi:10.1109/TSMC.1977.4309720

Fenton, N., & Neil, M. (2001). Making decisions: using Bayesian nets and MCDA. *Knowledge-Based Systems*, *14*(7), 307–325. doi:10.1016/S0950-7051(00)00071-X

Fildes, R., Goodwin, P., Lawrence, M., & Nikolopoulos, K. (2008). Effective forecasting and judgmental adjustments: an empirical evaluation and strategies for improvement in supply-chain planning. *International Journal of Forecasting*, *25*(1), 3–23. doi:10.1016/j.ijforecast.2008.11.010

French, S. (1986). *Decision Theory: An introduction of the Mathematics of Rationality*. Chichester: Ellis Harwood Limited.

French, S. (1996). Multi-attribute decision support in the event of a nuclear accident. *Journal of Multi-Criteria Decision Analysis*, *5*(1), 39–57. doi:10.1002/(SICI)1099-1360(199603)5:1<39::AID-MCDA109>3.0.CO;2-Q

French, S., & Geldermann, J. (2005). The varied contexts of environmental decision problems and their implications for decision support. *Environmental Science & Policy*, *8*(4), 378–391. doi:10.1016/j.envsci.2005.04.008

French, S., & Niculae, C. (2005). Believe in the Model: Mishandle the Emergency. *Journal of Homeland Security and Emergency Management*, *2*(1). doi:10.2202/1547-7355.1108

Godet, M. (1990). Integration of scenarios and strategic management: Using relevant, consistent and likely scenarios. *Futures*, *22*(7), 730–739. doi:10.1016/0016-3287(90)90029-H

Heckerman, D., Mamdani, A., & Wellman, M. (1995). Real-world applications of Bayesian networks. *Communications of the ACM*, *38*(3), 24–26. doi:10.1145/203330.203334

Heugens, P., & van Oosterhout, J. (2001). To boldly go where no man has gone before: integrating cognitive and physical features in scenario studies. *Futures*, *33*(10), 861–872. doi:10.1016/S0016-3287(01)00023-4

Hodgkin, J., Belton, V., & Koulouri, A. (2005). Supporting the intelligent MCDA user: A case study in multi-person multi-criteria decision support. *European Journal of Operational Research*, *160*(1), 172–189. doi:10.1016/j.ejor.2004.03.007

INFORMS. (2009). OR/MS Resource Collection. Retrieved October 1, 2009, from www.informs.org/Resources/Computer_Programs/.

Jakobson, G., Buford, J. & Lewis, L. (2008). Models of feedback and adaptation in multi-agent systems for disaster situation management (6943). Orlando, USA.

Kahneman, D., Slovic, P., & Tversky, A. (1982). *Judgment under Uncertainty: Heuristics and biases*. Cambridge: Cambridge University Press.

Kirkwood, C. (1992). An Overview of Methods for Applied Decision Analysis. *Interfaces*, *22*(6), 28–39. doi:10.1287/inte.22.6.28

Lempert, R., Groves, D., Popper, S., & Bankes, S. (2006). A General, Analytic Method for Generating Robust Strategies and Narrative Scenarios. *Management Science*, *52*(4), 514–528. doi:10.1287/mnsc.1050.0472

Mahmoud, M., Liu, Y., Hartmann, H., Stewart, S., Wagener, T., & Semmens, D. (2009). A formal framework for scenario development in support of environmental decision-making. *Environmental Modelling & Software*, *24*(7), 798–808. doi:10.1016/j.envsoft.2008.11.010

Merz, M., Hiete, M., & Bertsch, V. (2009). Multicriteria decision support for business continuity planning in the event of critical infrastructure disruptions. *International. Journal of Critical Infrastructures*, *5*(1/2), 156–174. doi:10.1504/IJCIS.2009.022854

Mingers, J., & Rosenhead, J. (2004). Problem structuring methods in action. *European Journal of Operational Research*, *152*(3), 530–554. doi:10.1016/S0377-2217(03)00056-0

Mustajoki, J., Hämäläinen, R., & Sinkko, K. (2007). Interactive computer support in decision conferencing: Two cases on off-site nuclear emergency management. *Decision Support Systems*, *42*(4), 2247–2260. doi:10.1016/j.dss.2006.07.003

Nakamura, K., Iwai, S., & Sawaragi, T. (1982). Decision Support Using Causation Knowledge Base. *Systems, Man and Cybernetics . IEEE Transactions on*, *12*(6), 765–777.

Nunamaker, J. (1997). Future research in group support systems: needs, some questions and possible directions. *International Journal of Human-Computer Studies*, *47*(3), 357–385. doi:10.1006/ijhc.1997.0142

O'Brien, F. (2004). Scenario planning--lessons for practice from teaching and learning. *European Journal of Operational Research*, *152*(3), 709–722. doi:10.1016/S0377-2217(03)00068-7

O'Hagan, A., & Oakley, J. (2004). Probability is perfect, but we can't elicit it perfectly. *Reliability Engineering & System Safety*, *85*(1-3), 239–248. doi:10.1016/j.ress.2004.03.014

Paprika, Z., & Kiss, I. (1985). Interactions in decision support systems: Division of labor in DSSs. *Engineering Costs and Production Economics*, *8*(4), 281–289. doi:10.1016/0167-188X(85)90053-9

Perrow, C. (1984). *Normal Accidents: Living with High-Risk Technologies*. New York: Basic Books.

Raskob, W., Gering, F., & Bertsch, V. (2009). Approaches to visualisation of uncertainties to decision makers in an operational Decision Support System. In *Proceedings of the 6th International Conference on Information Systems for Crisis Response and Management (ISCRAM 2009)*.

Ríos-Insua, D., & French, S. (1991). A framework for sensitivity analysis in discrete multi-objective decision-making. *European Journal of Operational Research, 54*(2), 176–190. doi:10.1016/0377-2217(91)90296-8

Roubelat, F. (2000). Scenario Planning as a Networking Process. *Technological Forecasting and Social Change, 65*(1), 99–112. doi:10.1016/S0040-1625(99)00125-0

Russell, S., & Norvig, P. (2003). *Artificial intelligence*. Upper Saddle River: Prentice Hall.

Schnaars, S. (1987). How to develop and use scenarios. *Long Range Planning, 20*(1), 105–114. doi:10.1016/0024-6301(87)90038-0

Schnaars, S., & Ziamou, P. (2001). The essentials of scenario writing. *Business Horizons, 44*(4), 25–31. doi:10.1016/S0007-6813(01)80044-6

Schoemaker, P. (1993). Multiple Scenario Development: Its Conceptual and Behavioral Foundation. *Strategic Management Journal, 14*(3), 193–213. doi:10.1002/smj.4250140304

Scholz, R., & Tietje, O. (2002). *Embedded Case Study Methods. Integrating Quantitative and Qualitative Knowledge*. Thousand Oaks: Sage Publications.

Shim, J., Warkentin, M., Courtney, J., Power, D., Sharda, R., & Carlsson, C. (2002). Past, present, and future of decision support technology. *Decision Support Systems, 33*(2), 111–126. doi:10.1016/S0167-9236(01)00139-7

Stewart, T. (2005). Dealing with uncertainties in MCDA. In Multiple Criteria Decision Analysis: State of the Art Surveys, International Series in Operations Research & Management Science (pp. 445-470). Springer.

Tietje, O. (2005). Identification of a small reliable and efficient set of consistent scenarios. *European Journal of Operational Research, 162*(2), 418–432. doi:10.1016/j.ejor.2003.08.054

Tress, B., & Tress, G. (2003). Scenario visualisation for participatory landscape planning—a study from Denmark. *Landscape and Urban Planning, 65*(3), 161–178. doi:10.1016/S0169-2046(02)00219-0

van der Heijden, K. (2007). *Scenarios: The Art of Strategic Conversation*. Chichester: Wiley.

von Winterfeldt, D., & Edwards, W. (1986). *Decision Analysis and Behavioral Research*. Cambridge: Cambridge University Press.

von Winterfeldt, D., & Fasolo, B. (2009). Structuring decision problems: A case study and reflections for practitioners. *European Journal of Operational Research, 199*(3), 857–866. doi:10.1016/j.ejor.2009.01.063

Watthayu, W., & Peng, Y. (2004). A Bayesian network based framework for multi-criteria decision making. In *Proceedings of the 17th International Conference on Multiple Criteria Decision Analysis* (pp. 6-11).

Weber, M. (1987). Decision making with incomplete information. *European Journal of Operational Research, 28*(1), 44–57. doi:10.1016/0377-2217(87)90168-8

Weistroffer, R., Smith, C., & Subhash, N. (2005). Multiple Criteria Decision Support Software . In Figueira, J., Greco, S., & Ehrgott, M. (Eds.), *Multiple Criteria Decision Analysis: State of the Art Surveys* (pp. 989–1018). New York: Springer.

ENDNOTE

[1] Diadem: Distributed information acquisition and decision-making for environmental management (funded by the European Union under the 7th Framework Programme; see http://www.ist-diadem.eu/)

Chapter 15
POP–C++ and Alpine3D:
Petition for a New HPC Approach

Pierre Kuonen
University of Applied Sciences of Western Switzerland, Switzerland

Mathias Bavay
WSL Institute for Snow and Avalanche Research SLF, Switzerland

Michael Lehning
WSL Institute for Snow and Avalanche Research SLF, Switzerland

ABSTRACT

In the developed world, an ever better and finer understanding of the processes leading to natural hazards is expected. This is in part achieved using the invaluable tool of numerical modeling, which offers the possibility of applying scenarios to a given situation. This in turn leads to a dramatic increase in the complexity of the processes that the scientific community wants to simulate. A numerical model is becoming more and more like a galaxy of various sub-process models, each with their own numerical characteristics. The traditional approach to High Performance Computing (HPC) can hardly face this challenge without rethinking its paradigms. A possible evolution would be to move away from the Single Program, Multi Data (SPMD) approach and towards an approach that leverages the well known Object Oriented approach. This evolution is at the foundation of the POP parallel programming model that is presented here, as well as its C++ implementation, POP-C++.

INTRODUCTION

In the developed world, there is a general trend of rising public expectations on protection from natural hazards. The goal is to be able to reduce the number of fatalities from natural hazards as well as contain the economic impact of such events. Moreover, long term understanding of potential trends in frequency and size of hazardous events

is also sought as necessary information for public policy planning (Marty et al., 2009).

In order to satisfy these requirements, various tools have been developed by the scientific community. These tools are designed to provide answers to questions such as "what impact will climate change have on water availability for a specific place", or "would a warmer climate lead to more frequent flooding", that is for predictions of general trends (*forecasting*, see for example Bavay et al., 2009). They are also used for helping

DOI: 10.4018/978-1-61520-987-3.ch015

analyze a current situation (*nowcasting*): knowing only the situation at a discreet set of points, what is the full picture? Using the information that is provided by a limited set of sensors, can we predict the danger level?

Damage by natural hazards has an obvious economic cost. But a less obvious cost has also to be taken into account: if being overly cautious reduces the direct cost of natural hazards, it however creates a potentially massive indirect cost (lost business because of closed roads or evacuated towns, investment in unnecessary security equipment, as well as general public discontent for unnecessary safety measures). This means that the tools used for natural hazards forecasting and nowcasting have to walk a fine line between too cautious and too optimistic, struggling to get as close to the truth as possible. A lot of recent research has tried to find optimal solutions for risk management in mountains (Lehning and Wilhelm, 2005; Bründl et al., 2009).

The goal of these nowcasting and forecasting tools is to predict a danger level or expected changes based on current data or already forecasted data. Two approaches are usually represented: the first one consists of building a statistical model on past data. The second one consists of trying to describe the details of the physical processes at play.

Usual Approach and Limitations

This approach is statistical as it is based on the assumption that past events would describe future events and behavior. Therefore, a statistical correlation is looked for between some data used as inputs (for example, measurements from weather stations) and the output data of interest (for example, a catchment outflow). The outcome is a purely statistical relationship between the input and output parameters that is of very low computational complexity. For example, the degree-day hydrological approach considers that a catchment's outflow depends on the time integral of positive air temperatures. As Ohmura (2001) has shown, this basically statistical approach has a sound physical basis, however.

Practically, this method requires a calibration phase for every new setup, by looking for statistical relations between the input and output parameters in a large enough dataset. This calibration has to be redone when looking at other input or output parameters, or when looking at another geographical location. Even if the investment required to build such a model is quite small, it has to be fully redone for each new application. Moreover, it requires that no statistically extraordinary event occurs. In such a case, since the relationship between the input and the output is based on normal, regular behavior, the model could simply not be applied. These models therefore appear ill suited for studying climate scenarios (which are characterized by a departure from the normal trends) or extreme events.

Phenomenological Approach and Limitations

There is another way to simulate how various inputs would impact the output: it is to simulate the physical processes that are taking place in between. This requires to carefully identify the relevant processes and their interactions and to model them. Some processes might be modeled using mostly statistical methods while some others might be split into even smaller processes (usually at a smaller scale). This reliance on some statistical methods might appear as a perversion of this approach (reverting back to the drawbacks described in the previous section). The rational is that a statistical description of a small scale process might be more robust than the statistical description of a larger scale set of processes. Therefore, such description should be reserved for well bounded processes. However, the general tendency is to replace these statistical descriptions by physical models during the evolution of the code.

This is indeed a key characteristic of phenomenological models: because modeling each of the relevant processes and determining their parameters (by direct measurements or by statistical fits) is a time consuming task, a given model might evolve over several years, backed up by experimental, theoretical and numerical studies and validated by experimental campaigns.

This highlights one of the major drawbacks of such an approach: the investment is high. Developing the model requires expertise, time and might spawn multiple secondary studies in order to gain better understanding of the individual processes. Moreover, the structure of the model is directly dictated by the fundamental structure of the phenomena that are modeled and its physical processes: as this model can be mostly seen more as an array of several, interacting process-models, the complexity of the overall model can rapidly grow. This is also matched by the computational complexity that can be very high.

On the other hand, such modeling has major benefits to offer: as it is based on the simulation of physical processes, which are usually quite independent of the geographical location, the overall model can be used unchanged in any new location (the only restriction being if the conditions are dramatically different and thus invalidate some of the hypothesis made for some of the processes). This also means that no new "calibration" has to be done before applying the model to new conditions. This is specifically relevant for climate change scenarios or the investigation of extreme events studies. But this also means that although the cost of setting up the model is quite high, then the cost of using the model in different conditions (different location, different climate, different output parameters) is often much smaller.

The high computing requirements issue is what is tackled in the coming sections. This issue can be mitigated by good programming but it will often not be enough and therefore parallel programming techniques have to be used in order to allow running the model on a parallel infrastructure (cluster or grid). The major issue has to do with the intrinsic complexity of the model: since the model's structure reflects the natural complexity of the processes that are simulated, the odds of the computational problem to be "embarrassingly parallel" are quite slim. Very often, the model will be made of highly parallel modules, running independently of each other, linked by purely sequential modules, alongside with partially coupled modules. This means that the traditional HPC approach, such as used by Message Passing Interface (MPI), does not match the structure of the processes that have to be simulated. Therefore, a new approach is required, that would deal with the heterogeneous nature of the model.

Need for a New DHPC Programming Paradigm

The community of High Performance Computing (HPC) is a very conservative community and moves only slowly to new programming models. Some people still think that the old Fortran programming language born in the 50s is the only viable option for programming large simulations programs.. For more than fifty years, programming models have evolved a lot and, to some extend, Fortran tried to follow this evolution but it falls short of offering all the powerful features of modern programming languages. Programming models become an even more critical issue if we consider parallel and distributed HPC (DHPC).

The first parallel programming model largely accepted by the HPC community in the late 1970s, was the "vectorial model". This model has largely dominated the development of DHPC software during almost 20 years (1970s to 1990s). Nevertheless, in the mid-1990s the first revolution occurred: it was the transition from the vectorial programming paradigm to the message passing programming paradigm. This revolution was a major difficulty for this community but it was largely driven and eased by the fact that major commercial constructors of vectorial machines (such as Cray) decided, rather brutally, to switch to the Massively Parallel Processing (MPP) model as well as by the emergence of the standard MPI library. Today, more than ten years after this revolution, MPI is still dominating the world of distributed HPC even if today supercomputer architectures are far away from the first MPP machine.

On the programming tools side, FORTRAN tends to disappear and to be slowly replaced by C. Nevertheless, even if the object oriented programming paradigm is now forty years old and widely accepted by the computer science community, it still struggles to gain acceptance in the DHPC community. However, the requirements for DHCP programming tools are currently evolving, mostly for two reasons: first, the architectures of supercomputer have evolved toward more heterogeneity and new alternatives have emerged to execute high performance distributed programs such as P2P or Grid architectures (Foster and Kesselman 1998). Second, the applications themselves are also becoming more and more heterogeneous. If traditional "flat parallel programs" as those generated by the *domain decomposition* approach still exist, they are more and more complemented by multi-physic or multi-simulation techniques. This is, for example, the case for code coupling where different simulations, each of them potentially using its own simulation method, have to collaborate to the simulation of a highly complex phenomenon.

These changes progressively create new needs as the Single Program Multiple Data (SPMD) model on which MPI is based does not contain the necessary concepts to address heterogeneous architectures and applications. As this change is progressive, it is difficult to predict when the inadequacies between the current programming tools and problems addressed in HPC will become important enough to overcome the reluctance of the HPC community to change its main parallel programming paradigm. In the mean time numerous research teams are proposing original alternatives or complements to MPI (Badia et al. 2003, Baude et al. 2006, Dünnweber et al. 2004, Pérez et al. 2003, Vanneschi 2002, Van Nieuwpoort et al. 2002). It is impossible to predict which of these alternatives will penetrate the HPC community and even if it would be one of the currently known ones, because the process of adopting a particular programming model is largely driven by non-technical or non-scientific considerations ranging from ideological opinions to pragmatic or economical reasons.

Fortunately there exists a community of researchers that addresses code coupling or multiphysic simulations and that did not have too much exposure to the traditional MPI programming paradigm. These communities are much more open to new programming approaches such as proposed by distributed object oriented models. This is the case of the Swiss Federal Institute for Forest, Snow and Landscape Research (WSL) and more specifically of its Institute for Snow and Avalanche Research located in Davos, Switzerland. This institute decided in 2003 to adopt a new object oriented parallel programming tool, called POP-C++, to implement a distributed version of its applications Alpine3D which aims at simulating alpine surface processes and in particular snow processes. The next section of this chapter will be dedicated to describe the structure of the Alpine3D model as an example of application requiring a new DHPC paradigm.

Then, the following section will be dedicated to the description of the POP-C++ tool and of its adequateness to address heterogeneous architectures and applications. In particular we will show that POP-C++ complements instead of rivaling MPI. This will allow a smooth transition from the MPI paradigm to a hierarchic paradigm as an elegant combination of message passing and distributed object oriented models.

Figure 1. Snow Water Equivalent as simulated by Alpine3D for the 43 km² of the Dischma Alpine catchment (Eastern Switzerland). Maximum SWE is 512 mm.

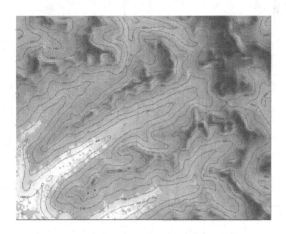

Case Study: The Alpine3D Model

As explained in the introduction, the aim of Alpine3D is to simulate the physical processes at the earth's surface in alpine regions with a particular focus on the snow cover. Practically, this means that given a digital terrain (with a Digital Elevation Model and a land use grid giving information about the type of land, e.g. pasture, forest, lake, etc) and using meteorological input such as typically measured by a set of weather stations or as provided by a meteorological model (air temperature, relative humidity, wind, precipitation, radiation), Alpine3D calculates a detailed state of the soil, snow (if present) and vegetation and predicts the spatial distribution of surface temperatures, snow cover or runoff generation (Lehning et al., 2006).

In the context of natural hazards, this means that issues like snowpack stability (for avalanches) or snow hydrology (for flooding or water/hydro power availability) can be tackled on a catchment scale by this model. For example, the model can generate Snow Water Equivalent spatial distribution (see figure 1) showing how much water is stored as snow in each cell of the simulated domain. This feeds a runoff model which in turn feeds the hydrological simulation of the catchment discharge. Because the model is physically based, this can be used to study the discharge changes induced by extreme events or climate change.

Another benefit of such a model is that it can be used to gain access to specific physical parameters at any location. For example, the analysis of an avalanche benefits from knowing the structure of the snowpack just before it triggered. An Alpine3D simulation allows simulating a virtual station that would output a detailed snowpack profile, which can then be analyzed for understanding the reasons of the avalanche itself. The Figure 2 shows the simulated snow depth evolution over several years at the Stillberg station, located at 2131 m in the Dischma catchment. No measurements from this station were used for the simulation, its role being solely of verification. A detailed snow profile then contains for each snow layer its microstructure information (grains types, grains sizes, etc), the temperature profile, the energy balance information, the amount of liquid water... This makes it possible to classify this profile like a regular field snow profile and therefore to evaluate the avalanche danger for this specific location.

Once again, the usage of a physically based modeling makes the model flexible enough to run on various geographical locations without a need for recalibration (within the validity of the model hypothesis themselves) as well as to provide otherwise hard to measure physical parameters (like the grain bond size).

In order to achieve these results, the overall modeling has been split into several key processes, that have been identified as:

1. The larger scale but still non-local contribution of the meteorological forcing is used as a boundary condition. This is typically provided by spatially distributing the measurements from the weather stations.
2. The local conditions will be affected by the local energy balance. A major component of the local energy balance is the radiation bal-

Figure 2. Snow heights as simulated by Alpine3D for the Stillberg station in the Dischma catchment (Eastern Switzerland) compared to measurements.

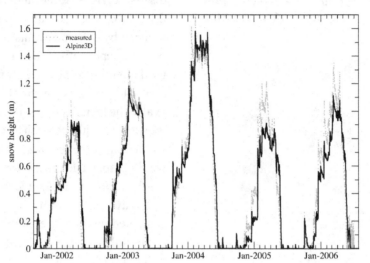

ance: an external source (the sun) provides energy through radiation that is spatially distributed according to the characteristics of the local terrain and the characteristics of the atmosphere. This includes effects such as shading or multiple reflections from (snow covered) terrain surfaces.

3. According to the local meteorological parameters and local energy balance, a snow cover may develop. This is turn affects the local energy balance as well as the local radioactive properties (albedo and long wave emission).

4. If snow is available and strong wind, some snow might be blown by the wind. Alternatively, during snow precipitation, a moderate wind is already sufficient to create a spatially varying snow distribution (Lehning et al., 2008). These snow transport processes depend on the precipitation rate, the properties of the existing snow surface (density, bonding) and the local wind speed.

5. Depending on the conditions in the snowpack, and again the energy and/or water input at the surface, some runoff might be generated. This water would then be routed

in the soil depending on the properties of soil and subsurface.

Therefore, the general structure of the model is as illustrated in Figure 3: a main module coordinates the various modules that each represents a set of physical processes.

MeteoIO

This module is implemented in a separate library. Its goal, as stated in the previous section, is to read meteorological forcing data and to create a regular grid of input parameters. For example, point measurements of weather stations have to be distributed onto the DEM's grid, filling every cell with reasonable values. This is done using statistical tools tailored to reflect some physical properties of each parameter. The basic element is a Inverse Distance Weighting (IDW) method that builds the local value as a weighted average of the point measurements, the weight being the inverse of the distance to each specific point measurement. Since some parameters are known to have a strong elevation dependence, a regression line can be computed on the data, allowing to reduce

Figure 3. Alpine3D modules and organization

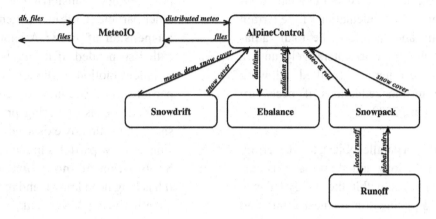

the data to a common elevation, do the spatial interpolation and then recalculating the values of each cell at its real elevation. The algorithms are therefore the following:

- for air temperature, elevation correction and interpolation
- for relative humidity, conversion to dew point temperature, elevation correction and IDW, then conversion back to relative humidity
- for the precipitations, a simple IDW
- for the wind, elevation correction and IDW on the wind velocity followed by a heuristic model tweaking the wind speed and direction depending on the terrain.

This module is also responsible for accessing the data through various means (files, databases, etc), selecting the data for the requested timestamp, filtering and – if necessary – re-sampling.

This module is being implemented in a parallel object, for cleanliness but also because though not computationally intensive, the IOs can introduce delays, specially when connecting to a database. So far, no finer scale parallelism is introduced into this module.

Energy Balance

This module is computing the radiation balance on the grid scale. It is computationally very intensive, both in term of memory and cpu. It is modeling several processes:

- Radiation coming directly from the sun, split between direct and diffuse radiation in the atmosphere
- Reflection (shortwave) and re-emission (long-wave) of the radiation by the terrain using a radiosity approach.

The algorithm works with two wavelength bands, distinguishing between long wave (mostly infrared) and short wave (mostly visible):

- Long wave radiation coming from the sky is uniformly distributed on the terrain (thermal radiation)
- Terrain re-emits some long wave radiation depending on its surface temperature, modifying the local radiation balance
- Cells having a direct view of the sun receive direct radiation.
- All cells receive diffuse radiation which is supposed to be isotropic.
- Terrain reflects some short wave radiation depending on its albedo and the view fac-

tor of the cells, modifying the local radiation balance. The calculation of the terrain reflections and therefore the total radiation distribution works iteratively until a convergence criteria is fulfilled, allowing multiple terrain reflections if necessary (Helbig et al., 2009).

This module is a parallel object, but the terrain radiation, both for long and short wave, is currently not parallelized within the energy balance module. This is planned for the near future and will increase the overall efficiency of the module.

Snow Distribution

This module is responsible for modeling the spatial distribution of snow as caused by wind – topography interactions (Raderschall et al., 2008; Lehning et al., 2008). A variety of processes need to be considered:

- **Preferential deposition of snow during precipitation:** This has recently been identified to be an important process in Alpine terrain (Lehning et al., 2008; Mott et al., 2008) and describes the process that wind features such as speed-up or separation (Raderschall et al., 2008) lead to an uneven deposition of snow during snow falls.
- **Erosion of an existing snow cover:** Depending on the density and the sintered state of the uppermost snow layers, a certain wind speed will be able to erode deposited snow (Clifton et al., 2006).
- **Saltation of snow:** Erosion of snow most typically leads to saltation, where particles follow ballistic trajectories over the surface (Doorschot and Lehning, 2002). Each particle receives enough kinetic energy from the wind or from the impact of another particle or just from rebound to start a trajectory and impact again the surface at the end, possibly remaining at the surface,

rebounding or transferring its energy to another particle (Clifton and Lehning, 2008).
- **Suspension of snow:** A snow particle is called suspended, if it largely follows the turbulent motion of the air flow. Saltating particles may become suspended at larger wind speeds or during precipitation of snow, there already exists a natural suspension of snow particles in the moving air.
- **Sublimation of snow during transport:** It has long been known and recently shown quantitatively (Wever et al., 2009) that suspended particles lead to significant mass loss in form of vapor transport back to the atmosphere. This process can remove a significant fraction of snow from the mountains, e.g. during Föhn events.

A Lagrangian saltation model (Clifton and Lehning, 2008) is used for the first process whereas the latter two processes are modeled by a continuum description in three dimensions using finite element methods (Lehning et al., 2008). This is computationally very demanding, but a parallel version has been developed (currently using standard libraries relying on MPI). This module is therefore implemented as a POP-C++ object that itself contains MPI elements.

Snowpack

The snowpack module is a one dimensional model of the soil – snow – vegetation column. It simulates the surface exchanges (mass and energy) of the column, the temperature profile in the column as well as the structure of the soil/snow column (grains, densities, water/air/ice content). This allows for example to determine the surface albedo and the runoff at the cell scale.

This module consists of the following models:

- a surface energy balance model using state of the art descriptions of turbulent heat transfer in the boundary layer (Lehning et al., 2002b; Stössel et al., 2009) and relying on the radiation balance from the radiation module (Helbig et al., 2009).

- a heat transport equation (modeled using finite elements) describes the heat fluxes including the heat transported by water vapor using an appropriate thermal conductivity (Lehning et al., 2002a).
- a description of snow settling using a microstructure based viscosity law for snow.
- a simple model of soil layers based on the soil density and thermal conductivity.
- a water transport model.
- a canopy model treating energy (radiative and turbulent heat exchange) and mass (evaporation, interception, transpiration) exchange between the atmosphere, the canopy and the underlying surface.

This module distributes cells among workers on various nodes (each worker being a parallel object), independently computing the cells in parallel before re-assembling the grid.

Runoff

The goal of the runoff module is to simulate the flow at a catchment outlet. It collects the runoff at the lower boundary of each cell (as calculated by the SNOWPACK module) and then fills various reservoirs. These reservoirs are storage modules for the fast surface flow (generated by rocky, impervious and frozen soil, in water-saturated and Hortonian runoff), interflow (usually dominant for alpine regions) and baseflow (slow and stable component in mountain catchments, which depends on the groundwater storage in the different aquifers and the amount of ground water recharge). The parameters of each reservoir have default values (Michlmayr et al., 2008) but could in principle be calibrated such as typically done for less complex runoff models (e.g. PREVAH, Zappa et al., 2009 or Viviroli et al., 2009).

This module being Fortran code, it is not parallelized. This is not an issue, since it is very inexpensive from a computational point of view. A new, more complex implementation is envisioned that would be a parallel object.

The multi-physics nature of Alpine3D makes it a good example of the new needs arising for DHCP. As numerical simulations become every day more prevalent in research, so does the complexity of the processes that the scientific community wants to simulate. This in turn pushes the traditional approaches to their limits and calls for new, innovative solutions. Among those, the POP-C++ programming model that had been chosen by the Alpine3D team for building its parallel implementation is described in the next section.

DISTRIBUTED HPC WITH THE POP PROGRAMMING MODEL

The Parallel Object Programming (POP) Model

POP-C++ is an implementation, as an extension of the C++ programming language (Nguyen and Kuonen 2007), of the POP programming model first introduced by Dr. Tuan Anh Nguyen in his PhD thesis (Nguyen 2004). The POP programming model is based on the very simple idea that objects are suitable structures to distribute data and executable codes over heterogeneous distributed hardware and to make them interact between each other.

Fundamentally this idea is not new. In the early 1990s this idea was already used by the Common Object Request Broker Architecture (CORBA) standard. Nevertheless CORBA focuses on interoperability by enabling software components written in multiple computer languages and running on multiple computers to work together. Performance is only a secondary issue for CORBA. For example nothing has been done in CORBA to allow the programmer to easily maximize parallelism in his application. CORBA also focuses on standardization issues. It aims at standardizing the method-call semantics between application objects that reside on different computers. Target applications of CORBA are distributed systems rather than HPC parallel or distributed applications. Anyway, even if the first version of CORBA is now almost twenty years old, it was never significantly used by the HPC community. Nevertheless the idea of using the already existing concept of objects as the foundation to distribute computation was already present in CORBA. The POP model expands this idea and adds the necessary functionality to allow for an easy development of coarse grain parallel high performance applications.

The object oriented paradigm has unified the concept of module and type to create the new concept of class. The next step introduced by the POP model is to unify the concept of class with the concept of task (or process). This is realized by adding to traditional "sequential" classes a new type of class: *the parallel class*. By instantiating parallel classes we are able to create a new category of objects we will call *parallel objects* in the rest of this chapter. Parallel objects are objects that can be remotely executed. They coexist and cooperate with traditional sequential objects during the application execution. Unlike MPI applications, which must follow the static SPMD execution model, the POP model allows usage of the more general Multiple Program Multiple Data (MPMD) execution model. New and different tasks, i.e. new parallel objects, can be dynamically created or destroyed during execution. As a consequence, the POP model is a programming model well adapted to heterogeneous and dynamic computing environments such as GRIDs. Parallel classes generalize the sequential class concept by keeping the good properties of the object-orientation paradigm, such as data encapsulation, inheritance and polymorphism, and adding new properties such as various method invocation semantics, the dynamic and transparent object allocation driven

by the high-level requirement descriptions as well as the possibility to share references to parallel objects. These new concepts are presented in the next sub-sections.

Remote Method Invocation Semantics in the POP Model

When unifying the concept of class and of task, the first issue that has to be solved is the question of the autonomic or non-autonomic behavior of parallel objects. Autonomous objects, referred to as *active objects*, are objects that have an independent internal activity, regardless of the fact that methods of the object have been called or not been called. In the traditional object oriented model, objects are inactive. This means that an object starts to be active only if a method of this object is called and that the activity of the object stops when the method terminates its execution. The activity will start again at the next method call. When creating the POP model, a major objective we pursued was to have this model as close as possible to the traditional sequential object oriented model. As a consequence it has been decided that, in the POP model, parallel objects will be inactive. This decision has a major impact on the ability of the POP model to have parallel executions. Indeed when a method is called, the caller object stops its activity to transfer it to the called object. The activity will be transferred again to the caller object when the execution of the method terminates. With this semantic we always have only one single flow of execution and, even if objects are located on different computers, there will never be parallel execution. To overcome this difficulty, the POP model introduces the *asynchronous method call*. Unlike synchronous method calls, which corresponds to the traditional way of calling a method, asynchronous calls do not wait for the end of the execution of the method to return control of the activity to the calling object, but immediately return the activity to the calling object. Asynchronous method calls allows the creation of multiple flows

Figure 4. The synchronous and asynchronous method calls

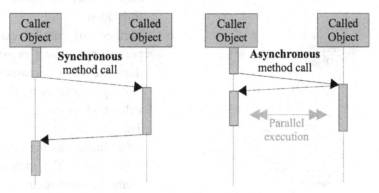

of activities and as a consequence the ability to have parallel execution. Figure 4 illustrates the synchronous and the asynchronous semantics for method calls.

The second question one has to answer is in which order methods have to be executed within parallel objects. The most common answer is to decide that methods are executed in the same order as the order calls arrived on the object. Nevertheless this approach appears not to be the best one as it does not allow prioritizing some calls over others. Having priorities between calls is an important feature to allow the control of parallelism.

The POP model defines three different semantics for method calls in order to manage the order of execution, i.e. the *concurrent semantic*, the *sequential semantic* and the *mutex semantic*. A concurrent method is a method which can be executed immediately when the call arrives on the object. This means that it can be executed concurrently with other on-going methods in the object. A sequential method is a method which can be executed only when all sequential method calls previously received by the object are terminated. This means, in particular, that sequential methods cannot be executed in concurrence with other sequential methods. Finally mutex methods are methods which can be executed only when all methods calls previously received, regardless of their semantic, have been executed. In addition, the execution of every method call received after

a mutex call, regardless of its semantic, is blocked until the mutex call has been executed. This is the only situation where the execution of a concurrent method call can be delayed. The exact definition of these three semantics is summarized below:

- An arriving *concurrent* method call can be executed immediately when it arrives, except if mutex calls are pending or executing. In the later case it is executed after completion of all mutex calls previously arrived.
- An arriving *sequential* method call is executed after completion of all sequential and mutex calls previously arrived.
- An arriving *mutex* method call is executed after completion of all calls previously arrived.

Figure 5 illustrates the order of the execution of methods calls on a parallel object regarding the semantic of the method.

Synchronous/asynchronous semantics and concurrent/sequential/mutex semantics are independent and can be combined. This leads to six different possible semantics for methods calls in the POP model:

- synchronous concurrent
- synchronous sequential
- synchronous mutex

Figure 5. Order of execution of methods regarding their semantic

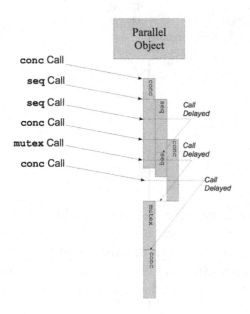

- asynchronous concurrent
- asynchronous sequential
- asynchronous mutex

The notion of parallel classes and of the six different semantics for method calls in parallel classes constitute the core of the POP model.

Due to the distributed nature of the POP model and to the fact that objects are supposed to be executed in different memory address spaces, some constraints apply to parallel classes in comparison with traditional sequential classes. These restrictions are the following:

a.   A parallel class cannot contain public attributes: This Constraint is a consequence that parallel objects are running in different memory address spaces and therefore cannot share data or pointers.

b.   A parallel class cannot contain class attributes (static attributes): Class attributes are attributes which are global to (shared by) all instances of the class. As parallel objects cannot share variables, class attributes are forbidden.

c.   Methods of parallel classes cannot have pointer parameters or return pointers: for the same reason as discussed above (a).

d.   An asynchronous method cannot return a value and cannot have output parameters: This is inherent to asynchronous method calls. Indeed as the control returns to the caller before the method terminates, the method cannot return results.

e.   Parallel classes can only inherit from parallel classes and sequential classes can only inherit from sequential classes: This constraint expresses the fact that sequential classes and parallel classes are classes of different natures. As a consequence, the hierarchy of sequential classes is separated from the hierarchy of parallel classes.

f.   Parallel objects can only be passed by reference to any methods, procedures or functions: This statement is more a nice feature of the POP model than a restriction. Indeed it allows to send the reference of a parallel object to another parallel object. This allows several parallel objects to share access to other parallel objects. Using this feature one can easily and dynamically build any topology of collaborating parallel objects.

Even if the POP model is very general and is not linked to any particular OO language, we will present, in the next section, its implementation as an extension of the C++ object oriented programming language, as it has been realized by the GRID & Ubiquitous Computing Group of the University of Applied Science of Fribourg, Switzerland.

The POP-C++ Programming Language

The POP-C++ programming language is an extension of the C++ programming language that implements the POP model. This extension is intentionally kept as close as possible to the standard C++ programming language. Our objective is that C++ programmers can easily learn POP-C++ and that existing C++ programs and libraries can be parallelized using POP-C++ with a minimal effort.

Only six new keywords have been added to standard C++ programming language. These new six keywords are the following:

- parclass: to allow the declaration of parallel classes
- async: to indicate that a method is asynchronous
- sync: to indicate that a method is synchronous
- conc: to indicate that a method is concurrent
- seq: to indicate that a method is sequential
- mutex: to indicate that a method is mutex

These six keywords are only used in declarations of parallel classes. All the rest of a POP-C++ program is written with pure C++ syntax.

Below is a first example of declaration of parallel class in POP-C++.

```
parclass MyParallelClass
{
  public:
    MyParallelClass();
// Default constructor
    ~MyParallelClass();
// Destructor
    sync conc int GetValue();
//. ..
    async seq void SetValue(int
v);  // some methods
  private:
    int theValue;
// a private attribute
};
```

As one can see, this parallel class declaration is very close to the declaration of a standard class in C++. The only differences are the usage of the keyword **parclass** instead of **class** and the addition of the indications of the semantic of the method calls. It also has to be mentioned that the constraints indicated in the previous section must be fulfill. For example the attribute **theValue** must be private.

A possible implementation of this class in POP-C++ could be:

```
MyParallelClass::MyParallelClass
(){theValue=0;}
MyParallelClass::~MyParallelCla
ss(){;} // nothing to do
int MyParallelClass::GetValue()
{return theValue;}
void
MyParallelClass::SetValue(int v)
{theValue=v;}
```

This implementation is exactly the same that would be done in pure C++. In fact the POP-C++ environment will use a standard C++ compiler to compile this implementation.

Now, suppose one writes the following **main** program:.

```
int main(int argc, char** argv)
{
  MyParallelClass ob;
  ob.SetValue(10);
  printf("The value is: %d\n",
ob.GetValue());
}
```

In this **main** program, the object **ob** is a parallel object because it is an instantiation of a parallel class but it is used exactly as a traditional sequential object. Nevertheless there is one important thing which makes the execution of the methods **SetValue** and **GetValue** very different from the execution of methods of a sequential object: it is the fact that, as the object **ob** is potentially located on a different computer, one has to transmit the values of the parameters. In this case the execution of **SetValue** requires that the value of the parameter **v** is sent to the remote object **ob** and the execution of the method **GetValue** requires that the returned value is sent from the remote object to the main program. For values of standard basic types of C++ such as **int**, **bool**, **float**, **char**, etc. there is no particular difficulty and POP-C++ is able to send these values in a totally transparent way for the programmer, as it is done in this example. If parameters are more complex data structures such as classes, the situation is more complicated because, a priori, POP-C++ is not able to know how to serialize any arbitrary complex data structures to transmit it on the network. It is the responsibility of the programmer to show to POP-C++ how to serialize the complex data structures he creates. As a consequence any class which should be transmitted as parameters of methods of parallel classes needs to be made *serializable*. In order to be serializable, a class must derive from the class **POPBase** provided in the standard POP-C++ environment and must implement the **Serialize** virtual method.

Below most important parts of the declaration of the **POPBase** class is provided:

```
class POPBase
{
  public:
    POPBase();
    ~POPBase();
    virtual void
Serialize(POPBuffer &buf, bool
pack);
. . .
};
```

Below is an example of a serializable class (declaration and implementation):

```
class SerializableClass: public
POPBase
{
```

```
public:
    SerializableClass();
    ~SerializableClass();
    virtual void
Serialize(POPBuffer &buf, bool
pack);
  private:
    int myInt;
    float myFloat;
};
SerializableClass::Serializabl
eClass(){myInt = 0; myFloat =
0.0;}
SerializableClass::~Serializable
Class(){;} // nothing to do
void SerializableClass::Serializ
e(POPBuffer &buf, bool pack)
{
  if (pack)  // How to serialize
the object
  {
    buf.Pack(&myInt,1);
    buf.Pack(&myFloat,1);
  }
  else       // How to de-seri-
alize the object
  {
    buf.UnPack(&myInt,1);
    buf.UnPack(&myFloat,1);
  }
}
```

The main difficulty resides in writing the **Serialize** method. Nevertheless the POP-C++ environment provides helpful tools for this task.

The signature of the **Serialize** method contains parameters of two different types:

1. **POPBuffer**: this class is provided by the standard environment of POP-C++. It is used to pack the serialized object to send it over the network.

2. **bool**: this parameter indicates if the **Serialize** method is called to package the object (to

send the objet) or to unpack the object (to receive the object). The value **true** means *to pack the object*.

It has to be mentioned that the **Serialize** method is automatically called by the POP-C++ environment when needed. The programmer only needs to define it. The **POPBuffer** class provides methods to pack and unpack all basic standard types of C++ as it can be seen on this example for the types **int** and **float**. The signature of the **Pack** and the **UnPack** methods of the **POPBuffer** class have the same parameters:

- a pointer on the first value to pack/unpack
- the number of values to pack/unpack

The second parameter enables the programmer to pack/unpack arrays of values. It is important to note that the pack and the unpack operations must be done in the same order when serializing and de-serializing the object.

Before being able to write efficient distributed programs with POP-C++ one needs to have the answer to one last important question:

When creating a parallel object (instantiate a parallel class), how the programmers can influence the choice of the remote machine on which the parallel object will be executed?

The next sub-section is dedicated to answer to this question.

Object Resources Description in POP-C++

In order to allow programmers to influence the choice of the remote machine for executing a particular instance of parallel class i.e. a given parallel object, the POP-C++ programming language provides the so-called *object descriptors*. Object descriptors use the fact that, in C++ as in all OO languages, there exists constructors which are used to instantiate objects. In POP-C++ objects, descriptors can be associated with constructors to help the constructor to find the appropriate remote computer to execute the parallel object it is constructing. Object descriptors allow specifying constraints on the remote machine to choose in terms of:

- Computing power of the machine.
- Memory size of the machine.
- Network bandwidth with the machine.
- Location of the machine.
- Protocol to use to communicate with the machine.
- The encoding method to send the data to the machine.

Each constructor of a parallel class can be associated with an object descriptor that is indicated directly after the constructor declaration, before the ";". The syntax is the following:

```
@{expression};
```

The **expression** contains one or a combination of the constraints presented above. In the example below we added object descriptors to the class **MyParallelClass** previously presented.

```
parclass MyParallelClass
{
  public:
    MyParallelClass()@{od.
url("localhost");};
    MyParallelClass(float p)@
{od.power(p);};
    MyParallelClass(int m)@{od.
memory(m), od.encoding(xdr);};
    sync conc int GetValue();
    async seq void SetValue(int
v);
  private:
    int theValue;
};
```

In this example, if the default constructor is used to construct the parallel object, it will be located on the same computer (**localhost**) than the object who requests for the parallel object instantiation. If the constructor with one **float** parameter is used (the second one) then the chosen machine will have at least **p** Mflops of available computing power. Finally if the constructor with one **int** parameter is used then the chosen machine will have at least **m** Mbytes of available memory and the data to send, to and from, the remote machine will be encoded using the **XDR** standard encoding method. More details on the usage of object descriptors can be found in the pages 11 to 12 of the "POP-C++: Parallel Object Programming C++, User and Installation Manual" (Nguyen et al. 2007).

In this section, we have described how to start writing distributed efficient POP-C++ programs. In the next sub-section we are going to present how these efficient POP-C++ distributed programs can effectively and smoothly integrate MPI code.

Coupling MPI Code with POP-C++ Code

As mentioned in the first section of this chapter, the MPI programming tool and the SPMD programming paradigm are today largely dominating the world of HPC programming. A lot of HPC applications and a lot of numerical algebra libraries (Balay et al. 1997) have been developed using this approach. In addition, for some applications, or parts of applications, MPI is very convenient tool. For these reasons we consider as important that any new HPC programming tool allows for an easy integration of MPI code.

To smoothly integrate MPI code in POP-C++ programs, the POP-C++ environment allows parallel objects to be MPI processes. These special parallel objects have the ability to communicate using the MPI paradigm as well as the POP-C++ remote method invocations. POP-C++ uses the template feature provided by C++ to implement MPI parallel objects.

In order to create MPI parallel objects in POP-C++, the built-in **POPMPI** template class has to be used. The declaration of the **POPMPI** template class is given below.

```
Template <class T> class POPMPI
{
public:
  POPMPI();              // Do
not create any MPI process
  POPMPI(int np);        // Cre-
ate np MPI process of type T
  ~POPMPI();             // De-
structor
  bool Success();        // Re-
turn true if MPI is started.
                         // Oth-
erwise, return false
  int GetNP();           // Get
number of MPI processes
  bool ExecuteMPI();     // Ex-
ecute method ExecuteMPI on all
                         // pro-
cesses
  inline operator T*();  //
type-cast to an array of par-
class T
};
```

The first thing to do for creating parallel MPI objects is to write the MPI code. To do so one has to create a parallel class that contains at least one method having the following signature:

```
async void ExecuteMPI();
```

This method is in fact the MPI process which has to be replicated to use the Single Program Multiple Data (SPMD) programming paradigm. This method will contain necessary calls to MPI primitives to communicate with the other MPI parallel objects. The class can also contain other user-defined methods which can be called from the POP-C++ program. Attributes of the class are, as usually in the OO model, accessible by all methods of the class and allow the different methods, namely the **ExecuteMPI** method and the other user-defined methods to share data. Attributes are typically used to store results obtained by the MPI process (the **ExecutMPI** method).

Here is an example of such a parallel class:

```
parclass TestMPI
{
  public:
    TestMPI();
    async seq void ExecuteMPI();
    async seq void SetVal(int
v);
    sync seq void GetVal();
  private:
    int val;
};

TestMPI::TestMPI()    { val=0; }
void TestMPI::ExecuteMPI()
{
  MPI_Status status;
  int myRank;
  MPI_Bcast(&val,1,MPI_INT, 0,
MPI_COMM_WORLD);
  MPI_Comm_Rank(MPI_COMM_WORLD,
&myRank);
  if (myRank > 0)
    MPI_Recv(&val, 1, MPI_INT,
0, MPI_ANY_TAG, MPI_COMM_WORLD,
&status);
}
void TestMPI::SetVal(int v)   {
val=v; }
int TestMPI::GetVal()   { return
val; }
```

One can notice the presence of the mandatory method **ExecuteMPI**. In this example the MPI process is very basic; it just broadcasts (primitive **MPI_Bcast**) the value of the attribute **val** from the MPI process having the rank 0. All the other processes only receive (primitive **MPI_Recv**) the broadcasted value and store it in their **val** attribute. One can also notice how cryptic - compared to POP-C++ code - the MPI code is to send an integer. The **TestMPI** class also contains other methods such as such as **SetVal** and **GetVal** which allows the rest of the POP-C++ program to communicate with the instances of this class.

When the parallel class containing the MPI code is created (for example the class presented above) the MPI parallel objects (built using the **POPMPI** built-in class) can be instantiated. This is done as shown below.

```
int main(int argc, char **argv)
{
  POPMPI<TestMPI> mpi(2); // In-
stantiate 2 parallel MPI objects
                             // us-
ing the class TestMPI
  mpi[0].SetVal(100);       // Set
100 for the attribute val of MPI
                             //
process 0
  printf("Values are: %d,
%d \n", mpi[0].Getval(),
mpi[1].GetVal());
  mpi.ExecuteMPI();         //Call
ExecuteMPI methods on all MPI
proc.
  printf("Values are:: %d,
%d\n",mpi[0].Get(), mpi[1].
Get());
}
```

In the main program one can observe that each MPI process can be addressed individually. Indeed **mpi[i]** identifies the ith MPI process. The instruction **mpi.ExecuteMPI()** starts the execution of the MPI program by launching the **ExecuteMPI** method on all created MPI processes.

In this example, all methods of the **TestMPI** class have been declared with the sequential (**seq**) semantic. This will guarantee that the methods will be executed in the object using the order of arrival of the calls. Other semantics are of course also possible. The only restriction is that the **ExecutMPI** method must be asynchronous (**async**).

Integrating MPI code in POP-C++ code allows easy coupling of several MPI programs. Finally, form the programming point of view, using MPI inside POP-C++ frees the programmer of writing the tedious parts concerning the initialization and the termination of MPI processes (**MPI_Init**, **MPI_Finalize**,...). This is automatically handled by the POP-C++ environment.

We are now able to write complex distributed programs integrating POP-C++ and MPI code. The last sub-section will show how all the power of the OO paradigm of the POP model considerably eases the development of distributed OO programs using complex data structures.

Advanced Distributed OO Programming with POP-C++

In this sub-section we are going to explore how to use some of the powerful features of the C++ Object Oriented language in conjunction with POP-C++ extensions. In fact the complexity of writing POP-C++ programs does not come from the POP model itself but from the C++ language, which can be in some situations rather tricky. To introduce this aspect of the distributed OO programming we are going to use a case study.

Let us suppose that your simulation code uses the vector class of the C++ Standard Template Library. Vectors are data structures commonly used in simulation codes which normally are based on linear algebra computation.

The **vector** class of the C++ Standard Template Library has the following declaration:

```
template < class T, class Al-
locator = allocator<T> > class
vector;
```

It takes two template parameters having the following meanings:

- **T**: Type of the elements.
- **allocator**: Type of the allocator object used to define the storage allocation model. By default, the allocator class template for type **T** is used, which defines the simplest memory allocation model and is value-independent.

In our case one will use the default value for the allocator. One assumes that one has defined a class **MySimulData** containing relevant data for our simulation and that one wants to store these data in vectors. The declaration will be as follows:

```
vector<MySimulData> myData;
```

The object **myData** is a vector of objects of the class **MySimulData**. Now, how can we send this data to a remote object; in other words:

```
        Can an object of class
vector<MySimulData> be the pa-
rameters of a method of a paral-
lel class ?
```

The answer is obviously NO because **vector<MySimulData>** is not a serializable class. It does not inherit from the **POPBase** class and, as a consequence, does not implement the **Serialize** method.

The best way to solve this problem is to build a serializable class from the **vector<MySimulData>** class. To do so we will use the possibility of having multiple inheritances in C++. One creates the class **VectorOfMySimulData** in the following way:

```
class VectorOfMySimulData:public
POPBase,public
vector<MySimulData>
{
  . . .
  virtual void
Serialize(POPBuffer &buf, bool
pack);
  . . .
}
```

The class **VectorOfMySimulData** inherits from the class **POPBase** and from the class **vector<MySimulData>** thus it can implement the **Serialize** method and it is a **vector<MySimulData>**.

The question is now: how to write the **Serialize** method for the class **VectorOfMySimulData** ? There are two different situations:

- we have access to the source code of the class **MySimulData** and we can modify it;
- we cannot modify the source code of **MySimulData**.

In the first case the best approach is to make the class **MySimulData** to inherit from the **POPBase** class and to implement the **Serialize** method for the **MySimulData** class. Doing this eases the writing of the method **Serialize** for the **VectorOfMySimulData** class. The code of this method is given below:

```
void VectorOfMySimulData::Serial
ize(POPBuffer &buf, bool pack)
{
  long vsize;          // used
to Pack/UnPack the vector size
```

```
  vector<MySimulData>::iterator
iter;
  if (pack)
  {
    vsize = size();
    buf.Pack(&vsize,1);     //
Pack the size of the vector
  }
  else
  {
    buf.UnPack(&vsize, 1); //
UnPack the size of the vector
    clear();
    resize(vsize);         // Re-
create a vector of the received
size
  }
  for (iter=begin();
iter!=end(); iter++)
    iter->Serialize(buf, pack);
// Pack/UnPack the data in

// the vector
}
```

To understand this code we have to know how to use the template class vector of the C++ Standard Template Library (http://en.wikipedia. org/wiki/Vector_(STL)). The serialize/de-serialize operation of the vector is done in two steps. First we Pack/UnPack the size of the vector and second we Pack/UnPack the values in the vector. In our case the the Pack/UnPack of the value is greatly eased by the fact that these values are instances of a serializable class. Thus we only have to call the **Serialize** method of the class **MySimulData** to Pack/UnPack these values. This is shown in bold in the program above.

In the second situation, where we cannot modify the source code of the **MySimulData** class, it gets a bit more complicated. Indeed, as the **Serialize** method is not available to Pack/UnPack data in the vector we will have to do it by fetching the different values that constitute the **MySimulData** class and to separately **Pack** or **UnPack** them. This is a bit trickier than using the **Serialize** method but remains at a reasonable complexity when compared to what would be required to send the same data structure with MPI.

Below is the skeleton of the Serialize method in the case where the source code of the **MySimulData** class can not be modified.

```
void VectorOfMySimulData::Serial
ize(POPBuffer &buf, bool pack)
{
  long vsize;          // used
to Pack/UnPack the vector size
  vector<MySimulData>::iterator
iter;
  if (pack)
  {
    vsize = size();
    buf.Pack(&vsize,1);      //
Pack the size of the vector
    for (iter=begin();
iter!=end(); iter++)
      . .. // Here we have to
Pack all the attributes
      . .. // of the MySimulData
class from the vector
  }
  else
  {
    buf.UnPack(&vsize, 1); //
UnPack the size of the vector
    clear();
    resize(vsize);          // Re-
create a vector of the received
size
    for (iter=begin();
iter!=end(); iter++)
      . .. // Here we have to
UnPack all the attributes
      . .. // of the MySimulData
```

```
class in the vector
  }
}
```

In both situations we are now able to create a parallel class which can have methods with parameters of the class **VectorOfMySimulData**. Below is an example of such a class.

```
parclass ParallelSimulation
{
  public:
    ParallelSimulation();
    ~ParallelSimulation();
    sync seq void
Simulate(VectorOfMySimulData
theData);
    . ..
};
```

In Figure 6 is represented, using the UML (Unified Modeling Language) formalism, the structure of the classes (pattern) that have been used to send a vector of objects of the class **MySimulData**, when the class **MySimulData** inherited from the class **POPBase**. On this diagram the parallel class is represented in dark grey whereas the class we want to transmit to a remote parallel object is represented in light grey. The other classes are represented in white.

In the case study presented above the main difficulty comes from the fact that we cannot make the class **vector<MySimulData>** serializable, because we cannot modify its code (since it is a class of the C++ Standard Template Library). In the case we can modify the source code of the template class, a better and simpler solution consists in making the template class directly inherit from the **POPBase** class. The corresponding pattern is represented in Figure 7.

Generally speaking it is always a good option to also make serializable the classes included in the classes we want to transmit to parallel objects. This is the case when the serializable class

Figure 6. Pattern used to send a vector of user defined data

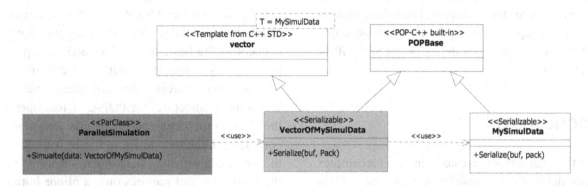

inherits from another class or contains an attribute of another class. We have seen this on the case study with the class **MySimulData**. This greatly helps to write the Serialize method of the class we have to transmit.

In these two examples the **Serialize** method of the class in light grey is as simple as:

```
void. . . .
.::Serialize(POPBuffer &buf,
bool pack)
{
  Data::Serialize(buf, pack);
}
```

In the case the class **Data** cannot inherit from **POPBase** (for example, because we do not have access to the source code of the class), then the **Serialize** method will be a little trickier.

In this section we presented the POP model and its implementation as an extension of the C++ programming language called POP-C++. It has been shown that POP-C++ complements MPI to develop heterogeneous and dynamic distributed programs. Using POP-C++ is straightforward for reasonably experienced C++ programmers. The majority of the code is pure C++, the declaration of parallel classes being the only addition, moreover relying on only six new keywords.

POP-C++ is a free software, it can be redistributed or modified under the terms of the GNU General Public License as published by the Free Software Foundation. The current stable version of POP-C++ is version 1.1.1. It is available on the "Grid & Ubiquitous Computing Group" web

Figure 7. Pattern used to send a data using template user-defined class

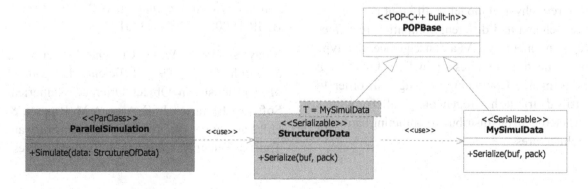

site (http://gridgroup.hefr.ch/popc/) and on the Source Forge Open Source software development web site (http://sourceforge.net/projects/popcpp/). A version 1.3 is under development and will be released soon.

CONCLUSION

In this chapter we presented Alpine3D, a complex parallel application simulating the process of snow cover at the earth's surface in alpine regions. By "complex" we mean applications making usage of several simulation methods, several physical models and several parallelization schema. We oppose this type of applications to what we refer as "flat parallel applications" using only one simulation method which can be homogeneously distributed on each cells of the domain that is simulated. For this type of applications the traditional SPMD programming paradigm, as provided by the MPI programming could be sufficient.

In Alpine3D three major phenomena have to be modeled:

1. the surface exchanges (mass and energy), the temperature profile and the structure of the soil/snow inside the snow bed.
2. the spatial distribution of snow as caused by wind.
3. the radiation balance due to the solar radiation.

Each of these three simulations requires a different physical model, a different parallelization schema and different computing resources (cpu and memory). As a consequence, this type of applications requires a new Distributed High Performance Computing paradigm in order to easily distribute heterogeneous parallel programs on heterogeneous distributed computing hardware architectures.

The POP-C++ programming tool based on the Parallel Object Oriented (POP) paradigm claims to fulfill these needs. By using the object concept as the basic brick of a distributed program and by allowing to easily and effectively distribute objects over distributed heterogeneous hardware architecture, the POP-C++ tool opens the door to the development of new and more complex simulations similar as those covered by the Alpine3D software. In addition, the usage of the POP-C++ tool requires only a minor initial effort for C++ programmers because the POP-C++ syntax and model is very close to C++. Finally, in order to allow the reutilization of MPI code, the POP-C++ programming environment offers a straightforward integration of MPI code inside POP-C++ programs.

We believe that recent and future simulation problems are, and will be, more and more complex in the sense that we discussed above. As a consequence, the need for a new HPC programming paradigm and tools going beyond the traditional SPMD model will become more and more important. The development of the Alpine3D application using the POP-C++ programming tool is a first step in the direction of such a new HPC programming paradigm.

REFERENCES

Badia, R.M., & Labarta, J. Sirvent, J., Pérez, J.M., Cela, J. M. Cela, & Grima R. (2003). Programming Grid Applications with GRID Superscalar. *Journal of Grid Computing, 1*(2). doi:10.1023/B:GRID.0000024072.93701.f3

Balay, S., Gropp, W. D., Curfman McInnes, L., & Smith, B. F. (1997), Efficient Management of Parallelism in Object Oriented Numerical Software Libraries. In E. Arge, A.M. Bruaset, & H.P. Langtangen (pp. 163-202). Modern Software Tools in Scientific Computing. Birkhauser Press

Baude, F., Baduel, L., Caromel, D., Contes, A., Huet, F., Morel, M., & Quilici, R. (2004). Programming, Composing, Deploying for the Grid. In Cunha, J. C., & Rana, O. F. (Eds.), *GRID COMPUTING: Software Environments and Tools*. Springer Verlag.

Bavay, M., Lehning, M., Jonas, T., & Löwe, H. (2009). Simulations of future snow cover and discharge in Alpine headwater catchments. *Hydrological Processes, 22*. doi:.doi:10.1002/hyp.7195

Bründl, M., Romang, H. E., Bischof, N., & Rheinberger, C. M. (2009). The Risk Concept and Its Application in Natural Hazard Risk Management in Switzerland. *Natural Hazards and Earth System Sciences, 9*, 801–813. doi:10.5194/nhess-9-801-2009

Clifton, A., & Lehning, M. (2008). Simulations of wind tunnel snow drift using a semi-stochastic model. *Earth Surface Processes and Landforms, 33*(14), 2156–2173. .doi:10.1002/esp.1673

Clifton, A., Rüedi, J.-D., & Lehning, M. (2006). Snow saltation threshold measurements in a drifting snow wind tunnel. *Journal of Glaciology, 52*(179), 585–596. doi:10.3189/172756506781828430

Doorschot, J., & Lehning, M. (2002). Equilibrium saltation: mass fluxes, aerodynamic entrainment and dependence on grain properties. *Boundary-Layer Meteorology, 104*(1), 111–130. doi:10.1023/A:1015516420286

Dünnweber, J., & Gorlatch, S. (2004). HOC-SA: A grid service architecture for higher-order components. In *IEEE International Conference on Services Computing, Shanghai, China* (pp. 288-294). IEEE Computer Society Press.

Foster, I., & Kesselman, C. (1998). *The Grid: Blueprint for a New Computing Infrastructure*. Morgan Kaufmann Publishers.

Helbig, N., Löwe, H., & Lehning, M. (2009). Radiosity approach for the shortwave surface radiation balance in complex terrain. *Journal of the Atmospheric Sciences.*.doi:10.1175/2009JAS2940.1

Lehning, M., Bartelt, P. B., Brown, R. L., Fierz, C., & Satyawali, P. (2002a). A physical SNOWPACK model for the Swiss Avalanche Warning Services. Part II: Snow Microstructure. *Cold Regions Science and Technology, 35*(3), 147–167. doi:10.1016/S0165-232X(02)00073-3

Lehning, M., Bartelt, P. B., Brown, R. L., Fierz, C., & Satyawali, P. (2002b). A physical SNOWPACK model for the Swiss Avalanche Warning Services. Part III: Meteorological Boundary Conditions, Thin Layer Formation and Evaluation. *Cold Regions Science and Technology, 35*(3), 169–184. doi:10.1016/S0165-232X(02)00072-1

Lehning, M., Löwe, H., Ryser, M., & Raderschall, N. (2008). Inhomogeneous precipitation distribution and snow transport in steep terrain. *Water Resources Research, 44*, W07404. .doi:10.1029/2007WR006545

Lehning, M., Völksch, I., Gustafsson, D., Nguyen, T. A., Stähli, M., & Zappa, M. (2006). ALPINE3D: A detailed model of mountain surface processes and its application to snow hydrology. *Hydrological Processes, 20*, 2111–2128. doi:10.1002/hyp.6204

Lehning, M., & Wilhelm, C. (2005). Integral Risk Management and Physical Modelling for Mountainous Natural Hazards. In Albeverio, S., Jentsch, V., & Kantz, H. (Eds.), *Extreme Events in Nature and Society* (p. 349). Springer.

Marty, C., Phillips, M., Lehning, M., Wilhelm, C., & Bauder, A. (2009). Climate change and natural hazards in the Grisons. *Schweizerischen Zeitschrift für Forstwesen, 160*, 201–209. doi:10.3188/szf.2009.0201

Michlmayr, G., & Lehning, M.Holzmann. H., Koboltschnig, G., Mott, R., Schöner, W., & Zappa. M. (2008). Application of Alpine3D for glacier mass balance and runoff studies at Goldbergkees, Austria. *Hydrological Processes*. doi:.doi:10.1002/hyp.7102

Mott, R., Faure, F., Lehning, M., Löwe, H., Hynek, B., & Michlmayr, G. (2008). Simulation of seasonal snow cover distribution for glacierized sites (Sonnblick, Austrian Alps). *Annals of Glaciology*, *49*, 155–160. doi:10.3189/172756408787814924

Nguyen, T. A. (2004). *An object-oriented model for adaptive high-performance computing on the computational GRID*. PhD Thesis no 3079, EPFL, Switzerland.

Nguyen, T. A., & Kuonen, P. (2007). Programming the Grid with POP-C++. [FGCS]. *Future Generation Computer Systems*, *23*(1), 23–30. doi:10.1016/j.future.2006.04.012

Nguyen, T. A., Pasin, M., & Kuonen, P. (2007). *POP-C++, Parallel Object programming C++: User and Installation Manual*. University of Applied Sciences of Western Switzerland, Fribourg.

Ohmura, A. (2001). Physical basis for the temperature-based melt-index method. *Journal of Applied Meteorology*, *40*, 753–761. doi:10.1175/1520-0450(2001)040<0753:PBFTTB>2.0.CO;2

Pérez, C., Priol, T., & Ribes, A. (2003). A Parallel CORBA Component Model for Numerical Code Coupling. [IJHPCA]. *International Journal of High Performance Computing Applications*, *17*(4), 417–429. doi:10.1177/10943420030174006

Raderschall, N., Lehning, M., & Schär, C. (2008). Fine scale modelling of the boundary layer wind field over steep topography. *Water Resources Research*, *44*, W09425. .doi:10.1029/2007WR006544

Stössel, F., Guala, M., Fierz, C., Manes, C., & Lehning, M. (Manuscript submitted for publication). On the micrometeorology of surface hoar on mountain snow covers. *Water Resources Research*.

Van Nieuwpoort, R. V., Maassen, J., Hofman, R., Kielmann, K., & Bal, H. E. (2002). Ibis: an efficient Java-based grid programming environment. In Proceedings of joint ACM-ISCOPE conference on Java Grande table of contents, Seattle, Washington, USA (pp.18- 27).

Vanneschi, M. (2002). The programming model of ASSIST, an environment for parallel and distributed portable applications. *Parallel Computing*, *28*(12), 1709–1732. doi:10.1016/S0167-8191(02)00188-6

Viviroli, D., Zappa, M., Gurtz, J., & Weingartner, R. (2009). An introduction to the hydrological modelling system PREVAH and its pre- and post-processing-tools. *Environmental Modelling & Software*, *24*(10), 1209–1222. .doi:10.1016/j.envsoft.2009.04.001

Wever, N., Lehning, M., Clifton, A., Rüedi, J.-D., Nishimura, K., & Yamaguchi, S. (2009). Verification of moisture budgets during drifting snow conditions in a cold wind tunnel. *Water Resources Research*, *45*. .doi:10.1029/2008WR007522

Zappa, M., Pos, F., Strasser, U., Warmerdam, P., & Gurtz, J. (2003). Seasonal water balance of an Alpine catchment as evaluated by different methods for spatially distributed snowmelt modeling. *Nordic Hydrology*, *34*, 179–202.

KEY TERMS AND DEFINITIONS

Multi-Physic: Refers to simulations consisting of multiple physical models cooperating to the phenomenon of interest.

Alpine Surface Processes: Refers to the interaction between the atmosphere and the surface (soil and cryosthere) in alpine environment.

Water Availability: It is the study of catchments discharge depending on snow melt, precipitations and climate. Distributed HPC: Distributed High Performance Computing refers to high performance programs running on distributed hardware, i.e, network of computers with no shared memory.

Heterogeneous Distributed Hardware: Network of computers without shared memory where computers can be of different nature in terms hardware and/or operating systems.

Multiple Program Multiple Data (MPMD): Refers to parallel or distributed programs where tasks or processes can be different. MPMD is usually opposed to the SPMD (Single Program Multiple Data) where all processes are identical.

Distributed Object Oriented Model: Programming model using the object paradigm to distribute data and executable code over distributed hardware.

Serializable: A data structure is serializable if a function allowing to concatenate all the data structure inside a continuous memory space has been defined.

Chapter 16

Sensor and Computing Infrastructure for Environmental Risks:
The SCIER System

Odysseas Sekkas
University of Athens, Greece

Dimitrios V. Manatakis
University of Athens, Greece

Elias S. Manolakos
University of Athens, Greece

Stathes Hadjiefthymiades
University of Athens, Greece

ABSTRACT

The SCIER platform is an integrated system of networked sensors and distributed computing facilities, aiming to detect and monitor a hazard, predict its evolution and assist the authorities in crisis management for hazards occurring at Wildlife Urban Interface (WUI) areas. The goal of SCIER is to make the vulnerable WUI zone safer for the citizens and protect their lives and property from environmental risks. To achieve its objective, SCIER integrates technologies such as: (1) wireless sensor networks for the detection and monitoring of disastrous natural hazards, (2) advanced sensor data fusion and management for accurately monitoring the dynamics of multiple interrelated risks, (3) environmental risk models for simulating and predicting the evolution of hazardous phenomena using Grid-computing. In this chapter we present the key software components of the SCIER system architecture, namely the sensor data fusion component and the predictive modeling and simulation component.

DOI: 10.4018/978-1-61520-987-3.ch016

INTRODUCTION

The tendency for the development of extensive WUI (Wildlife Urban Interface) areas is a relatively new phenomenon. This refers to all types of areas where forests, water bodies, and rural lands interface with homes, other buildings and infrastructures, including first and secondary home areas, industrial areas and tourist developments (Stewart, 2007). The related problems that it generated, especially with regards to increasing fire and flood risks, started becoming noticeable only in the 1990s. The rapid development of WUI areas is the result of pollution and overpopulation of city centers that grew in the 1970s. However, in many cases, the rapid development of such WUIs was unplanned, or poorly planned. Settlements were built without efficient road networks, and homes and other buildings were developed in or near areas that form the flood plain of water catchments. Often there is no provision for routes of escape in case of a disaster.

The *Sensing and Computing Infrastructure for Environmental Risks* (SCIER) system constitutes an integrated sensing and computing platform capable of delivering to the authorities and the citizens valuable real time information regarding natural hazards that may affect the WUI. SCIER aims at providing the functionality needed for detecting, monitoring and forecasting the hazard's evolution. Sensors deployed in the region monitor environmental parameters (e.g., temperature, humidity, wind direction and speed) and feed the data to predictive models running in the computing infrastructure. The SCIER platform builds upon existing technical expertise and recent progress in the areas of sensors, communications, Grid computing, Geographical Information Systems (GIS), data fusion and predictive modeling. Indeed, the information produced by the SCIER platform can in many cases be a key factor in the effective fighting of the hazard's consequences. SCIER predicts the evolution of the main phenomenon as well as the risks associated with any

secondary phenomena it may trigger. Furthermore, for the people living in vulnerable WUI areas, it addresses their needs for security and reliable alerting services. Finally, SCIER provides Civil Protection Authorities with a tool for the effective management of crisis situations caused by natural hazards.

Related Work

In this section we briefly discuss prior research activities on natural hazard detection and monitoring. Most of them deal with fire detection and make use of temperature and humidity sensors, smoke detectors and infrared cameras. In (Chen, 2003) a fire-detection system is proposed based on multi-sensor technology and neural networks. The sensed contextual data includes environmental temperature, smoke density and CO density. In (Pehrsson, 2000) and (Pehrsson, 2003), the authors present a system that is based on various types of sensors and use neural networks. However such systems require the use of training data and most of them are evaluated indoors where the weather conditions are fully controllable and, surely, completely different in comparison with those observed in a WUI. A system for wildfire monitoring using a wireless sensor network (WSN) that collects temperature, relative humidity and barometric pressure is described in (Doolin, 2005). The authors in (Calle, 2006) and (Sivathanu, 1996) propose systems based on infrared (IR) technology for the detection of fires. Furthermore, (Kucuk, 2008) and (Kosucu, 2009) have proposed solutions in which sensors are deployed from an aircraft. In (Hefeeda, 2007) the authors propose a distributed k-coverage algorithm to balance the load across all deployed sensor nodes. However, these systems use either in-field or out-field sensors, thus rendering them vulnerable to false alarms. In addition, aerial or satellite images are frequently used for outdoor fire detection and monitoring. In (Mandel, 2007) the authors present a system architecture which attempts

Figure 1. SCIER system architecture

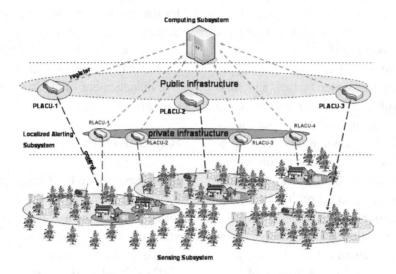

to integrate recent sensing measurements (from satellite spectral image data and other sensors) with simulation based predictive modelling into a closed loop system. The sensor data are used to calibrate the predictions of the simulation model in order to minimize the evolution prediction error. These methods are known as Dynamic Data Driven Application System (DDDAS) techniques and are recently of great interest to the scientific community. Satellite based monitoring is used for detecting forest fires in (Zhanqing, 2001). However, the scan period and the low resolution of satellite images make such method incapable for early (real-time) fire detection.

SCIER System

The SCIER platform is a complex system which integrates technologies and techniques from different scientific fields. It is customary that the architecture of such a large-scale integrated system is visualized by a vertical, bi-directional flow-chart divided into different layers. Each layer performs a specific set of activities. Neighboring layers contribute to their common interface so that all bilateral transactions are reliable and safe. In the SCIER case, we identify three (3) architectural layers: the Sensing Subsystem, the Localized Alerting Subsystem and the Computing Subsystem.

Sensing Subsystem

The purpose of the SCIER sensing subsystem is the monitoring of the environmental parameters that are relevant to the assessment of a natural hazard. In the WUI two kinds of sensors are deployed: Citizen Owned Sensors (COS), installed by land/home owners in fixed and registered locations in private areas, and Publicly Owned Sensors (POS), installed by state authorities in fixed and known locations in public areas. The sensor nodes (Figure 2) are energy efficient and form a multi-hop,

Figure 2. Sensor nodes used in SCIER system

Figure 3. Distribution of the nodes on the field

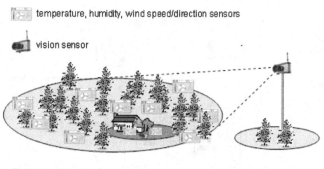

The figure is not to scale.

self-organized, robust Wireless Sensor Network transmitting the raw measurements to a sink node via appropriate routing protocols. They are enclosed in appropriately designed (temperature and water resistant) plastic boxes which isolate them from the environment but do not affect the radio propagation. Outside the box is only the actual sensor (sensing part).The transmission range of the nodes is adjustable and can reach up to 250 m in open space areas and up to 100 m in a forest zone.

In SCIER the following categories of nodes are identified.

- **In-field sensor nodes.** They carry two or three sensors, and they are responsible for measuring and transmitting values of meteorological parameters (e.g., temperature and humidity). Since they are battery-powered, these nodes have high energy-consumption constraints. They transmit their readings towards the data sinks, following the multi-hop organization of the WSN. The WSN can be dynamically deployed which means that nodes can be added or removed at run-time according to the circumstances.
- **In-field sensor nodes - cluster heads.** They are similar to the sensor nodes except that they are more powerful and can support more complex sensors (e.g. wind

measurements) as well. They propagate the information to the data sinks.
- **Out-of-field vision sensors ("smart cameras").** They report data from the scene under surveillance.
- **Data sinks.** They relay the information from/to the sensor nodes towards/from the SCIER's higher layers.

The use of out-field vision sensors that monitor the same area adds an extra feature to the system which contributes to the effective detection of a hazardous natural phenomenon. Vision sensors, in the case of a fire, transmit information about smoke or flame probability corresponding to locations on which the camera focuses and which are different from the location of the vision sensor. They provide a very stable representation of the scene under uncontrolled illumination conditions. Vision sensor nodes are fixed nodes, most likely installed on poles or citizens' homes. A monitored area could be covered by one or more vision sensors. Figure 3 depicts the distribution of the nodes on the monitored field.

Localized Alerting Subsystem

The Localized Alerting Subsystem (LAS) includes the Local Alerting Control Unit (LACU) which comprises a computing element of primary importance to the SCIER system. Each LACU controls

Figure 4. The Localized Alerting Subsystem of SCIER

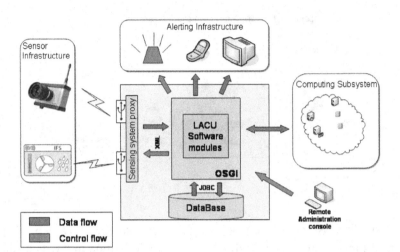

a network of sensors, receives input from them, and executes fusion algorithms on the received data. Multiple LACUs are deployed in the area that needs to be monitored for potential emergency situations arising from environmental hazards, such as forest fires or floods. Using LACUs can be extended to handling different environmental risks by adjusting the type of sensors and fusion algorithms used. The LACUs self organize into a network, where each node (a LACU) has certain functional autonomy, but can also be controlled by the Computing Subsystem, in certain cases. In SCIER, two types of LACUs are identified (Figure 1):

- Public LACU (P-LACU). It is installed and operated by public authorities and controls a wireless network of Public Owned Sensors.
- PRivate LACU (R-LACU). It controls Citizen Owned Sensors and is installed by individuals in order to protect their private properties.

The LACU is a device which mediates between the Computing Subsystem and the Sensing Subsystem. It is a simple computing unit, which runs specific software developed in the context of the SCIER project. This software is based on the OSGI framework (OSGI Alliance, 2009) and is highly modular enabling to load/unload or update components on demand. The basic software components comprising LACU (Figure 4) have the following roles:

- Control of the WSN
- Acquisition of data from the underlying WSN. Currently supported sensors include temperature, humidity, wind speed & direction sensors, pluviometers and vision sensors assessing smoke and flame probability
- Administration of the system, which can be performed either locally or remotely
- Execution of flood/fire detection algorithms that assess the severity of the readings flowing in the system and produce alerts
- Support for various alerting components (visible, acoustic, SMS messaging) used to provide notifications when potential emergencies are detected
- Communication with external computing elements (i.e., the Computing Subsystem)

Computing Subsystem

The Computing Subsystem (CS) is largely based on a GIS where the fused sensor information is stored, processed and visualized. Multiple environmental models of different time scales are used in order to establish an accurate tracking (and simulation) of the hazardous phenomenon. The risk models are executed in the CS, providing estimates on the spreading of the risk and the expected impact and arrival time for settlements, villages and farms. Advanced computing infrastructure (e.g., a Grid setup) can be used to run "what-if" scenarios, thus investigating the consequences of potential changes in key environmental parameters such as the wind speed and direction. The Grid infrastructure offers the capability of parallel simulations which allow the exploration of the effects of potential changes in such parameters.

The main functionalities of Computing Subsystem are:

- Collect and store sensor-measurements from the area of interest.
- Perform data-fusion-algorithms to assess the level of risk.
- Trigger a simulation in case of a perceived real alarm, i.e. (a) retrieve geographical data from the GIS Database on the terrain layout of the area of interest, (b) generate different slightly perturbed scenarios on the wind speed and wind direction for the area of interest, (c) submit to the Grid parallel simulations, one for each scenario, (d) retrieve results and pass them to the GIS module for visualization on a reference map

Fusion Process in SCIER

In SCIER, as discussed in previous sections, data derived from various sensors are used for the detection and monitoring of hazardous events. Such sensors include temperature, humidity, wind speed/direction, pluviometry, soil moisture and vision sensors. The volume of raw data that are generated requires an effective post-processing in order to decide on the occurrence (or not) of an event. Thus, sensor fusion techniques which process and assess the data and reason about an event are adopted.

In the case of fire detection one requires measurements derived from in-field sensors that are deployed in the area or out-field sensors which monitor from a distance the area. In the SCIER system temperature, humidity sensors (in-field) and vision sensors (out-field) are used. The last are based on a high-dynamic range contrast camera in which the contrast representation of a scene can be used (through appropriate algorithms) to detect smoke or flames and estimate/generate a probability (confidence level) on this event. These two categories of sensors are combined in a two-level fusion scheme (Zervas, 2009), (Zervas, 2007) thus improving the reliability of the system in fire detection. At the first level of fusion (data fusion) we adopt the cumulative sum technique (Gombay, 2005), (Page, 1954) for fusing data from in-field sensors and assign a probability of fire occurrence in each of them. At the second level fusion (information fusion), probability values about fire events from the first level are combined through evidential theory with probability values about fire events from the vision sensor. The adoption of such a scheme has the advantage of early fire detection and, simultaneously, eliminates any false alarms in case of no fire occurrence. Figure 5 depicts the whole work flow regarding the aspect of the fusion process (data and information) in SCIER.

First Level Fusion

In each LACU readings from in-field sensors are constantly processed through a sensor data fusion procedure in order to detect any significant change in the environmental state. For instance, consider the event in which the normal ambient

Figure 5. Fusion process in SCIER

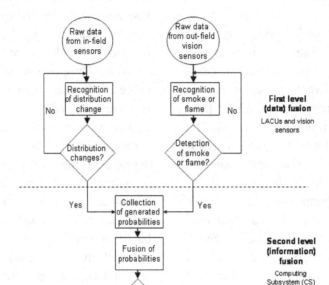

temperature increases in an abnormal way. This could indicate the occurrence of a possible fire event. The system regularly monitors the data distribution that is generated over time. If a change in data distribution is detected, this is reflected on a specific "metric". Such "metric" postulates a translation of the impact of the evidence to a certain amount of belief on a current hypothesis (the fire event hypothesis in our case). Such probability is calculated for each sensor.

In an analogous data fusion procedure, for each out-field vision sensor, any significant change in the contrast or the luminance of the monitored scene is translated, through specific algorithms, to a probability of fire (smoke or flame). According to the luminance of the environment which depends on the time of the day the vision sensor which divides the scene into tiles, generates the appropriate probability for each tile. Smoke probability in the daylight and flame probability at night. Each sensor Si (temperature, humidity, vision sensor) produces a confidence level c_i for fire detection in its monitoring area and reports this

value to the Computing Subsystem (CS) where the second level fusion takes place. The confidence levels indicate an order in probability of positive detection in the sense that a higher confidence level implies a higher probability of fire.

Second Level Fusion

In the second level fusion process vision sensor (camera) data and data coming from LACUs are combined. Each single fusion process will be based on data for a single camera tile together with data from the sensors that this camera tile oversees. In those cases where a camera tile does not oversee any sensor(s), or a/any sensor(s) is/ are not overseen by a camera, a degenerate fusion process will be carried out taking into account the probabilities of a single camera tile or any sensor(s) respectively.

Upon reception of the confidence levels c_i, $i=1,...,M$ the fusion process at CS evaluates a discriminant function $f(c_1, c_2, c_3 ...,c_M)$ and compares it with a threshold t, to decide if a fire

Figure 6. Second level fusion in SCIER

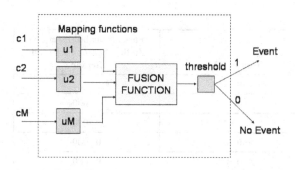

event is present or not. The fusion methods may treat the influence of the confidence levels differently and this is shown schematically in Figure 6 where mapping functions are used to scale the confidence levels.

Confidence levels are combined through Dempster–Shafer (DS) evidential theory (Shafer, 1976). In DS process for each sensor we need the basic probability assignments $m(F)$, $m(no - F)$ and the unsigned probability mass $m(F \cup no - F)$. These quantities sum to one, thus only two of them need to be specified. The mass $m(F)$ represents the belief in fire detection, $m(no - F)$ the belief in the no–fire case and $m(F \cup no - F)$ represents the uncertainty of the sensor. Given the confidence level of a sensor, c_i, and the uncertainty u_i, the mapping function can be:

$$m_i(F) = (1 - u_i)c_i$$
$$m_i(noF) = (1 - u_i)(1 - c_i)$$
$$m_i(F \cup no - F) = u_i$$

The Dempster – Shafer algorithm for fusing two probability masses is

$$m_{12}(F) = \frac{1}{1 - K}[m_1(F)m_2(F) + m_1(F)m_2(F \cup no - F) + m_1(F \cup no - F)m_2(F)]$$

$$m_{12}(no - F) = \frac{1}{1 - K}[m_1(no - F)m_2(no - F) + m_1(no - F)m_2(F \cup no - F) - m_1(F \cup no - F)m_2(no - F)]$$

$$m_{12}(F \cup no - F) = \frac{1}{1 - K}[m_1(F \cup no - F)m_2(F \cup no - F)]$$

where

$$K = m_1(F)m_2(no - F) + m_1(no - F)m_2(F)$$

For fusing the probabilities of three sensors, one has first to combine two sensors to obtain $m_{12}(F)$, $m_{12}(no - F)$ and $m_{12}(F \cup no - F)$ and then use Dempster-Shafer rule to obtain $m_{123}(F)$, $m_{123}(no - F)$ and $m_{123}(F \cup no - F)$. Once exhausting the sensors we are left with the basic probability masses:

$$m_{123\cdots M}(F), \; m_{123\cdots M}(noF) \; and \; m_{123\cdots M}(F \cup no - F)$$

For detection purposes we can use either the support $m_{123\cdots M}(F)$ and compare it to a threshold t, or the plausibility $m_{123\cdots M}(F) + m_{123\cdots M}(F \cup no - F)$ and compare it to a threshold t, or even the average of these two.

Examples of the Fusion Process

Scenario 1. In this scenario we combine the maximum probability induced by the in-field sensors (in order to minimize the false alarm rate) with the fire detection probability of the vision sensor. We assume three different fire probabilities for the vision sensor (0.1, 0.5, 0,9), each one fused with three different probabilities (0.1, 0.5, 0.8), inferred by the LACU using in-field sensor measurements. Table 1 shows the combination results of the DS algorithm. As it is observed, unless both fused probabilities exceed the value

Table 1. DS algorithm on fusing the probability regarding the fire event of an in-field sensor with the vision sensor

In field Sensor 1 $m_1(F)$	Vision Sensor $m_2(F)$	Conflict (K)	Fused Probabilities $m_{12}(F)$ In field Sensor 1 + Vision Sensor
0.1	0.1	0.18	0.0122
0.1	0.5	0.50	0.1000
0.1	0.9	0.82	0.5000
0.5	0.1	0.50	0.1000
0.5	0.5	0.50	0.5000
0.5	0.9	0.50	0.9000
0.8	0.1	0.74	0.3077
0.8	0.5	0.50	0.8000
0.8	0.9	0.26	0.9730

of 0.5, the final probability is kept in relatively small values. Thus, a malfunctioning sensor is not able by itself to trigger a fire alarm. Moreover high probability values of both sensors will enhance our confidence for a fire event.

Scenario 2. In this scenario we combine the fused probabilities obtained in the first scenario with the fire detection probability of another in-field sensor (Sensor 3). For the latter, we assume two values (0.2 and 0.6) and the results are depicted in Table 2 and Table 3 respectively. As it is observed from the entries of Table 2 (rows 3 and 5), the value 0.5 when combined with the small probability of the in-field sensor i.e., 0.2, yields the lowest value, that is 0.2.

On the contrary, as it is observed from Table 3 (rows 3 and 5), the value 0.5 when fused with a higher probability, i.e., 0.6, yields the maximum value, that is 0.6. If all constituent probabilities are greater than 0.5 (last two rows of Table 3) then our belief for a fire event is reinforced as it is indicated by the high values of the final fused probability.

Table 2. DS fusion algorithm with a second in-field sensor (probability 0.2)

Fused Probabilities $m_{12}(F)$ In field Sensor 1 + Vision Sensor	In field Sensor 3 $m_3(F)$	Conflict (K)	Fused Probabilities $m_{123}(F)$
0.0122	0.2	0.2073	0.0031
0.1000	0.2	0.2600	0.0270
0.5000	0.2	0.5000	0.2000
0.1000	0.2	0.2600	0.0270
0.5000	0.2	0.5000	0.6000
0.9000	0.2	0.7400	0.6923
0.3077	0.2	0.3846	0.1000
0.8000	0.2	0.6800	0.5000
0.9730	0.2	0.7837	0.9000

Table 3. DS fusion algorithm with a second in-field sensor (probability 0.6)

Fused Probabilities m_{12}(F) In field Sensor 1 + Vision Sensor	In field Sensor 3 m_3(F)	Conflict (K)	Fused Probabilities m_{123}(F)
0.0122	0.6	0.5975	0.0182
0.1000	0.6	0.5800	0.1429
0.5000	0.6	0.5000	0.6000
0.1000	0.6	0.5800	0.1429
0.5000	0.6	0.5000	0.6000
0.9000	0.6	0.4200	0.9310
0.3077	0.6	0.5384	0.4000
0.8000	0.6	0.4400	0.8571

FIRE FRONT EVOLUTION SIMULATION - GRID WORKFLOW

The fusion algorithms, described in the previous section, decide whether the event constitutes a real threat or it is a false alarm. In the former case the SCIER Computing Subsystem initiates a simulation in the Grid infrastructure consisting of several parallel runs. Each run is based on a different set of input parameters and computes:

- The expected evolution of the fireline for up to 180 minutes after fire detection. The Fire Spread Engine (FSE) software program developed by Technoma S.A. (EUFIRELAB, 2006) is used for this purpose. The FSE is a computer application which estimates the fire front expansion on surface forest fuels, using spatial data about the topography, moisture content, type of the fuel and dynamic environmental parameters, such as the wind vector field. The software is based on the Rothermel-Frandsen theoretical model and uses algorithms similar to those found in the BEHAVE system (Andrews, 1986). A cellular-automata algorithm is used for the estimation of the fire spread. However, the major limitation of all fire spread simula-

tors (including the FSE) is that their accuracy decreases as the span of the prediction time increases. An acceptable prediction time span for the FSE is usually up to 2-3 hours after fire detection.

- In addition to the fire front line, a simulation run also estimates the time-evolving temperature field it induces as it moves through an area. For that purpose we use the *Temperature Field Modeling* (TFM) software component that we have developed (Manolakos, 2008).

At the end of a simulation, the different temperature fields estimated by the parallel TFM components are compared to the real temperature field measured at the deployed sensor locations in order to infer the posterior probabilities of the different simulated fire front scenarios given the real sensor data. Algorithms for matching (real to hypothesized scenarios) and scoring scenarios probabilistic similarity have been developed for this purpose. The larger the number N of simulated fire front evolution scenarios, the more accurate the estimated posterior probabilities of the scenarios given (sufficient data collected from) the real event. Since fire spread is a very complex phenomenon influenced by many diverse factors, one can hope to estimate accurately the probability

Figure 7. Components of the SCIER forest fire simulation workflow. Most components run inside the Grid. The matching and scoring components currently are not GRIDified.

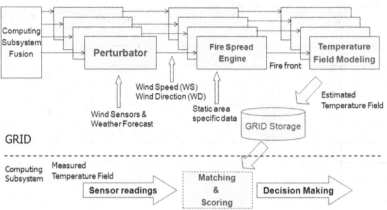

of arrival of a fire line to a specific area only if an effective sampling mechanism of the very large parameter space is in place (Bianchini, 2005). The parallel processing capabilities provided by the Grid infrastructure, help us complete the simulation of a large number of alternative scenarios in a reasonable amount of time, compared to a serial execution.

The workflow of a wildfire evolution simulation includes a set of components whose interaction is shown in Figure 7. The workflow starts with the generation of the different wind speed and direction fire scenarios to be simulated. This is the task of a component called the "Perturbator". The *Perturbator* uses current wind sensor measurements and wind weather forecasts for the next few hours to estimate the mean and extend of wind parameters and then produces, by random sampling from a statistical distribution (Gaussian or uniform), a wind speed and wind direction combination (WS, WD) to be simulated.

A wind scenario ((speed and direction combination generated by the Perturbator) is supplied as input to an FSE component instance which estimates the corresponding fire front's spatio-temporal evolution for the next 180 min. The number of simulated scenarios depends on the number of different wind speed and direction com-

bination pairs (WS, WD) generated and is a user controllable parameter. Each scenario corresponds to a simulation run and all the runs included in the same simulation job are executed in parallel in the Grid. In addition to the wind speed and direction dynamic data, the FSE needs as input data files related to the ground morphology (slope, aspect), moisture, fuel type etc. These files are specific to a geographic area of interest, do not change from run to run (static data) and need to be available before a Grid simulation can be launched.

The execution of an FSE component instance generates the data needed to launch subsequently a *Temperature Field Modeling* (TFM) component instance (Manatakis, 2010, Manolakos 2008). This component takes as input the output files produced by the FSE (time of arrival of the fire line at each geographical cell, fire flame length at time of arrival) and the corresponding wind speed and direction files and produces *in-silico* an estimate of the temperatures that the in-field temperature sensors located close to the fire front line are expected to "feel". Temperature estimates are updated every time the front line is advanced, i.e. every minute. Therefore, if a simulation includes N (WS,WD) runs, these will simulate in parallel N different fire front line evolution scenarios and produce their respective N temperature fields. If

Figure 8. Fire front evolution estimation

the FSE is simulated say for 180 minutes, then the corresponding temperature field produced by the TFM component basically corresponds to a "thermal video" with 180 successive "temperature frames" and as many pixels per frame as the deployed SCIER sensors in the field. The TFM component is the most computationally demanding in the Grid workflow. At each simulation time step its time complexity increases with the product of the number of cells newly affected by the fire and the number of sensors (monitoring points) existing close to the fire front line.

To rank order the simulated scenarios in terms of how probable they are, we compare periodically the temperature fields, estimated in-silico by the different Grid TFM runs, to the set of real sensor measurements (temperature readings) as they become available from the SCIER sensing subsystem after the confirmation of a true fire event. The process that assesses the degree of similarity of the real temperature sensor data to the in-silico produced temperature estimates is called field *matching*. It is followed by the *scoring* process which estimates how probable is each one of the simulated fire front evolution scenarios based on the results of the matching and in light of the already collected sensor measurements. Scoring also produces a map that shows the probability that each geographical cell will be affected by the fire

for the next up to 180 minutes after fire detection. This may be estimated using either all N simulated scenarios or the $M < N$ most probable ones identified by the scoring process. The matching and scoring processes can be iterated periodically to improve the scoring accuracy by exploiting more WSN collected temperature data, as they become increasingly available from the SCIER sensing subsystem as the fire progresses. The matching and scoring SCIER software components run outside the Grid (in the CS server) because their computational time is not substantial. However if N becomes very large, a Grid implementation may also be considered for their implementation. The visualization of a typical scenario regarding the fire front evolution is depicted in Figure 8.

RESULTS AND DISCUSSION

To evaluate the efficiency of the proposed GRID computing workflow, we have performed extensive simulations with different numbers of runs (ranging from 5 to 110). Each run corresponds to a specific forest fire scenario which simulates 180 min of fire front evolution and temperature field generation for a fire that is started at a specific ignition point in the area of Stamata, a WUI community outside Athens, Greece. Stamata has been selected as one of the SCIER project field trial areas and a WSN with 20 temperature sensor nodes and 2 vision sensors has been deployed in order to be able to test and evaluate the overall SCIER system's behavior and performance under real world conditions during the summer period, when the risk for fire events occurring is higher. All the simulated fire front evolution scenarios run in one of the sites of the national GRID infrastructure, where $P = 40$ CPU slots (*specint*2000=1367) have been reserved for SCIER simulations. The adopted middleware used was *glite* version 3.0 (Hellas Grid, 2008).

Figure 9. (a) Burned area (in hectares), under different wind speed and wind direction conditions (b) Number of runs completed at each time step (min) of the grid execution time.

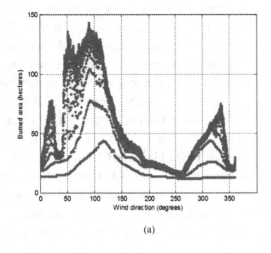

(a)

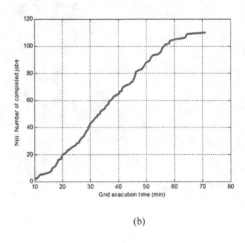

(b)

Simulation Setup

For the FSE component to work the area of Stamata has been organized as a geographical grid with 501x501 square "cells", each cell having a side of 6 m. The static input files (produced by field experts using GIS systems and data bases) provide to the FSE information about the ground morphology (slope, aspect), the fuel type and fuel moisture (dead and live) which dominates at each cell.

In order to simulate a fire spread evolution scenario, apart from the aforementioned static input files, the FSE needs information about the prevailing wind speed and direction conditions. Running the FSE with different wind speed and direction input parameters provides us with different fire front evolutions scenarios. In Figure 9(a) we observe for a given wind direction, the dependence of the burned area (in hectares) on the different wind speeds is ranging from 1 m/sec to 42 m/sec. Since the execution times of the FSE and TFM components depend on the burned area, which may vary greatly with the wind speed and direction for the same geographical area and ignition points, we cannot predict a priori the execution time of a simulation run.

For the simulation experiments whose results are reported here, the Perturbator component used as current wind speed and direction the values 10 m/sec, 270 degrees and as weather forecasted values 20 m/sec, 360 degrees respectively. To generate different wind speed and wind direction scenarios the Perturbator sampled a normal distribution with mean value the mid-range and standard deviation the half-difference between forecasted and current values. The number of deployed sensors, whose temperature vs. time curves have to be estimated by the TFM component for each scenario, was 20 matching the number of sensors which have actually been deployed in the Stamata test site.

Evaluation Method

Let $N(t)$ denote the number of parallel runs included in a simulation that have been completed in the Grid by time t (min) after the simulation's initiation. Let $T_s(t)$ be the aggregate serial execution time of all these runs if they were to be executed serially (the one after the other) in a single Grid worker node. Let $T_q(t)$ be the aggregate Grid overhead time spent collectively by all the $N(t)$ runs in scheduling queues and other non-execution Grid stages. Let $S(t)$ denote the *speedup* factor

Figure 10. (a) Solid and dotted curves show the aggregate serial computational time Ts(t) and the aggregate time in queue Tq(t) respectively. (b) Computational speedup using the Grid.

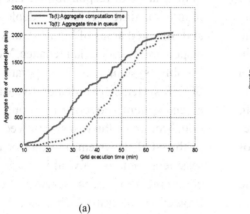

(a)

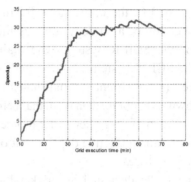

(b)

which has been realized using the Grid by time t. It can be computed as $S(t) = T_s(t) / t$. Finally let $E(t)$ denote the grid running *efficiency*, measuring deviation from the ideal linear speedup at time t and computed as $E(t) = S(t) / p$, where p is the number of slots allocated to the execution of the $N(t)$ parallel runs.

Experimental Results and Discussion

The experimental results presented next correspond to a simulation consisting of 110 runs (i.e. sumulating 110 different fire front evolution scenarios) executed using 40 reserved CPU slots provided by a single HellasGrid site (Hellas Grid, 2008). Figure 9(b) shows the number of completed $N(t)$ jobs as a function of parallel grid time t. We observe that it takes 71 min to complete all 110 jobs, i.e. the effective grid throughput is approximately ~1.54 jobs/min, as opposed to ~0.054 job/min in a single worker node machine.

Figure 10(a) shows two curves, the solid and dotted curves correspond to the aggregate computation time $T_s(t)$ and the aggregate scheduling and in-queue time $T_q(t)$ respectively, of all the runs completed by time t. As time progresses,

we observe that the aggregate time spent by all completed runs collectively in the queue increases (as expected) and tends to reach the aggregate serial time of the runs. Two factors contribute to the rate (slope) of $T_q(t)$, one is the rate of the number of completed jobs $N(t)$ (shown in Figure9(b)) and the other is the time the corresponding jobs spent waiting in the queue. For $t < 30$ min, the rate of $T_q(t)$ is lower as this part of the curve corresponds to the first bunch of jobs that are scheduled to CPU slots immediately and therefore spend minimum time in the queue. Also, as shown in Figure 9(b), the rate of $N(t)$ for $t > 60$ min decreases, which explains the decrease of the rate of $T_q(t)$, as at this phase the rate of $N(t)$ becomes the dominating factor. Figure 10(b) shows the grid realizable speedup as grid execution time increases. We observe that when the number of parallel runs exceeds the number of available CPU spots the speedup factor levels off, but is sustained at ~30 which is a very respectable steady state value considering that we are using $p=40$ CPU slots i.e. the maximum theoretical speedup is equal to 40. This indicates that for a large number of runs (scenarios) parallel computation can be very efficient, exceeding the 75% efficiency level.

EXPLOITATION

The definition of the spatial scale and the main protected entity will be the cornerstone for the design and execution of a strategy towards the exploitation of the SCIER system. The main objective of the SCIER system implemented in a protection cell (a settlement, a small housing area, a small town) is to obtain data from that cell, monitor the evolution of fire, offer better protection, and perform more effective deployment of forces in the affected areas.

The product to be designed as marketable should be based on what has been developed in the project, namely network of wireless sensors, prediction models, spatial data processing and communication technologies. It should follow a scalable and inter-operable approach, in this way the initial deployment of the SCIER components will serve as a system seed around which new nodes can be added to the initial protection cell. New cells can be aggregated to support co-operative monitoring among settlements meening that the sensors of remote cells can be used locally, thus extending the power and efficiency of all sensors deployed in the area. The design of the system should count on private homeowners or industrial installations, that may also request more specific and dense network of sensors for their own purposes, but that can be used within the network to protect other properties and infrastructures as well. For the definition of the concept, a protection cell will include also a buffer zone (e.g. 0.5 km) and, depending on the pattern and density of houses/vegetation, other buffer areas inside the settlement where sensors should be deployed. Experience shows that, in case of WUI detected fires, the protection should focus mainly on the first ring of houses that are normally more exposed to the incoming risk.

The SCIER system can be adopted by local authorities like municipalities and prefectures for the protection of varying scales of land against fires. Such authorities are strongly interested in detecting incidents at the earliest possible time and issue early warning alarms to the people living in their areas of responsibility. The IT personnel of municipalities/prefectures deploys the system at designated areas of high risk and assumes the responsibility for its maintenance. Moreover, such a public system may interoperate with private SCIER systems in a collaborative way in order to extend system's coverage and optimize its operation. Moreover, the vast sensor feed accumulated in the SCIER system and the local danger assessment could be of vital importance to central government emergency services, such as the Civil Protection Authority. Such authority should obtain accurate information on the development of hazardous phenomena and centrally orchestrate the risk management tasks (e.g., allocate fire fighting units, dispatch ambulances etc).

FUTURE WORK

As a future work, we propose the enhancement of the implemented fusion algorithms with alternative combination rules, e.g., (Yager. 1987) and the adoption of the Fuzzy Set theory to deal with uncertainty, imprecision and incompleteness of the underlying data. Further validation through trials with controlled flaming or smoldering fire should be conducted to quantify parameters such as thresholds, false alarm rates and fusion weights. As mentioned in previous sections, 40 CPU slots have been reserved for SCIER purposes in the GRID infrastructure, therefore the waiting time of a simulation job is more predictable. However, considering that SCIER jobs are triggered rarely but unexpectedly by non-periodic events and that sustained CPU reservations may not be practical in large-scale deployments, we are investigating novel grid resources scheduling models that could provide a viable alternative. In addition to the notion of grid CPU slots, which typically have a one to one correspondence with the operating system CPU slots of the worker nodes, this model introduces to grid middleware the concept of *virtual grid CPU slots*. Virtual grid CPU slots are made available only to specific job types with soft real time requirements and short deadlines. According to this model SCIER jobs will be neither delayed due to other jobs nor queued while waiting for a grid CPU slots to be released.

CONCLUSION

In this chapter we have presented the SCIER system which focuses on the detection and monitoring of environmental risks. In its current form, the SCIER system/architecture can deal with forest fires and flash floods, with emphasis on the protection of the WUI areas. Throughout the chapter, we elaborated on the fire hazard case. Detection of fires in SCIER is performed through a multi-sensor infrastructure integrating wind-speed, wind-direction, temperature, humidity and smart vision sensors. To cope with all these different types of sensors and deliver alarms with increased accuracy and confidence a layered fusion scheme has been adopted. Different sensor feeds are processed in the two layers of the fusion scheme. On the lower layer, the statistical behavior of sensor data is constantly assessed. On the higher layer, D-S theory of evidence is adopted in order to mix the indications coming from the lower layer and the out-of-field vision sensors. We provide examples to clearly illustrate the adopted scheme. Apart from the SCIER fusion model, we also discuss the advanced simulation architecture developed in the context of the project. Specifically, we have developed an end-to-end grid workflow which involves the parallel execution of many alternative fire evolution scenarios. The workflow has been evaluated in the area of Stamata, Greece. The analysis of our experimental results demonstrates the potential speedup that can be delivered by the grid infrastructure to the application. Parallel simulations allow the investigation of the effects of perturbations in critical environmental variables and the joint assessment of probabilistic scenarios to derive meaningful predictions and corresponding crisis management plans before the disaster strikes.

REFERENCES

Andrews, P. L. (1986). *BEHAVE: fire behavior prediction and fuel modeling system- BURN subsystem. Part 1. GTR-INT-194.* USDA Forest Service.

Bianchini, G., Cortés, A., Margalef, T., & Luque, E. (2005). S2F2M - Statistical System for Forest Fire Management, Atlanta, USA. In *Proceedings of the Int. Conf. on Computational Science,* ICCS Book of Abstracts.

Calle, A., Casanova, J. L., & Romo, A. (2006). Fire Detection and Monitoring using MSG Spinning Enhanced Visible and Infrared Imager (SEVIRI) Data. *Journal of Geophysical Research, 111,* G04S06. doi:10.1029/2005JG000116

Chen, S., Bao, H., Zeng, X., & Yang, Y. (2003). A Fire Detecting Method based on multi-sensor Data Fusion. In IEEE Systems Man and Cybernetics, 4.

Deliverable, E. U. F. I. R. E. L. A. B. D-03-06. (n.d.). Retrieved September 9, 2008, from http://www.eufirelab.org

Doolin, D. M., & Sitar, N. (2005). *5765.* Wireless Sensors for Wildfire Monitoring. In Smart Structures and Materials.

Gombay, E., & Serban, D. (2005). An adaptation of Pages CUSUM test for change detection. *Periodica Mathematica Hungarica, 50,* 135–147. doi:10.1007/s10998-005-0007-7

Hefeeda, M., & Bagheri, M. (2007). Wireless Sensor Networks for Early Detection of Forest Fires. Pisa, Italy. In *Proceedings of the 1st International Workshop on Mobile Ad hoc and Sensor Systems for Global and Homeland Security (MASS-GHS),* Pisa, Italy.

Hellas Grid. (n.d.). Retrieved September 10, 2008, from www.hellasgrid.gr.

Kosucu, B., Irgan, K., Kucuk, G., & Baydere, S. (2009). FireSenseTB: A Wireless Sensor Networks TestBed for Forest Fire Detection. In the *5th ACM International Wireless Communications and Mobile Computing Conference (IWCMC'09), Leipzig, Germany*.

Kucuk, G., Kosucu, B., Yavas, A., & Baydere, S. (2008). FireSense: Forest Fire Prediction and Detection System using Wireless Sensor Networks. In the *4th IEEE/ACM International Conference on Distributed Computing in Sensor Systems (DCOSS'08), Santorini Island, Greece*.

Manatakis, D. V., Manolakos, E. S., Xanthopoulos, G., Roussos, A., & Viegas, D. (2010). In-silico estimation of the temperature field induced by a moving fire – Predictive modeling and validation using prescribed burn data. In the *VI International Conf. on Forest Fire Research*, Coimbra, Portugal.

Mandel, J., Beezley, J., Bennethum, L., Chakraborty, S., Coen, J., & Douglas, C. L, Hatcher, J., Kim, M., & Vodacek, A. (2007). A Dynamic Data Driven Wildland Fire Model. Computational Science - ICCS 2007 (pp. 1042-1049).

Manolakos, E. S., Manatakis, D., & Xanthopoulos, G. (2008). Temperature Field Modeling And Simulation of Wireless Sensor Network Behavior During a Spreading Wildfire, Lausanne – Switzerland. In *Proceedings of the European Signal Processing Conference*, EUSIPCO.

OSGi Alliance. (2009). http://www.osgi.org.

Page, E. S. (1954). Continuous Inspection Schemes. *Biometrika, 41*, 100–115.

Rose-Pehrsson, S. L., & Hart, S, J., Street, T. T., Williams, F. W., Hammond, M. H., Gottuk, D. T., Wright, M. T., & Wong, J. T. (2003). Early warning fire detection system using a probabilistic neural network. *Fire Technology, 39*(2), 147–171. doi:10.1023/A:1024260130050

Rose-Pehrsson, S. L., Shaffer, R. E., Hart, S. J., Williams, F. W., Gottuk, D. T., Strehlen, B. D., & Hill, S. A. (2000). Multi-criteria Fire Detection Systems Using a Probabilistic Neural network. *Sensors and Actuators. B, Chemical, 69*(3), 325–335. doi:10.1016/S0925-4005(00)00481-0

Shafer, G. (1976). *A mathematical theory of evidence*. Princeton, NJ: Princeton Univ. Press.

Sivathanu, C., & Tseng, L. K. (1996). *Fire Detection Using Near-IR Radiation and Source Temperature Discrimination*. National Institute of Standards and Technology. Annual Conference on Fire Research.

Stewart, S. I., Radeloff, V. C., Hammer, R. B., & Hawbaker, T. J. (2007). Defining the Wildland Urban Interface. *Journal of Forestry, 105*, 201–207.

Yager, R. R. (1987). On the Dempster-Shafer Framework and New Combination Rules. *Information Sciences, 41*, 93–137. doi:10.1016/0020-0255(87)90007-7

Zervas, E., Mpimpoudis, A., Anagnostopoulos, C., Sekkas, O., & Hadjiefthymiades, S. (2009). *Multisensor Data Fusion for Fire Detection and Monitoring*. International Journal on Multi-Sensor, Multi-Source Information Fusion.

Zervas, E., Sekkas, O., Hadjiefthymiades, S., & Anagnostopoulos, C. (2007). Fire Detection in the Urban Rural Interface through Fusion techniques, Pisa, Italy. In *Proceedings of the 1st International Workshop on Mobile Ad hoc and Sensor Systems for Global and Homeland Security (MASS-GHS)*, Pisa, Italy.

Zhanqing, L., Khananian, A., Fraser, R., & Cihlar, J. (2001). *Automatic Detection of Fire Smoke using Artificial Neural Networks and Threshold Approaches applied to AVHRR Imagery*. IEEE Trans. On Geoscience and Remote Sensing.

Chapter 17
A Personalized Forest Fire Evacuation Data Grid Push Service:
The FFED–GPS Approach

Eleana Asimakopoulou
University of Bedfordshire, UK

Nik Bessis
University of Bedfordshire, UK

Ravikanth Varaganti
University of Bedfordshire, UK

Peter Norrington
University of Bedfordshire, UK

ABSTRACT

Much work is under way in disaster reduction and emergency management towards the utilization of information and communication technologies (ICT) and the design of relevant services associated with risk management towards sustainable development and livelihood. Recent forest fires occurred in Southern Europe, caused environmental destruction and a number of fatalities. The effective and efficient production of forest fire evacuation plans requires decisions based on integrated data from heterogeneous and distributed sources that change over time very quickly. Recent ICT advances suggest the need for further work in the advanced evacuation systems area. We are particularly interested of how to automatically inform potential victims about the most relevant evacuation routes in the most-timely fashion so they can escape a forest fire safely. With this in mind, this chapter describes the concepts, architecture and implementation of the Personalized Forest Fire Evacuation Data Grid Push Service using data push and next generation grid technologies.

DOI: 10.4018/978-1-61520-987-3.ch017

Figure 1. Average Burnt Area in 2000-2006 and Total Burnt Area in 2007

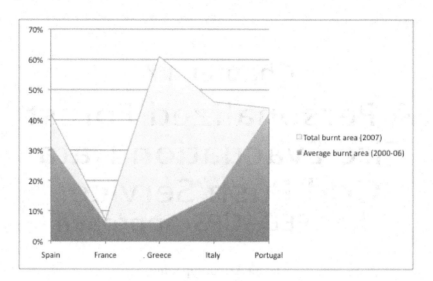

INTRODUCTION

Recent forest fires in Southern Europe caused serious environmental destruction and a number of fatalities. In such cases, emergency response is very important and the effective and efficient management of response operations requires a number of information and communication technologies (ICT) and relevant collaborative computer-based systems to assist and bring together the intellectual and physical resources of different authorities and the sharing of accurate information in a timely manner. However, several obstacles arise in the design and implementation of such services (Bessis and Asimakopoulou, 2009).

In their report for the European Parliament, Bassi and Kettunen (2008) discussed the significant impact of forest fires in the environment, economy and society, as well as that the most affected state between 2000 and 2006 was Portugal, followed by Spain, Italy, Greece and France. In 2007, the situation was completely reversed. Greece recorded a maximum percentage of burnt area followed by Italy. The situation re-occurred in August 2009 with many fires in Greece, France and Portugal. At the time of writing there was

little data available to present relevant to the 2009 destructions. Figure 1 shows the average burnt area between 2000 and 2006 and the total burnt area in 2007.

In previous works, we (Bessis and Asimakopoulou, 2008; 2009 and Bessis et al., 2009a; 2009b) have proposed the development of a grid technologies based service architecture encompassing a number of platform independent features such as support for real time data, resource access and integration from heterogeneous and distributed sources, flexible policies, expert input, assessment processes and simulation tools. To address these issues, this chapter describes the conceptualization, development and implementation of a personalized 'forest fire evacuation data-based grid push service' (FFED-GPS) as a means to automate the information integration, access, assessment and matchmaking processes. This in turn, will be used for the production of effective personalized evacuation routes in a timely manner, which can be made available to potential victims and thus, to timely receive and follow these personalized evacuation instructions.

With this in mind, the contributions of the chapter are to: (1) present a brief background review including a case scenario for illustrating service requirements and as a way to discuss possible ways for improving the current situation, as well as offer a brief discussion of how the FFPD-GDS will fit with other relevant manifestations, as well as present any assumptions towards our service development; (2) offer an overview of relevant technologies, as well as describe the data pull and push matchmaking processes; (3) discuss the development approach and theoretical underpinnings including the provision of concept flow diagrams, methods and proposed service architecture; (4) present scenarios to describe our algorithms and thus, detail our implementation method in the form of a client application that is accessible via a web browser and a mobile device. We finally conclude with some experimental results and recommend future steps.

BRIEF BACKGROUND REVIEW

Natural phenomena are normal, unavoidable and necessary planetary actions, however they may cause instability to the human societies, often with serious losses in lives and environmental resources. Disaster management is the discipline responsible to mitigate, prepare for, response to and recover from disasters with the ultimate goal to save lives, property and the environment. In managing disasters and in particular during the response phase, it is apparent that a number of teams and individuals from multiple, geographically distributed organizations are required to communicate, cooperate and collaborate in order to take appropriate decisions and actions (Graves, 2004; Otten et al., 2004). In particular, disaster managers and other experts are in close cooperation towards the design and implementation of services associated with risk management, early warning systems, disaster response and recovery to achieve sustainable development and livelihood.

Amongst other natural disasters forest fires is an extreme phenomenon with rapid increase lately and it usually causes extensive damage to the local population, property, and local and global environment. A forest fire can be caused either by natural causes, such as hot days, or lightning, or by careless people and accidents.

The European Union (EU) considers forest fires as a frequent event in the Mediterranean region, particularly during the summer season. Forest fires destroy large-scale areas of trees and other types of vegetation.

During 2007 various forest fires occurred across the Mediterranean region. According to European Civil Protection, 'in 2007 Greece experienced the worst year on record for forest fires. Extremely hot and dry weather conditions in Greece, combined with strong winds led to a disastrous upsurge of forest fires and wildfires. That year Greece requested assistance four times through the Monitoring Information Centre of the European Schedule Commission during the months of June, July and August. The total burnt area in 2007 amounted to 268,834 hectares, of which 180,000 burnt between the 24 and 30 August 2007'. In 2009, the total burnt area in Greece from forest fires between 22 and 25 August was more than 150,000 hectares. Fortunately, during the 2009 events in Greece there were no people killed (despite the ecological destruction and the many houses destroyed) but during the 2007 events 5,392 people were affected and 67 of them killed – some of them fire fighters – while 'the damage has been calculated at 1,750,000 US$' (EM-DAT, 2008).

In addition to the large number of fires all over the country that divided emergency management units into smaller groups, other conditions making their work difficult included the geological characteristics of the area, which consists of numerous mountains and valleys, the high temperature, which did not allow relevant aircrafts to operate continuously and the chaotic situation caused by overstressed residents of the affected areas, who could not logically follow the emergency services' instructions. Further to these, TV channels, which in such situations are a source of information for the locals and a visual source of information for relevant authorities, could not approach all areas closely and further to this the flow of news limited the timely arrival of information for a particular area. In turn, this resulted in people trapped by the fire. Some of them were killed, while for some others emergency services lost vital time to save them. The EU has experienced similar situations in Portugal (2005), and Italy and Croatia (2008).

Apart from the fact that people require regular training and consequently respond maturely to the orders provided by relevant emergency co-ordination units and their rescue services, most fatalities occurred because they had been surrounded by the raging fire. On the other hand, forest fires will undoubtedly occur if drought and consecutive heat waves during summer periods continue. These circumstances in turn suggest the need for work in the area of advanced warning systems during the response phase of a forest fire in progress. In turn, many scholars in the area have focused on developing a number of technological platforms to manage natural hazards including forest fires, as well as to develop systems for a safer European citizen. For example, SCIER (Sensor and Computing Infrastructure for Environmental Risks)–a 2006 IST project involving the Czech Republic, France, Greece, Spain, Switzerland, Portugal and the UK – aims to develop an integrated system consisting of sensors, networking and computing infrastructure, in order to detect, monitor, predict and assist in crisis management of natural hazards or accidents in areas where forests and rural lands interface with homes, other buildings and infrastructures. The robust approach taken in the SCIER project is very similar to the overall approach taken by others in the field.

Bessis and Asimakopoulou (2008; 2009) extended these approaches to include the importance of how to automatically inform potential victims of the most relevant evacuation routes in the timeliest fashion so they can escape safely from a forest fire in progress. Recent evidence shows that in many

instances where people have been trapped by the fire, some of them were killed instantly; while for others, emergency services lost vital time to save them. It is therefore crucial in such cases, for both emergency services and the public to know the most effective evacuation route. If this were known, emergency services could redirect people and safely evacuate the area and then solidly focus on the fighting of the fire alone. On this basis, our chapter introduces an advanced warning concept, which involves the simulation of a forest fire in progress, assessment of the hazard it poses, and transmission of a warning which broadcasts available evacuation routes – to nearby inhabitants – ahead of any significant danger using advanced ICT. In brief, our aim and contribution here is to extend and complement others' attempts, and thus produce a personalized evacuation method serving the overarching EU initiative for a safer European citizen.

Case Scenario: Improving Current Practice

Southern Europe is an area with high vulnerability to forest fires due to its morphology and to the weather conditions it experiences, especially during the summer periods. The area is characterized by a rich morphology, which includes many mountains, valleys, forests, hills, and rivers, temperatures that can reach 45-47oC, and very strong winds that can last for several days during the summer.

Let's assume that a number of forest fires occurred in a particular region like the one described above, which accommodates a number of cities, towns and villages. The fires have caused disruptions to some road network connections and many people have been trapped in different areas. We must emphasize that during the summer period there are also holidaymakers who are not necessarily familiar with alternative road network connections. In this scenario, the disaster management stakeholders are responsible for working col-

Figure 2. Current emergency response decision making process

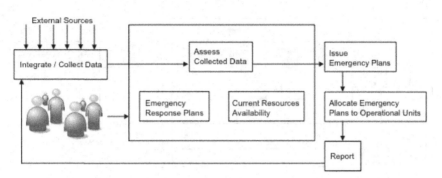

laboratively in order to collect information about the affected area and the phenomenon, to assess it and to decide and issue emergency response plans in order to control the fire, as well as to safely evacuate the area and to protect the infrastructure of the region and the local environment.

Collection of such information is a continuous process and must take place via the integration of real-time alerting systems and other distributed and heterogeneous sources such as scientific instruments, geographical information systems, weather stations and satellites, mobile technologies carried by operational units located in the affected area, TV channels and finally calls made by victims requiring assistance. Disaster managers, and other experts collect and assess this data and by taking into consideration existing emergency plans, policies and availability and status of physical emergency resources, make decisions and form an emergency plan, which is issued and allocated to the relevant operational units. This is shown in figure 2.

In our previous works, we have suggested the use of current digital and wireless technology that is embedded in cell telephony and capable of reading, identifying and clustering groups of victims based on their exact positioning (cell area 1, cell area 2, etc) and thus, pushing this information to the relevant parties. Our assumption here is that other scholars' works, including findings from the SCIER project, suggest that sensors can be placed across the fires and real time satellite images of

the area can be received by disaster managers as tools to continuously read and monitor current fire activity. Data received could be analyzed in real-time and fed into a number of collaborative decision support systems encompassing a simulation tool, which could forecast projected fire path based on parameters like surround materials (sensitivity factor), wind speed, temperature, moisture, oxygen levels and other atmospheric readings. Another simulation tool could take the projected fire path simulation results and combine them with geographical maps towards the identification of possible local evacuation routes. The expectation is that this information would be critical for disaster specialists in order to make informed decisions and timely broadcast and push a number of currently suitable projected evacuation plans that are directly relevant to specific groups of victims based on their identical position across the region.

We therefore suggest that current cell and wireless technology that is capable of receiving and transmitting signals from/to dispersed locations could be used to read, broadcast and stream different digital signals encompassing different images (evacuations plans), which could be pushed to relevant groups based on their exact positioning. The method will clearly serve as a victim-focused timely source of information alleviating concerns with regard to victims receiving a generalized flow of images and news that are best limited to presently unrelated evacuation plans. Figure 3

Figure 3. Improved emergency response decision making process

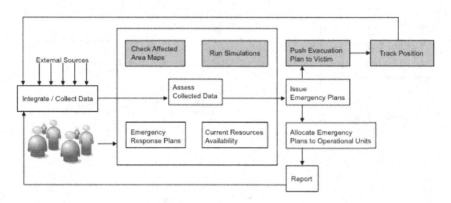

illustrates our proposed data flow encompassing additional parts – of critical importance – to the existing processes shown in fig. 2 as an improvement to the current practice. In this chapter, we take as a given that there is an effective and efficient platform such as the one researched from the SCIER project, enabling us to built upon it and, as suggested in our previous works, combine their projected forest fire simulation results with geographical maps and potential victims exact positioning towards the identification of possible local evacuation routes.

OVERVIEW OF RELATED TECHNOLOGIES

Recent advances in networking and digital resource integration resulted in the concept of grid technology. The most important standard that has emerged recently within the grid technology community is the OGSA-DAIS (Open Grid Services Architecture – Data, Access and Integration Services), which as a data integration specification addresses the ability to allow users to specify 'what' information is required without having to provide detailed instructions on 'how' or 'from where' to obtain the information. Further developments in the area have resulted (Bessis, 2009) in producing services to automatically

'keep' users 'informed' of latest, relevant, specific changes about data that are registered within the grid application in which users have access to. In terms of standards, grid environments share the same protocols with web services (XML, WSDL, SOAP, UDDI) and their main difference is that web services typically provide stateless, persistent services whereas a grid provides state-full, transient instances of objects. Recent developments include the development of lightweight semantic service descriptions (LWSSD) such as those produced from the recently emerged WSMO, SAWSDL or OWL-S specifications.

In general, there are two models, namely the Pull and Push models, for a client to retrieve data from a data source. In the Pull model, a user always initiates a search by specifying search parameters. The Push approach as originally discussed in 2003 (Bessis, 2009) suggests that every time a data provider commits a new data entry in the data source, a trigger will cause an automatic search between the records stored in a subscriber's specified search parameters to identify possible, relevant matches. If there are any retrieved data, these are pushed automatically to the subscriber. In this model, users as subscribers are required to subscribe once and are not required to initiate the search function every time new data are committed within a data source. In brief, user-tailored Push technology enables a decision maker to specify

Figure 4. FFED-GPS conceptual model

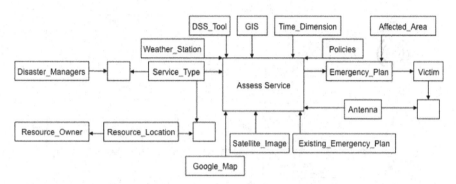

a set of parameters into a subscriber form and as those conditions match specific parts of the newly updated and/or created content, such information are 'pushed' to the subscribed decision maker. More information is available from Bessis (2009).

Overall, Push can streamline the delivery of user-tailored, specific information to a variety of users via the Internet or a web browser. There are various technologies including Web 2.0, AJAX, WS-Notification and/or WS-Eventing, supporting the development of pure push or push-like mechanisms. It is therefore the viewpoint here that user-tailored Push technology addresses data consistency, which is an important property as it determines how 'relevant', 'accurate' and 'fresh' the data is. Updates within a distributed environment are more frequent compared to a centralized one and therefore, there is a need for updates occurring at the back-end to be migrated to other sites in the network so that all the copies of relevant and up-to-date data are synchronized to maintain consistency as well as preserving personalization of individuals across the stakeholder community.

Its applicability within this particular scenario rests on the aspect of grid technology being capable in providing disaster management teams with an infrastructure allowing seamless and flexible collaborative remote access to various resources (including data, computational power, software, satellites or any other resource instrumentation, which could virtually connected to/from a per-

sonal computer) that are stored in (and/or managed via) multiple autonomous, distributed and heterogeneous computers. That is to say, disaster management teams are now capable of remotely assessing and managing disastrous situations through the utilization of available collaborative tools in a highly effective and efficient manner. In addition to this, our view here is that keeping all interested parties, including disaster managers and potential victims informed of fresh and relevant changes that are directly relevant to their interest will contribute to an efficient personalized and accurate forest fire evacuation service.

FFED-GPS ARCHITECTURE

Figure 4 illustrates the back-end of our forest fire evacuation data grid push service (FFED-GPS) in the form of a conceptual model. It illustrates some of the entities shown in Figure 3 that are required during an emergency response decision-making process. Our conceptual model is centred on six core entities, namely Disaster_Manager, Service_Type, Assess_Service, Antenna, Evacuation_Plan and Victim. We appreciate that there are other entities, which are not shown and others, which are included in our model, however these are considered of less importance when describing our concept, although they are of critical importance in terms of the overall model's functionality. There

Figure 5. FFED-GPS process flow architecture

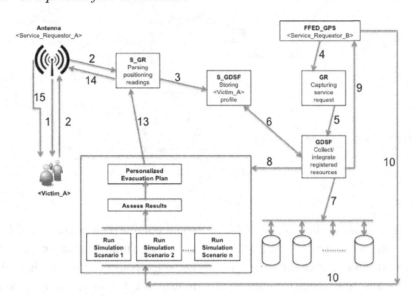

are also a number of weak entities (shown as blank boxes) representing the resolution between entities having a many-to-many relationship. Arrowheads denote many relationships whilst optionalities and relationship names are not denoted on the diagram in order to reduce complexity.

Let's assume that a Disaster_Manager invokes one or many Service_Type(s), which as an entity provides disaster managers with access to various dispersed resources via the Assess_Service(s) entity. Such resources include but are not limited to heterogeneous data and tools collected from a number of entities like the Weather_Station, GIS (Geographical Information System), Satellite_Image(s), Google_Map(s) and Time_Dimension.

The Assess_Service entity is considered to be the heart of the system as it will allow a disaster manager to forecast the projected fire path based on parameters such as surround material (sensitivity factor), wind speed, temperature, moisture, oxygen levels and other atmospheric readings. The entity recalls the action of 'start collecting and integrating' a fresh image of the aforementioned dispersed data. It is assumed that it also invokes the same action when an alerting situation occurs. It also provides access to the Antenna entity, which

continuously takes readings about the victims' exact positioning via their registered mobile phones (cell area 1, cell area 2, etc). At this stage it is important to emphasize that the proposed system is a support mechanism, which does not aim to replace existing practices. The Assess_Service(s) entity also provides access to available Existing_Emergency_Plan(s) and various decision support systems, data mining and simulation tools (DSS_Tools), which are important to the disaster managers' decision-making process. This will also assess the projected fire path and combine it with geographical maps (from satellite images, GIS and/or Google Maps) towards the identification of possible evacuation routes. It is anticipated that these will be informed with readings taken from the Antenna entity, which as mentioned earlier will provide exact positioning readings of victims. On this basis, disaster managers can produce relevant and personalized evacuation plans, which can then be issued or pushed to relevant parties such as operational units or victims.

Our FFED-GPS assumes that a victim affected by a forest fire in progress is the owner of a wireless device such as a mobile phone. Figure 5 illustrates the FFED-GPS service architecture in the form of a process flow chart. This layer demonstrates service functionality and offers the flexibility to a disaster manager to manually or to automatically control, access, integrate and monitor the flow and communications between dispersed data, which are made available via various entities.

Let's assume that <Victim_A> requests the identification of a safe evacuation plan. The request is saved as a <victim_profile> in the FFED_GPS's grid register, the <GR> and moves to the <GDSF>, the grid data service factory, which takes care of collecting and integrating data such as exact positioning readings and tools from dispersed and heterogeneous resources such as the Antenna, a GIS, Google Maps, DSS tools, etc, which must also register with <GDSF> prior to their availability. Once matches are found, the <GDSF> will make results available through the <Sink_GDS>, a grid data service encompassing all aggregated results found from the available grid data services. The FFED-GPS is now able to run simulations and assess various evacuation paths for producing a personalized to <Victim_A> evacuation plan. Once the plan meets the requirements, this is made available at the <S_GDSF>, the sink, which will push evacuation details automatically to the registered and subscribed <Victim_A>. These are shown as Actions 1-15 in Figure 5, including all aforementioned steps.

Action 1: The Antenna as a "Service Requestor A" establishes connection with <Victim_A>.

Action 2: The Antenna as a "Service Subscriber A" starts capturing <Victim_A>'s exact positioning readings and passes them to the <S_GR>.

Action 3: <S_GR> sends a request to the Subscriber Grid Data Services Factory <S_GDSF> to create the <Victim_A> profile.

Action 4: The FFED_GPS as a "Service Requestor B" requests to register their query to access services via the <GR>.

Action 5: The <GR> requests <GDSF> starting collecting/integrating resources registered to it.

Action 6: The <S_GDSF> sends the request of collecting and matchmaking process to the <GDSF>.

Action 7: The <GDSF> sends requests to registered resources.

Action 8: The <GDSF> returns results to the <Sink_GDS>.

Action 9: The <GDSF> informs the FFED_GPS to access the relevant results.

Action 10: The FFED-GPS runs 'what-if' simulation scenarios via the <Sink_GDS>.

Action 11: The FFED-GPS assesses simulation results via the <Sink_GDS>.

Action 12: The FFED-GPS produces a personalized evacuation plan via the services available from the <Sink_GDS>.

Action 13: The <Sink_GDS> pushes the personalized evacuation plan to the <S_GDSF>.

Action 14: The <S_GDSF> pushes the personalized evacuation plan to <Victim_A>.

Action 15: The Antenna maintains connection with <Victim_A>'s wireless device for monitoring and tracking the victim's positioning and movement (loops actions 1-3, once <Victim_A> moves to a safe place and according to the personalized evacuation plan, S_GDSF deletes the <Victim_A> profile).

FFED-GPS DEVELOPMENT APPROACH

To develop the FFED-GPS, we logically divided the forest area into four equal areas known as quadrants. Each quadrant is sub-divided into cells. The victim's mobile device sends the current location to the server via the Cell Telephony Antenna, which in turn runs a geo-code to determine and capture the latitude and longitude of the victim's current position.

Figure 6 shows the four logical quadrants used to define the forest. We assume that in each quadrant there is a safe point (S) in the far end point, which a victim has to reach. There are also two end diagonal points (E), which are used to

Figure 6. A forest area divided into quadrants and cells

define the far end safe point. Finally, we also use geo-coding for positioning the exact latitude and longitude of the fire in the relevant quadrant: cell. Our development revolves around the concept of the Cartesian coordinate system with a point P defined by the (x, y) coordinates, which are comparable to the latitude (x) and longitude (y) of a point P. Our rationale involves a program which starts and executes a load() method, which in return calls two methods namely the placeEndPoints() and drawGraph() methods. After executing both methods, the flow comes to an end.

First of all, victims are placed in the graph (the forest) randomly at several positions. There can be any number of victims. Then, fire is placed randomly at some point imitating as if fire started at that point. Before a fire is placed, it has to be set with a direction parameter that indicates the direction in which the fire would progress. Then, we start simulating the model. Each time we run the simulation, a rescue() method is called, which in turn calls the methods of setFire(), getFire() and assignDirections(). The rescue() method allows us to identify the positioning of a victim. For example, if a victim lies in the first quadrant, the rescueXPYP() is called. Depending on the invoked

rescue() method, a relevant drawGraph() method is used, which invokes Cartesian coordinates for positioning victim and fire movement along the relevant quadrant. Next, we produced a getFire() method, which checks and calls methods based on the wind direction. For example, if the wind direction is east, then it calls the getFireEast() method, which in turn calls the getFireEastIntensity() method; if the wind direction is west, it calls the getFireWest() method, which in turn calls the getFireWestIntensity() method and so on. The intensity methods simulate the victims feeling the heat intensity of the fire progressing in that particular direction when the victim is nearer to fire. Similarly, if the wind direction is southeast, relevant methods are invoked in a similar fashion.

The starting function of the complete program is the load() method. This is called once the page is loaded. This is specified in the <body> tag with an onLoad option. In this method, everything is initialized related to the map object (using the GMap2 class). Once the map object is created, markers are created and are placed on the map. The center of the forest area is selected and its (x, y) values are passed to GLatLng class to create and store the value of the point P. The center of the map is set to P(x, y) using setCenter() method. A marker object is created by the class GMarker and placed at the point specified as the center of the map by using the addOverlay() method. Other methods such as placeEndPoints() and drawGraph() are called inside this method.

Events are registered for the map and the markers that have been placed on the map. Thus, when we click on the map, the Click event responds and places a particular type of marker. This marker can be a fire marker or a victim marker. If a fire marker is placed, then the createFireMarkers() method is called passing the starting point of the fire. If a victim marker is placed, then the createVictimMarkers() method is called passing the current location of the victim.

The rescue() method is called each time a "Simulate" or "Rescue Me" button on the web page is pressed. An array named victim_start_points[] is initialized with no values. In this method, wind direction values are assigned to the wind direction variable by calling the assignDirections() method. The points of the fire markers are placed in the array fire_points[]. Now, using a FOR loop, every point in the array fire_points[] is passed to the getFireIntensity() method to get the intensity of the fire in a specified direction (east, west, etc). When used with fire progressing in non-cardinal compass directions (for example, southeast), the methods of setFire() and getFire() are called instead of method getFireIntensity().

In our simulation testbed, the position of the victim changes constantly. Hence, two arrays named victim_start_points[] and victim_end_points[] are used. The array victim_start_points[] is used to store the current position of the victim and the array victim_end_points[] is used to store the next position of the victim. Each time this method is called, both arrays are initialized with no values i.e. it is made empty. A FOR loop is used to copy the elements of the array victim_end_points[] into the array victim_start_points[]. Once the elements are copied, victim_end_points[] is initialized to hold no values i.e. an empty array again. Now, the points in the array victim_start_points[] are checked to see in which quadrant the points fall. In this particular scenario, a victim in a particular quadrant tries to escape using an execution plan to reach a safe point in that particular quadrant. Hence, each time the system is simulated, new points are copied into the victim_start_points[] from the victim_end_points[]. Each time this method is called, the new points are checked with the quadrant in which they match. In a real life scenario, a forest area would be very large and thus, a victim cannot run to escape through a safe point of another quadrant. Based on the quadrant it matches, the corresponding quadrant rescue method is invoked.

In developing this prototype, several assumptions are currently built in, namely:

1. A victim is made to move diagonally towards the safe point of that particular quadrant. Here, the concept of displacement is taken into consideration so as to evacuate the victim as soon as possible over shortest distance.

2. When a victim is near to the XY boundary and his next position is out of X boundary or Y boundary, then he is made to move along the boundary so as to reach the safe point. The victim might automatically become safe when he is out of the boundary, if the fire does not continue in the same direction, but the plan would be to create a rescue team that would save the victims nearer to boundary or at a place surrounded by fire. The rescue teams (i.e. Rescue_Team entities) would be created in a later development phase.

3. The speed of both victim and fire is selected randomly and made constant for any type of victim throughout the execution.

4. The direction of the fire is constant throughout the simulation. Thus, the code presentation used here employs readable, pseudocode-style method names (etc.) containing compass point designations.

5. Neither fires nor victims can cross the quadrant boundaries. These boundaries function in this prototype as natural or man-made barriers, such as rocky areas, chasms, rivers or firebreaks. In a more advanced implementation these would be accounted for wherever as features within the map, capable of blocking movement of fires or people. Clearly, for people without specialist knowledge or equipment, these features represent areas of risk of entrapment and must be avoided in the absence of ways of passing through them with less risk than of meeting the fire.

6. Fire started in one quadrant has no effect on the evacuation of victims of another quadrant. The complete development of the prototype is concentrated only on the evacuation plan produced by the disaster

managers and forwarded to victim's mobile phone assuming that other rescue forces would be available afterwards as this process is running. It means that, before other rescue forces try to save victims, this service would save almost all people.

The Algorithm for Quadrant 1

For a given victim point in this quadrant, eight end points are calculated. This is done by calculating the last point obtained in generating points in both x and y axes in both positive and negative directions. The last point obtained in generating points in positive x direction is named as vsafe1. The last point obtained in generating points in negative x direction is named as vsafe2 and so on. Now, the other four end points are calculated by taking the (x, y) values of the four points obtained. The top right corner point of the victim bounds is obtained by taking latitude of the point vsafe3 and longitude of the point vsafe1 and is named as vsafe5. The top left corner point of the victim bounds is obtained by taking latitude of the point vsafe3 and longitude of the point vsafe2 and is named as vsafe6 and so on. When the simulation starts, for each and every step, the value of the point vsafe5 is compared with the boundary of positive X and positive Y plane.

If there is no obstruction in the safe path, then the victim is made to move in diagonal direction by adding vsafe5 to the array victim_end_points[]. If vsafe5 lies outside Y plane, then the next point would be vsafe3 and the victim is made to move north along Y plane by adding vsafe3 to the array victim_end_points[]. Similarly, If vsafe5 lies outside X plane, then the next point would be vsafe1 and the victim is made to move east along X plane by adding vsafe1 to the array victim_end_points[].

Single Direction Fires

Let us consider that the direction of wind is south. This makes the fire spread southwards. Now, again there are two cases. In the first one, the victim is marked to the right of the fire marker point. In this case there is no need to worry, as the fire will spread south, and the victims can escape easily to the right side. In the second case, the victim is marked to the left of the fire marker point. Now, as the fire is moving in southwards, the victim is supposed to escape to the north. This is achieved as follows. Each time the rescue() method is called, it checks whether the victim's next step (point) lies in the first quadrant, and if it lies, it calls the rescueXPYP() method. Here, the victim bounds are calculated for each new step of the victim. As it is the first quadrant, the default next step would be the point vsafe5 obtained from the previous value. At each simulation step, it is checked whether the point vsafe5 lies in the bounds map_xpyp_bounds or not. The map_xpyp_bounds is a variable that stores the boundary information of the first quadrant. If the point lies in the bounds map_xpyp_bounds, then the intersection of victim_bounds and fire_bounds_xpyp will be checked; if not, the next value will be the point vsafe3 otherwise, the next value will be vsafe6. Here, vsafe6 is chosen as the victim safe point as the direction of the fire is southwards and the victim has to move north. Regardless whether it is vsafe3 or vsafe6, there is a chance that the victim point may cross the top boundary. So, if this is the case, the new point would change to vsafe1 making the victim move eastward. The next method is getFireSouth(). It is used to populate the arrays storing starting values of fire with end points and initializing the arrays with end points as empty arrays. Fire markers placed in first quadrant with south direction are stored in the arrays fire_points_xpyp_south_start[] and fire_points_xpyp_south_end[]. Fire markers placed in second quadrant with south direction are stored in the arrays fire_points_xnyp_south_start[] and fire_points_xnyp_south_end[] and so on. If there are fire markers in the first quadrant with south direction then, using a FOR loop, the contents of the array fire_points_xpyp_south_end[] are copied into the array fire_points_xpyp_south_start[] and the array fire_points_xpyp_south_end[] is initialized with no elements i.e. an empty array. Now, the method getFireIntensitySouth() is called by passing the elements of array fire_points_xpyp_south_start[] for its length number of times in a FOR loop.

Variable Direction Fires

Winds may flow in different directions in different parts of an area. The direction of wind flow will drastically affect the direction of progression of the fire. Our rationale of how a victim will safely evacuate the area follows next.

First of all, we have to select a wind direction (south, west, etc) before we place a fire marker in the map. When the simulation starts, it is expected that each fire continues to move in its respective direction in steps. As described earlier, the victim's bounds are also calculated. The direction of the victim depends on the direction of the fire ahead. If the fire ahead is moving in the easterly direction, then the victim tries to escape to the west. If the fire ahead is in the westerly direction, then the victim tries to escape eastwards and so on. In single direction fires, the single array called fire_points[] of the rescue() method is used as a parameter in the method getFireIntensity(). The latter method is called as many times as the number of fire markers. In contrast, with multiple directions fires, a single method called getFire() calls all four methods namely, getFireEast(), getFireWest(), getFireSouth() and getFireNorth(). Similarly, we could invoke a getFireSouthWest() if there was a fire from a south-west direction. The aforementioned four variables namely east, west, south and north are used to indicate the number of fire markers declared in the corresponding directions. When the fire markers are placed to spread in east direction, variable east is incremented by one. When the fire markers are placed to spread in west direction, variable west is incremented by one

and so on. If the value of variable east is greater than zero (i.e. the fire markers with east direction are placed in the map) the method getFireEast() is called. Similarly, if the value of variable west is greater than zero, the method getFireWest() is called and so on.

Let's now assume that the one of the wind directions is east. This makes the fire spread and carry in east direction. Now, again there are two cases. In the first one, victim is marked above the fire marker point. In this case there is no need to worry as the victim is in a safe place and can escape easily. In the second one, the victim is marked below the fire marker point. Now, as the fire is moving in the eastward direction, the victim is supposed to escape to the west. This is achieved as follows.

Each time the rescue() method is called, it checks whether the victim's next step (point) lies in the first quadrant, and if there is, it calls the rescueXPYP() method. Here, the victim bounds are calculated for each new step of the victim. As it is the first quadrant, the default next step would be the point vsafe5 obtained from the previous value. At each simulation step, it is checked whether the point vsafe5 lies in the bounds map_xpyp_bounds or not. The map_xpyp_bounds is a variable that stores the boundary information of the first quadrant. If the point lies in the bounds map_xpyp_bounds, then the intersection of victim_bounds and fire_bounds_xpyp will be checked. If it does not, the next value will be the point vsafe5, otherwise the next value will be vsafe6. Here, vsafe6 is chosen as the victim safe point as the direction of the fire is east and the victim has to move west. Before moving to the point, the program checks whether fire_bounds_xpyp contains the point vsafe6 or not. If it is not, the next point added to the array victim_end_points[] is vsafe6 or vsafe2. Regardless of whether it is vsafe2 or vsafe6, there is a chance that the victim point may cross the positive y axis. So, if this is the case, the new point would be vsafe3 and this point makes the victim move northwards once it tries to cross the positive y axis.

The next method employed is getFireIntensityEast(). This method is used to increase the intensity of fire in the easterly direction. This is done by calculating the end points (the locations described in the form of (x, y) in both positive and negative x and y directions. The end point obtained by calculating the last point on positive x axis is called fend1. The end point obtained by calculating the last point on negative x axis is called fend2. The end point obtained by calculating the last point on positive y axis is called fend3. The end point obtained by calculating the last point on negative y axis is called fend4. The top rightmost point is obtained by taking the values of latitude of the point fend3 and longitude of the point fend1 and is named fend5. The top leftmost point is obtained by taking the values of latitude of the point fend3 and longitude of the point fend2 and is named fend6 and so on. Now, the intensity of the fire in the easterly direction is calculated using the GLatLngBounds class by passing the points fend7 and fend5 and storing the result in the fire_bounds variable. The point is checked for whether it lies in the bounds of the first quadrant or not. If the given point lies in the first quadrant, then the calculated bounds are stored in the array fire_bounds_xpyp_east[]. Similarly, if the given point lies in the second quadrant, then the calculated bounds are stored in the array fire_bounds_xnyp_east[] and so on. The next point is calculated in the east direction. This is done by increasing the longitude value of the given point. If the new point lies in the first quadrant then it is stored in the array fire_points_xpyp_east_end[]. Similarly, if the new point lies in the second quadrant then it is stored in the array fire_points_xnyp_east_end[] and so on. To conclude, if there was a south-west wind direction, getFire() would call jointly the

getFireSouth() and getFireWest() methods using the aforementioned rationale.

FFED-GPS IMPLEMENTATION

The implementation of the mobile-based client application is implemented from Varaganti (2009) and is based on J2ME technologies. In this, the victim's current position is sent to the grid system. The grid system then produces an evacuation plan and forwards it to the victim's mobile device. The plan consists of the direction in which a victim should move to safely escape the fire and reach the safe point. The grid system sends these values via an XML document. The victim's mobile would access this document and manipulate the data via assessment with Static Google Maps and then display it on the device.

For the FFED-GPS prototype, everything was written in a class that extends the J2ME standard MIDlet class. This also involved the implementation of two interfaces including the CommandListener and Runnable.. The CommandListener interface contains a method named commandAction() that takes the reference variables of classes Command and Displayable as parameters. This method is used to produce an action when a command event occurs. The Runnable interface contains a method named run() that is used to create a separate thread that runs in parallel to the current method in which it is called. When the run() method is called on by an object of Thread class, it executes the start() method on that object.

When the emulator starts, it executes the constructor EvacuateMIDlet() which initializes the Display, Form, Command, and ImageItem objects. The code uses classes Image and ImageItem for displaying the static maps image on the mobile device. It uses KXMLParser for parsing and manipulating XML data. The Command and

Figure 7. Simulation results

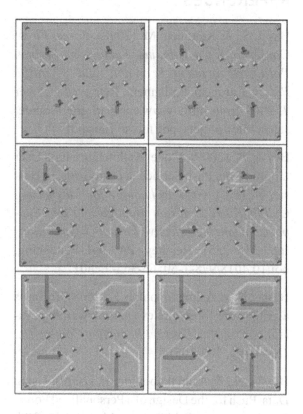

ImageItem objects are added to the Form objects and the Form object is added to the Display object. Two main methods used in the mobile implementation are readXML() and rescue(). readXML() is called inside constructor so as to get the initial and current position of the victim in forest and display it on mobile device.

The readXML() method is used to read the current position of the victim from the file victim.xml. This method is called at the time of initialization and every time the victim presses the "Rescue" button on her mobile device. The values of both latitude and longitude are known from the file and are used for displaying the victim marker on the mobile device. The rescue() method is then used to read the next position of the victim from the file rescue.xml. This method is called every time the victim presses the "Rescue" button on her mobile device. The values of both latitude and longitude

are known from the file and are used for displaying the path to the next point on the mobile device. Finally, the events generated when the "Rescue" button is pressed on the mobile device are handled by the commandAction() method, which in turn starts a thread that displays a static map on the mobile device. A static map is displayed by sending a string to the HttpConnection object which contains the locations at which markers have to be placed, different paths between different markers, and the color and size of the markers.

Testing and Simulation Results of the FFED-GPS Prototype

Using the prototype, we have produced twenty simulation scenarios. Some of these are shown in Figure 7.

A white-box test with five non-computer literate users was employed and it was found to be of full satisfaction in terms of the FFED-GPS prototype functionality and working order. In particular, during the simulation experiments, we were able to note full functionality:

- Even if the fire was progressing uniformly along any direction in any quadrant at a same time
- With sixteen victims receiving escape directions from the fire in any quadrant at the same time
- Even if victims attempted to cross a quadrant boundary
- With all sixteen victims successfully reaching the safe end points in good time
- With all victims of a particular quadrant reaching the safe point in that particular quadrant
- With all victims escaping from the fire before they were close to the fire
- With victims escaping in the opposite direction to the direction of progression of fire where possible

Users also commented that a voice feature would have been advantageous. Other comments include the inclusion of multilingual support and touch screen functionality.

CONCLUSION

In this chapter we have focused on describing the concept, design and implementation of the FFED-GPS, that is a data-push grid service enabling potential victims of a forest fire in progress to safely evacuate the area concerned. In particular, we demonstrated the way in which cell digital signal technology could be used within a grid aware infrastructure as a method to locate groups of victims and push them evacuation routes based on their exact positioning. To achieve this, we developed a number of methods and algorithms to calculate and monitor the dynamic positioning of fires and potential victims. We also used Cartesian coordinate system and geo-coding APIs available from Google Maps. The end result, a web-based client application prototype that is accessible from a mobile device, was tested and found to be of full satisfaction in terms of its functionality and working order.

The prototype development can easily be extended to take better account of the movement of both fires and victims, in terms of variability of speed, mobility and direction, noting for example that fires can jump obstacles or split, and that human groups may become dispersed. The terrain will affect anticipated movement, though this may be different from actual movement, and proper account must be taken of the position and assumed nature of natural or man-made regions and structures. With regard to fires, a fuller account must be developed of their intensity and longevity, as this affects the risk to victims when approaching a forest fire. Other future steps include prototype integration with other collaborative platforms and its testing in a simulation environment.

REFERENCES

Bassi, S., & Kettunen, M. (2008). *Forest Fires: Causes and Contributing Factors in Europe. Policy Department Economic and Scientific Policy*. European Parliament. Retrieved October 7, 2009 from http://www.europarl.gr/ressource/static/ files/projets_pdf/forest_fires.pdf

Bessis, N. (2009). Model Architecture for a User tailored Data Push Service in Data Grids. In Bessis, N. (Ed.), *Grid Technology for Maximizing Collaborative Decision Management and Support: Advancing Effective Virtual Organizations* (pp. 235–255). Hershey, PA: IGI Publishing. doi:10.4018/978-1-60566-364-7.ch012

Bessis, N., & Asimakopoulou, E. (2008). Towards a grid aware forest fire evacuation warning system. In *Proc. Int. Disaster Reduction Conference, Davos* (pp. 102-105).

Bessis, N., & Asimakopoulou, E. (2009). Using Data-Push for the Design of a Personalized Forest Fire Evacuation Grid Service. IADIS, 25th-28th February, Barcelona (pp. 59-68).

Bessis, N., Asimakopoulou, E., & Conrad, M. (2009b). A Mathematical Analysis of a Data-Grid Push Service for Disaster Management Response Scenarios. *The International Conference of Computing in Engineering, Science and Informatics* (ICC2009, sponsored by ACM and IEEE), 2nd-4th April, Fullerton, California.

Bessis, N., Asimakopoulou, E., & Varaganti, R. (2009a). The Implementation of a Personalized Forest Fire Evacuation Data Grid Service. IADIS, 19th-21st November, Rome.

Graves, R. J. (2004, May). *Key technologies for emergency response*. Paper presented at the International Community on Information Systems for Crisis Response (ICSCRAM2004) conference, Brussels.

Hatjefthymiades, S., Marias, G. F., & Priggouris, I. (2006). *SCIER: Sensor and Computing Infrastructure for Environmental Risks*. Retrieved October 7, 2009 from http://62.103.200.120/projects/scier2/site/Assets/D01_Project_Presentation_v01.pdf

Otten, J., Heijningen, B., & Lafortune, J. F. (2004, May). *The virtual crisis management centre. An ICT implementation to canalise information.* Paper presented at the International Community on Information Systems for Crisis Response Management (ISCRAM2004), Brussels.

Varaganti, R. (2009). *Personalized Forest Fire Evacuation Grid System*. Unpublished MSc Thesis, University of Bedfordshire, UK.

Compilation of References

Aalst, M. K. (2006). The impacts of climate change on the risk of natural disasters. Disasters, 30(1), 5–18. doi:10.1111/j.1467-9523.2006.00303.x

Abhyankar, M. (2006). Education, knowledge, innovation-building a culture of safety and resilience. An Indian Experience. In W. Almman et al. (Ed.), IDRC Davos 2006 (Vol. 2, Extended Abstracts, pp. 13-16). Davos, Switzerland: Swiss Federal Research Institute.

Adam, B., Beck, U., & van Loon, J. (Eds.). (2000). The risk society and beyond: Critical issues in social theory. London: Sage.

Adger, W. N. (2006). Vulnerability. Global Environmental Change, 16, 268–281. doi:10.1016/j.gloenvcha.2006.02.006

Alessio, M., & Vicuña, J. (2003). Reestablecimiento de la normalidad a través de la gerencia de la logística de catástrofes de origen natural. Trabajo especial de grado. Venezuela: Universidad Metropolitana.

Alexander, D. (1997). The study of natural disasters, 1977-1997: Some reflections on a changing field of knowledge. Disasters, 21(4), 284–304. doi:10.1111/1467-7717.00064

Alexander, D. (2005). An interpretation of disaster in terms of changes in culture, society and international relations. In R.W. Perry & E.L. Quarantelli (Eds.), What is a disaster? New answers to old questions (pp. 25-38). Philadelphia: XLibris.

Alexander, D. E. (2006). Globalization of disaster: trends, problems and dilemmas. Journal of International Affairs, 59(2), 1-22. Retrieved August 3, 2009, from http://www.policyinnovations.org/ideas/policy_library/data/01330/_res/id=sa_File1/alexander_globofdisaster.pdf

Almeida de, A. T. (2005). Modelagem multicritério para seleção de intervalos de manutenção preventiva baseada na teoria da utilidade multiatributo. Pesquisa Operacional, 25(1), 69–81.

Amaya, J. (2005). Fracasos y falacias de la educación actual. Ciudad de México, México: Editorial Trillas.

American Hospital Association. (2009). Fast Facts on US Hospitals. Retrieved on May 11, 2009 from http://www.aha.org/aha/resource-center/Statistics-and-Studies/fast-facts.html

Andrews, P. L. (1986). BEHAVE: fire behavior prediction and fuel modeling system- BURN subsystem. Part 1. GTR-INT-194. USDA Forest Service.

Angerou, C., Ciborra, C., & Land, F. (Eds.). (2004). The Social Study of Information and Communication Technology: Innovation, Actors and Contexts. New York: Oxford University Press.

Anil, K. S. (2003). Development of an Integrated Disaster Management System in India: Importance of Reliable Information. Paper Presented at the International Conference on Total Disaster Risk Management. 2-4 December 2003.

Annan, K. (1988). International decade for natural disaster reduction (Report of the Secretary-General). Report A/43/723.

Annan, K. (2005). In larger freedom: towards development, security and human rights for all (Report of the Secretary-General). Report A/59/2005, paragraph 66.

Anthony, R. N. (1987). Contabilidad Gerencial. Barcelona: Ediciones Orbis Colección Biblioteca de la Empresa Vol. I. N° 21. (English, 1974, Programmed learning aid for management accounting. Homewood, Illinois: Learning Systems).

Araz, C., Ozfirat, P. M., & Ozkarahan, I. (2007). An integrated multicriteria decision-making methodology for outsourcing management. Computers & Operations Research, 34(12), 3738–3756. doi:10.1016/j.cor.2006.01.014

Archibald, T. W., Black, D., & Glazebrook, K. D. (2009). An index heuristic for transshipment decisions in multi-location inventory systems based on a pairwise decomposition. European Journal of Operational Research, 192(1), 69–78. doi:10.1016/j.ejor.2007.09.019

Arellano, N. E. (2008). Internet Connect with Facebook generation, expert urges Canadian companies. Itbusiness.ca, Business Advantage through Technology. Retrieved April 1, 2009, from http://www.itbusiness.ca/it/client/en/home/News.asp?id=48910

Arellano, N. E. (2008). Internet Connect with Facebook generation, expert urges Canadian companies. Itbusiness.ca, Business Advantage through Technology. Retrieved April 1, 2009, from http://www.itbusiness.ca/it/client/en/home/News.asp?id=48910

Arnold, J. P. S. (2006). Disaster Management – A Hand Book for NGOs. Chennai: TNVHA.

Asamblea Nacional. (2009). Ley de Gestión Integral de Riesgos Socionaturales y Tecnológicos. Gaceta Oficial de la República Bolivariana de Venezuela Extraordinaria N° 39.095 del 9 enero de 2009.

Ascough, J., Maier, H., Ravalico, J., & Strudley, M. (2008). Future research challenges for incorporation of uncertainty in environmental and ecological decision-making. Ecological Modelling, 219(3-4), 383–399. doi:10.1016/j.ecolmodel.2008.07.015

Ashlin, A., & Ladle, R. J. (2007). 'Natural disasters' and newspapers: Post-tsunami environmental discourse. Environmental Hazards, 7(4), 330–341. doi:10.1016/j.envhaz.2007.09.008

Asimakopoulou, E., Sagun, A., Anumba, C. J., & Bouchlaghem, N. M. (2006). Use of ICT during the Response Phase in Emergency Management in Greece and the United Kingdom. International Disaster Reduction Conference, Davos, Switzerland.

Auf der Heide, E. (2009). The media: Friend and Foe. In Disaster Response: Principles of Preparation and Coordination. Retrieved April 13, 2009, from http://orgmail2.coe-dmha.org/dr/DisasterResponse.nsf/section/10?opendocument&home=html

Badia, R.M., & Labarta, J. Sirvent, J., Pérez, J.M., Cela, J. M. Cela, & Grima R. (2003). Programming Grid Applications with GRID Superscalar . Journal of Grid Computing, 1(2). doi:10.1023/B:GRID.0000024072.93701.f3

Baklanov, A., Sørensen, J., Hoe, S., & Amstrup, B. (2006). Urban meteorological modelling for nuclear emergency preparedness. Journal of Environmental Radioactivity, 85(2-3), 154–170. doi:10.1016/j.jenvrad.2005.01.018

Balay, S., Gropp, W. D., Curfman McInnes, L., & Smith, B. F. (1997), Efficient Management of Parallelism in Object Oriented Numerical Software Libraries. In E. Arge, A.M. Bruaset, & H.P. Langtangen (pp. 163-202). Modern Software Tools in Scientific Computing. Birkhauser Press

Barab, S. A., Thomas, M., Dodge, T., Carteaux, R., & Tuzun, H. (2005). Making learning fun: Quest Atlantis, a game without guns. Educational Technology Research and Development, 53 (1), 86-108. Retrieved January 28, 2009 from http://inkido.indiana.edu/research/onlinemanu/papers/QA_ETRD.pdf.

Barker, K., & Haimes, Y. Y. (2009). Assessing uncertainty in extreme events: Applications to risk-based decision making in interdependent infrastructure sectors. *Reliability Engineering & System Safety, 94*(4), 819–829. doi:10.1016/j.ress.2008.09.008

Barrientos, Y. (2006a). Campaña educativa comunidad-escuela para la mitigación de riesgos socionaturales asociados a las cuencas de los ríos Osorio y Piedra Azul, estado Vargas, Venezuela. Informe Técnico N° 3. Caracas, Venezuela. FONACIT-UPEL. Documento Inédito.

Barrientos, Y., Iztúriz, A., García, A., & Ruíz, S. (2006 b). Instructional and Methodological Strategies for Learning about Natural Hazards and Socionatural Risk at Elementary School at Vargas State, Venezuela. In W. Almman et al. (Ed.), IDRC Davos 2006 (Vol. 3, Extended Abstracts, pp. 696-697). Davos, Switzerland: Swiss Federal Research Institute.

Barton, A. H. (2005). Disaster and collective stress. In R.W. Perry & E.L. Quarantelli (Eds.), What is a disaster? New answers to old questions (pp.125-152). Philadelphia: XLibris.

Basher, R. (2006). Global Early Warning Systems for Natural Hazards: Systematic and People-centred, Philosophical Transactions of the Royal Society, (364), 2167-2182.

Bassi, S., & Kettunen, M. (2008). Forest Fires: Causes and Contributing Factors in Europe. Policy Department Economic and Scientific Policy. European Parliament. Retrieved October 7, 2009 from http://www.europarl.gr/ressource/static/ files/projets_pdf/forest_fires.pdf

Baucells, M., & Sarin, R. (2003). Group decisions with multiple criteria. *Management Science, 49*(8), 1105–1118. doi:10.1287/mnsc.49.8.1105.16400

Baude, F., Baduel, L., Caromel, D., Contes, A., Huet, F., Morel, M., & Quilici, R. (2004). Programming, Composing, Deploying for the Grid . In Cunha, J. C., & Rana, O. F. (Eds.), GRID COMPUTING: Software Environments and Tools. Springer Verlag.

Bauer, H. (2001). Wahrscheinlichkeitstheorie. Berlin: de Gruyter.

Bavay, M., Lehning, M., Jonas, T., & Löwe, H. (2009). Simulations of future snow cover and discharge in Alpine headwater catchments. Hydrological Processes, 22. doi:.doi:10.1002/hyp.7195

Bayrak, T. (2007). Performance Metrics for Disaster Monitoring Systems. In B. Van de Walle, P. Burghardt & C. Nieuwenhuis (Eds.), Proceedings of the 4th International Conference on Information Systems for Crisis Response and Management ISCRAM2007 (pp. 125-132). Newark, NJ.

BBK. (2005). Opportunities for Public Safety in Germany. Bonn, Germany: German Federal Agency for Public Safety and Disaster Preparedness BBK.

Beard, A. N. (1999). Some ideas on a systemic approach. *Civil Engineering and Environmental Systems, 16*, 197–209. doi:10.1080/02630259908970262

Beasley, J. E. (2002). O.R.-Notes. Inventory Control. Retrieved October 2008, from http://people.brunel.ac.uk/~mastjjb/jeb/or/invent.html

Beck, U. (1992). Risk Society: Towards a new modernity. London: Sage.

Beck, U. (2009). World at risk. Cambridge: Polity.

Belozercovsky, C., & Sensel, D. (2002). Sistema de apoyo a la toma de decisiones en caso de una catástrofe natural. Trabajo especial de grado. Venezuela: Universidad Metropolitana.

Belton, V., & Stewart, T. (2002). Multiple Criteria Decision Analysis. An Integrated Approach. Boston: Kluwer Academic Publishers.

Ben-Haim, Y. (2000). Robust rationality and decisions under severe uncertainty. *Journal of the Franklin Institute, 337*(2-3), 171–199. doi:10.1016/S0016-0032(00)00016-8

Ben-Mena, S. (2000). Introduction aux méthodes multicritères d'aide à la decision. Biotechnol. Agron. Soc. Environ, 4(2), 83–93.

Berger, E. (2008, Sept 13). Hurricane Ike: Targeting Florida? Houston Chronicle. Retrieved Sept 16, 2008, from http://blogs.chron.com/sciguy/archives/2008/09/hurricane_ike_t_1.html.

Berke, P. R., Kartez, J., & Wenger, D. (1993). Recovery after disaster: achieving sustainable development. Disasters, 17(2), 93–108. doi:10.1111/j.1467-7717.1993.tb01137.x

Bertelsen, O. W., & Bodker, S. (2003). Activity Theory . In Caroll, J. M. (Ed.), HCI Models Theories, and Frameworks: Toward A Multidisciplinary Science (pp. 291–324). San Francisco: Morgan Kaufmann. doi:10.1016/B978-155860808-5/50011-3

Bertsch, V. (2008). Uncertainty Handling in Multi Attribute Decision Support for Industrial Risk Management. Karlsruhe.

Bertsch, V., Geldermann, J., & Rentz, O. (2007). Preference Sensitivity Analyses for Multi-Attribute Decision Support. In Operations Research Proceedings 2006 (411-416).

Bertsch, V., Geldermann, J., Rentz, O., & Raskob, W. (2006). Multi-criteria decision support and stakeholder involvement in emergency management. International Journal of Emergency Management, 3(2/3), 114–130. doi:10.1504/IJEM.2006.011163

Bessis, N. (2009). Grid Technology for Maximizing Collaborative Decision Management and Support: Advancing Effective Virtual Organizations. Hershey, PA: IGI Publishing.

Bessis, N. (2009). Model Architecture for a User tailored Data Push Service in Data Grids . In Bessis, N. (Ed.), Grid Technology for Maximizing Collaborative Decision Management and Support: Advancing Effective Virtual Organizations (pp. 235–255). Hershey, PA: IGI Publishing. doi:10.4018/978-1-60566-364-7.ch012

Bessis, N., & Asimakopoulou, E. (2008). Towards a grid aware forest fire evacuation warning system. In Proc. Int. Disaster Reduction Conference, Davos (pp. 102-105).

Bessis, N., & Asimakopoulou, E. (2009). Using Data-Push for the Design of a Personalized Forest Fire Evacuation Grid Service. IADIS, 25th-28th February, Barcelona (pp. 59-68).

Bessis, N., Asimakopoulou, E., & Conrad, M. (2009b). A Mathematical Analysis of a Data-Grid Push Service for Disaster Management Response Scenarios. The International Conference of Computing in Engineering, Science and Informatics (ICC2009, sponsored by ACM and IEEE), 2nd-4th April, Fullerton, California.

Bessis, N., Asimakopoulou, E., & Varaganti, R. (2009a). The Implementation of a Personalized Forest Fire Evacuation Data Grid Service. IADIS, 19th-21st November, Rome.

Bianchini, G., Cortés, A., Margalef, T., & Luque, E. (2005). S2F2M - Statistical System for Forest Fire Management, Atlanta, USA. In Proceedings of the Int. Conf. on Computational Science, ICCS Book of Abstracts.

Black, R. (2006). 'Clear' human impact on climate. British Broadcasting Corporation NEWS. Retrieved May 3, 2006, from http://news.bbc.co.uk/go/pr/fr/-/2/hi/science/nature/4969772.stm

Blumer, J. G., & Katz, E. (1974). The uses of mass communications: Current perspectives on gratifications research. Beverly Hill, CA: Sage.

Boin, A. (2005). From crisis to disaster. In R.W. Perry & E.L. Quarantelli (Eds.), What is a disaster? New answers to old questions (pp. 153-172). Philadelphia: XLibris.

Bolívar, C., & Ríos, P. (1990). El uso de la informática en la educación. Investigación y Postgrado, 5(2), 59–91.

Bonfadelli, H. (2001). Medienwirkungsforschung. Grundlagen und theoretische Perspektiven. Konstanz, D: UVK Verlagsgesellschaft mbH.

Bos, T. (2005). The impact of using virtual reality technology to train for law enforcement critical incidents. Journal of California Law Enforcement, 39(2), 5–14.

Bossard, C., Kermarrec, G., Buche, C., & Tisseau, J. (2008). Transfer of learning in virtual environments: A new challenge. Virtual Reality (Waltham Cross), 12(3), 151–161. doi:10.1007/s10055-008-0093-y

Bottorell, A. (2006). The Common Alerting Protocol: An Open Standard for Alerting, Warning and Notification. In B. Van de Walle & M. Turoff (Eds.), Proceedings of the 3rd International ISCRAM Conference (pp. 497-503). Newark, NJ.

Botts, M., Percivall, G., Reed, C., & Davidson, J. (2008). OGC® Sensor Web Enablement: Overview and High Level Architecture (LNCS 4540, pp. 175-190). Berlin/Heidelberg: Springer.

Boyd, D. M., & Ellison, N. B. (2007). Social network sites: Definition, history, and scholarship. Journal of Computer-Mediated Communication, 13(1), article 11. Retrieved August 3, 2009 from http://jcmc.indiana.edu/vol13/issue1/boyd.ellison.html

Boykoff, M.T. & Rajan, R.S. (2007). Signals and noises. Mass-media coverage of climate change in the

British Broadcasting Corporation (BBC). (2006a, March 16). Quenching Mexico City's Thirst. BBC NEWS. Retrieved March 16, 2009, from http://news.bbc.co.uk/go/pr/fr/-/2/hi/americas/4812352.stm

British Broadcasting Corporation (BBC). (2006b, April 8). Quake survivors 'still need aid'. BBC NEWS. Retrieved April 8, 2009, from http://news.bbc.co.uk/go/pr/fr/-/2/hi/south_asia/4890252.stm

Bruckner, J., Keys, D., & Fisher, J. (2004). The Acute Exposure Guideline Level Program: Applications of Physiologically Based Pharmacokinetic Modeling. Journal of Toxicology and Environmental Health. Part A., 67(8-10), 621–634. doi:10.1080/15287390490428017

Bründl, M., Romang, H. E., Bischof, N., & Rheinberger, C. M. (2009). The Risk Concept and Its Application in Natural Hazard Risk Management in Switzerland. Natural Hazards and Earth System Sciences, 9, 801–813. doi:10.5194/nhess-9-801-2009

Bucher, H. G. (2002). Crisis communication and the internet: Risk and trust in a global media. First Monday, 7(4). Retrieved May 15, 2009, from http://outreach.lib.uic.edu/www/issues/issue7_4/bucher/index.html

Bucher, H.-J. (2003). Internet und Krieg: Informationsrisiken und Aufmerksamkeitsökonomie in der vernetzten Kriegskommunikation. In M. Lö (Ed.), Krieg als Medienereignis II: Krisenkommunikation im 21. Jahrhundert (pp. 275–296). Wiesbaden, D: VS Verlag für Sozialwissenschaften.

Buckle, P. (2005). Disaster: Mandated definitions, local knowledge and complexity. In R.W. Perry & E.L. Quarantelli (Eds.), What is a disaster? New answers to old questions (pp. 173-200). Philadelphia: XLibris.

Bui, T., & Lee, J. (1999). An agent-based framework for building decision support systems. Decision Support Systems, 25(3), 225–237. doi:10.1016/S0167-9236(99)00008-1

Bunn, D., & Salo, A. (1993). Forecasting with scenarios. European Journal of Operational Research, 68(3), 291–303. doi:10.1016/0377-2217(93)90186-Q

Burton, I., Saleemul, H. L. B., Pilifosova, O., & Schipper, E. L. (2002). From impacts assessment to adaptation priorities: the shaping of adaptation policy. Climate Policy, 2(2-3), 145–159. doi:10.1016/S1469-3062(02)00038-4

Butler, J., Jia, J., & Dyer, J. (1997). Simulation techniques for the sensitivity analysis of multi-criteria decision models. European Journal of Operational Research, 103(3), 531–546. doi:10.1016/S0377-2217(96)00307-4

Byrne, M., & Whitmore, C. (2008). Crisis informatics. IAEM Bulletin, February 2008, 8.

Calle, A., Casanova, J. L., & Romo, A. (2006). Fire Detection and Monitoring using MSG Spinning Enhanced Visible and Infrared Imager (SEVIRI) Data. Journal of Geophysical Research, 111, G04S06. doi:10.1029/2005JG000116

CAPART. (2006). Retrieved from Available at http://capart.nic.in/scheme.vrc.pdf

Cardós, M., & García-Sabater, J. P. (2006). Designing a consumer products retail chain inventory replenishment policy with the consideration of transportation costs. International Journal of Production Economics, 104(2), 525–535. doi:10.1016/j.ijpe.2004.12.022

Careem, M., Silva, C. D., Silva, R. D., Raschid, L., & Weerawarana, S. (2006). Sahana: Overview of a disaster management system. In IEEE International Conference on Information and Automation.

Carey, J. (2003). The functions and Uses of Media during the September 11 Crisis and Its Aftermath. In M.A. (Ed.), Crisis Communications: Lessons from September 11 (pp. 1-17). Lanhalm, MD: Rowman & Littlefield Publishing Group.

Caribbean Information and Communications Technology Community. (2004). Facilitating Effective Disaster Management in the Caribbean. Retrieved May, 2009, from http://www.devnet.org.gy /documents/Caribbean-disasterbrief-Final.pdf

Casse, P. (1982). Training for the multicultural manager: a practical and cross-cultural approach to the management of people. Washington, DC: SIETAR International.

Center for Disease Control and Prevention (CDC). (2006). Crisis and emergency risk communication (CERC): Pandemic influenza. "Checklist: Scientific risk communication or the public". Retrieved October 10, 2007 from http://www.bt.cdc.gov/erc/panflu/

Centro Nacional para el Mejoramiento de la Enseñanza de la Ciencia (CENAMEC). (1996). Rompecabezas de Placas Tectónicas. Caracas, Venezuela: Colsum.

Chang, Ch., & Lo, T. Y. (2009). On the inventory model with continuous and discrete lead time, backorders and lost sales. Applied Mathematical Modelling, 33(5), 2196–2206. doi:10.1016/j.apm.2008.05.028

Charnes, A., & Cooper, W. W. (1961). Management models and industrial applications of linear programming. New York: John Wiley and Sons.

Chaudhary, A. G. (2004): Convergence: Globalisation, Localization and New Communication Technologies In P.J (Ed.), Mass Media in Transition: An International Compendium (pp.11-24). Athens: Athens Institute for Education and Research.

Chaudhary, A. G. (2004): Convergence: Globalisation, Localization and New Communication Technologies In P.J (Ed.), Mass Media in Transition: An International Compendium (pp.11-24). Athens: Athens Institute for Education and Research.

Chen, Ch. (2001). A fuzzy approach to select the location of the distribution center. Fuzzy Sets and Systems, 118(1), 65–73. doi:10.1016/S0165-0114(98)00459-X

Chen, S., Bao, H., Zeng, X., & Yang, Y. (2003). A Fire Detecting Method based on multi-sensor Data Fusion. In IEEE Systems Man and Cybernetics, 4.

Chen, Y. F., Rebolledo-Mendez, G., Liarokapis, F., Freitas, S., & Parker, E. (2008). The use of virtual world platforms for supporting an emergency response training exercise. In Proceedings of the 13th International Conference on Computer Games: AI, Animation, Mobile, Interactive Multimedia, Educational & Serious Games.

Clark, J., & Dede, C. (2005, April). Making Learning Meaningful: An Exploratory Study of Using Multi-User Environments (MUVEs) in Middle School Science. Paper presented at the meeting of the American Educational Research Association, Montreal, Canada.

Clark, J., Stephens, M., & Fengler, W. (2005). Indonesia: Rebuilding a better Aceh and Nias (Six Month Report). Washington, Jakarta / Indonesia. Retrieved June 25, 2005, from http://www.reliefweb.int/rwarchive/rwb.nsf/db900sid/KHII-6DU466?OpenDocument

Clark, R., & Elvira, C. (2003). Organización de la comunidad en caso de desastres de origen natural: Un sistema de apoyo. Trabajo especial de grado. Venezuela: Universidad Metropolitana.

Clavaud, M., & Navarro, R. (2006). Distribución de la población a los centros asistenciales después de una catástrofe, dirigida al estado Miranda. Trabajo especial de grado. Venezuela: Universidad Metropolitana.

Clifton, A., & Lehning, M. (2008). Simulations of wind tunnel snow drift using a semi-stochastic model. Earth Surface Processes and Landforms, 33(14), 2156–2173. .doi:10.1002/esp.1673

Clifton, A., Rüedi, J.-D., & Lehning, M. (2006). Snow saltation threshold measurements in a drifting snow wind tunnel. Journal of Glaciology, 52(179), 585–596. doi:10.3189/172756506781828430

Cohen, E. L., Ball-Rokeach, S. J., Jung, J.-Y., & Kim, Y.-C. (2003). Civic Actions after September 11: A Communication Infrastructure Perspective. In M.A, No (Ed.), Crisis Communications: Lessons from September 11 (pp. 31-43). Lanhalm, MA: Rowman & Littlefield Publishing Group.

Collier, N., Doan, S., Kawazoe, A., Goodwin, R. M., Conway, M., & Tateno, Y. (2008). BioCaster: detecting public health rumors with a Web-based text mining system. Bioinformatics (Oxford, England), 24, 2940–2941. doi:10.1093/bioinformatics/btn534

Collins, H. (2009). Emergency managers and first responders use twitter and facebook to update communities. Retrieved August, 7, 2009, from http://www. emergencymgmt.com/safety/Emergency-Managers-and-First.html

Colombo, A. G., & Vetere Arellano, A. L. (2002). Dissemination of lessons learnt from disasters. NEDIES workshop. Italy: Ispra.

Comes, T., Hiete, M., Wijngaards, N., & Kempen, M. (2009). Integrating Scenario-Based Reasoning into Multi-Criteria Decision Analysis. In Proceedings of the 6th International Conference on Information Systems for Crisis Response and Management (ISCRAM 2009). Gothenburg.

Comfort, L. K., Dunn, M., Johnson, D., Skertich, R., & Zagorecki, A. (2004). Integrating Real-Time Information in to Disaster Management: The IISIS Dashboard. In D. Ma & T. Pl (Eds.), Disasters and Society - from Hazard Assessment to Risk Reduction (pp. 227-235). Berlin, D: Logos Verlag.

COMSYS. (2007). The COMSYS VSAT Report. Retrieved January 2, 2008, from http://www.comsys. co.uk/vsat_rep.htm

Cooke, R. M., & Goossens, L. H. J. (2004). Expert judgement elicitation for risk assessments of critical infrastructures. Journal of Risk Research, 7(6), 643–656. doi:10.1080/1366987042000192237

Cosgrave, J. (1996). Decision making in emergencies. Disaster Prevention and Management, 5(4), 28–35. doi:10.1108/09653569610127424

Cottle, S. (2009). Global Crisis Reporting: Journalism in the Global Age. Maidenhead, UK: Open University Press.

Covello, V. T., McCallum, D. B., & Pavlova, M. T. (1988). Effective Risk Communication. New York: Plenum Press.

Crowe, A. (2008). A closer look at emergency management websites. Crisis Response Journal, 4(3), 44.

Cuny, F. C. (1994). Disasters and development. Oxfam. Dallas, TX: Oxford University Press.

Currion, P. (2005). An ill wind? The role of accessible ITC following hurricane Katrina. Retrieved February, 10, 2009, from http://www.humanitarian.info/itc-and_katrina

Cutter, S. L. (1996). Vulnerability to environmental hazards. Progress in Human Geography, 20, 529–539. doi:10.1177/030913259602000407

D'Agostino, F., Farinelli, A., Grisetti, G., Iocchi, L., & Nardi, D. (2002). Monitoring and Information Fusion for Search and Rescue Operations in Large-scale Disasters. In Proceedings of the Fifth International Conference on Information Fusion. Annapolis, MD: Omnipress, USA.

Dalal, P. (2007). Cyber law in India needs rejuvenation. Indian Attorney.

Dandoulaki, M., & Andritsos, F. (2007). Autonomous sensors for just in time information supporting search and rescue in the event of a building collapse. International Journal of Emergency Management, 4(4), 704–725. doi:10.1504/IJEM.2007.015737

Dausey, D. Aledort, J., & Lurie, M. (2005). Tabletop exercises for pandemic influenza preparedness in local public health agencies. U.S. Department of Health and Human Services Office of the Assistant Secretary for Public Health Emergency Preparedness. Retrieved April 15, 2009 from http://www. pandemicflu.gov/plan/states/tr319.html.

de Cárdenas, E. Sánchez, T. de., & Quintero, N de. (1990). Revisión de los programas existentes para la Educación Básica con el objeto de incluir en ellos el material actualizado sobre riesgos naturales y prevención sísmica. Paper presented at meeting III Congreso Venezolano de Geografía, Universidad de Los Andes, Mérida, Venezuela.

De Groeve, T., & Eriksson, D. (2005). An Evaluation of the performance of the JRC Earthquake Alert Tool. DG Joint Research Centre of the European Commission. Retrieved June 30, 2009, from http://dma.jrc.it/services/gdas/Performance_of_Earthquake_Alert_Tool.pdf

de las Naciones Unidas, O. Estrategia Internacional para la Reducción de Desastres., UNICEF., Federación Internacional de las Sociedades de la Cruz Roja y Media Luna y Ayuda Humanitaria de la Comisión Europea. (2005). Campaña Mundial: La Reducción de Desastres Empieza en la Escuela. Consulta: 2006. Enero 22. Disponible en: http//www.eird.org

De Marchi, B. (1990). Assessing People's Information Needs about Major Accident Hazards: Improving Knowledge for a Better Response . In Gow, H. B. F., & Otway, H. (Eds.), Communicating with the Public about Major Accident Hazards. New York: Elsevier Science Pub.

Decker, K. C., & Holtermann, K. (2009). The role of exercises in senior policy pandemic influenza preparedness. Journal of Homeland Security and Emergency Management, 6(1), 32. doi:10.2202/1547-7355.1521

Dede, C. (1995). The evolution of constructivist learning environments: Immersion in distributed virtual worlds. Educational Technology, 35, 46-52. Retrieved January 28, 2009 from http://www.virtual.gmu. edu/ss_pdf/constr.pdf.

Deliverable, E. U. F. I. R. E. L. A. B. D-03-06. (n.d.). Retrieved September 9, 2008, from http://www.eufire-lab.org

Dey, B. (2006). Building a culture of safety and resilience towards natural hazards through education. In W. Almman et al. (Ed.), IDRC Davos 2006 (Vol. 2, Extended Abstracts, pp. 134-136). Davos, Switzerland: Swiss Federal Research Institute.

Dilekli, N., & Rashed, T. (2007). Towards a GIS data model for improving the emergency response in the least developing countries: Challenges and opportunities . In Van de Walle, B., & Carlé, B. (Eds.), Proceedings ISCRAM2007 (pp. 57–62).

Donath, J. (2007). Signals in social supernets. Journal of Computer-Mediated Communication, 13(1), article 12. Retrieved August 3, 2009, from http://jcmc.indiana.edu/vol13/issue1/donath.html

Dong, Y., Carter, C. R., & Dresner, M. E. (2001). JIT purchasing and performance: an exploratory analysis of buyer and supplier perspectives. Journal of Operations Management, 19(4), 471–483. doi:10.1016/S0272-6963(00)00066-8

Doolin, D. M., & Sitar, N. (2005). 5765. Wireless Sensors for Wildfire Monitoring. In Smart Structures and Materials.

Doorschot, J., & Lehning, M. (2002). Equilibrium saltation: mass fluxes, aerodynamic entrainment and dependence on grain properties. Boundary-Layer Meteorology, 104(1), 111–130. doi:10.1023/A:1015516420286

Dos Reis, A. (2009, agosto 11). Videojuego dispara para un clic contra la violencia. El Nacional (p. 6). ISRD-IDRC- Prevention Web. (2008). Virtual disaster risk reduction library. CD-ROM.

Dos Santos, A., & Rojas, J. (2006). Logística para reestablecer el ambiente de infraestructura en una empresa después de una catástrofe. Trabajo especial de grado. Venezuela: Universidad Metropolitana.

Dove, K. (2007). Emergency management information systems: Application of an intranet portal for disaster training and response. An examination of emerging technologies in a local emergency operations center. (Unpublished dissertation). Pepperdine University. 256 pages.AAT3293112.

Dreyfus, H., & Dreyfus, S. (1986). Mind over Machine: The power of human intuition and expertise in the era of the computer. Oxford: Basil Blackwell Pub.

Duffy, T. M., & Cunningham, D. J. (1996). Constructivism: Implications for the design and delivery of instruction . In Jonassen, D. H. (Ed.), Educational Communications and Technology (pp. 170–199). New York: Simon & Schuster Macmillan.

Duffy, T. M., & Jonassen, D. H. (1991). Constructivism: New implications for instructional technology? Educational Technology, 31(5), 7–11.

Dünnweber, J., & Gorlatch, S. (2004). HOC-SA: A grid service architecture for higher-order components. In IEEE International Conference on Services Computing, Shanghai, China (pp. 288-294). IEEE Computer Society Press.

Dunwoody, S., & Peters, H. P. (1992). Mass media coverage of technological and environmental risks: a survey of research in the United States and Germany. Public Understanding of Science (Bristol, England), 1(2), 199–230. doi:10.1088/0963-6625/1/2/004

Durbach, I., & Stewart, T. (2003). Integrating scenario planning and goal programming. Journal of Multi-Criteria Decision Analysis, 12(4-5), 261–271. .doi:10.1002/mcda.362

Eakin, H., & Luers, A. L. (2006). Assessing the vulnerability of social - environmental systems. Annual Review of Environment and Resources, 31, 365–394. doi:10.1146/annurev.energy.30.050504.144352

Earle, P. S., & Wald, S. J. (2006). Rapid post-earthquake information and assessment tools from the U.S. Geological Survey National Earthquake Information Center. In B. Van de Walle & M. Turoff (Eds.), Proceedings of the 3rd International ISCRAM Conference Newark, NJ (USA), May 2006

Economic Commission for Latin America and the Caribbean (ECLAC). (1991). Manual for estimating the socio-economic effects of natural disasters. United Nations. Retrieved from http://www.reliefweb.int/rw/lib.nsf/db900SID/LGEL-5E2CLJ?OpenDocument

Edwards, W. (1977). How to Use Multiattribute Utility Measurement for Social Decisionmaking. IEEE Transactions on Systems, Man, and Cybernetics, 7(5), 326–340. doi:10.1109/TSMC.1977.4309720

Egeland, J. (2006). Opening Address. In Proceedings of Third International Conference on Early Warning, March 27-29, 2006, Bonn, Germany.

Eglese, R., Maden, W., & Slater, A. (2006). A road Timetable TM to aid vehicle routing and scheduling. Computers & Operations Research, 33(12), 3508–3519. doi:10.1016/j.cor.2005.03.029

Ehrgott, M., & Gandibleux, X. (Eds.). (2002). Multiple criteria Optimization: State of the art annotated bibliographic surveys. Kluwer Academic Publishers.

Elgamal, A., Yan, L., Fraser, M., Lu, J., & Conte, J. P. (2005). Large- Scale Simulation and Data Analysis. In Proceedings of the 2005 ASCE International Conference on Computing in Civil Engineering, Cancun, Mexico.

Entman, R. M. (1993). Framing: Toward Clarification of a Fractured Paradigma. The Journal of Communication, 43(4), 51–58. doi:10.1111/j.1460-2466.1993.tb01304.x

Erich, J. (2008). Virtually REAL. Fort Atkinson . EMS Product News, 16(5), 24.

Erwin, S. (2001). Virtual metropolis underpins emergency response trainer. National Defense, 86(576), 42–32.

ETSI. (2006). Emergency Communications (EMTEL); Requirements for communications from authorities/organizations to individuals, groups or the general public during emergencies. ETSI TS 102 182 V1.2.1. Retrieved June 30, 2009, from http://portal.etsi.org.

European Commission. (2009). IMSK –Integrated Mobile Security Kit, European Research Framework Program 7. Retrieved June 30, 2009, from http://ec.europa.eu/enterprise/security/doc/fp7_project_flyers/imsk.pdf

Fahland, D., Gläßer, T. M., Quilitz, B., Weißleder, S., & Leser, U. (2007). HUODINI – Flexible information integration for disaster management . In Van de Walle, B., Burghardt, P., & Nieuwenhuis, C. (Eds.), Proceedings IS-CRAM2007 (pp. 255–262).

Farnham, S., Pedersen, E. R., & Kirkpatrick, R. (2006). Observation of Katrina/Rita Groove deployment: Addressing social and communication challenges of ephemeral groups. In B. Van de Walle & M. Turoff (Eds.) Proceedings of the 3rd International ISCRAM Conference, Newark, NJ (USA), May 2006 (pp. 39-49).

FCC. (2007). Review of the Emergency Alert System. Report of the Federal Communications Commission. Washington, D.C.: Federal Communications Commission.

Federal Emergency Management Agency (FEMA). (2009). Emergency Management Institute (EMI). Home Page. Retrieved May 11, 2009 from http://training.fema.gov.

Federal Emergency Management Agency (FEMA). (2009a).The National Exercise Simulation Center (NESC) (2009). Retrieved February 6, 2009 from http://www.fema.gov/news/newsrelease.fema? id=47280.

Federal Emergency Management Agency (FEMA). 2007. NIMS Compliance Metrics Terms of Reference. Retrieved January 29, 2008 from http://www.fema.gov/pdf/emergency/nims/comp_met_terms.pdf

Fenton, N., & Neil, M. (2001). Making decisions: using Bayesian nets and MCDA. Knowledge-Based Systems, 14(7), 307–325. doi:10.1016/S0950-7051(00)00071-X

Fernández, A. M. (1998a, February 3). Gobierno pudo evitar tragedia de Ica si la declaraba en estado de emergencia. La República (p. 5).

Fernández, A. M. (1998b, February 5). Ica, ciudad devastada. La República (p. 24).

Fildes, R., Goodwin, P., Lawrence, M., & Nikolopoulos, K. (2008). Effective forecasting and judgmental adjustments: an empirical evaluation and strategies for improvement in supply-chain planning. International Journal of Forecasting, 25(1), 3–23. doi:10.1016/j.ijforecast.2008.11.010

Fischhof, B. (1995). Risk Perception and Communication Unplugged: Twenty Years of Process. Risk Analysis, 15(2), 137–145. doi:10.1111/j.1539-6924.1995.tb00308.x

Fischhoff, B. (2006). Bevaviorally realistic risk management . In Daniels, R. J., Kettl, D. F., & Kunreuther, H. (Eds.), On risk and disaster: Lessons from hurricane Katrina (pp. 78–88). Philadelphia: University Pennsylvania Press.

Fisher, H. W. (1998). The role of the new information technologies in emergency mitigation, planning, response and recovery. Disaster Prevention and Management, 7(1), 28–37. doi:10.1108/09653569810206262

Foster, I., & Kesselman, C. (1998). The Grid: Blueprint for a New Computing Infrastructure. Morgan Kaufmann Publishers.

Freeman, P. K., Martin, L. A., Mechler, R., Warner, K., & Hausmann, P. (2002). Catastrophes and development: Integrating natural catastrophes into development planning (Working papers series No. 4). Washington, DC: World Bank.

Freifeld, C. C., Mandl, K. D., Reis, B. Y., & Brownstein, J. S. (2008). HealthMap: global infectious disease monitoring through automated classification and visualization of Internet media reports. *Journal of the American Medical Informatics Association*, 15, 150–157. doi:10.1197/jamia.M2544

Freire, P. (2007). Pedagogy of the Oppressed. New York: Continuum.

French, S. (1986). Decision Theory: An introduction of the Mathematics of Rationality. Chichester: Ellis Harwood Limited.

French, S. (1996). Multi-attribute decision support in the event of a nuclear accident. *Journal of Multi-Criteria Decision Analysis*, 5(1), 39–57. doi:10.1002/(SICI)1099-1360(199603)5:1<39::AID-MCDA109>3.0.CO;2-Q

French, S., & Geldermann, J. (2005). The varied contexts of environmental decision problems and their implications for decision support. *Environmental Science & Policy*, 8(4), 378–391. doi:10.1016/j.envsci.2005.04.008

French, S., & Niculae, C. (2005). Believe in the Model: Mishandle the Emergency. *Journal of Homeland Security and Emergency Management*, 2(1). doi:10.2202/1547-7355.1108

Fussel, H. M. (2007). Vulnerability: a generally applicable conceptual framework for climate change research. *Global Environmental Change*, 17(2), 155–167. doi:10.1016/j.gloenvcha.2006.05.002

Gallopin, G. C. (2006). Linkages between vulnerability, resilience, and adaptive capacity. *Global Environmental Change*, 16, 293–303. doi:10.1016/j.gloenvcha.2006.02.004

Galtung, J., & Ruge, M. H. (1965). The structure of foreign news. The Presentation of the Congo, Cuba and Cyprus crisis in four Norwegian newspapers. *Journal of Peace Research*, 2, 64–91. doi:10.1177/002234336500200104

Gamboa, D., & Peña, D. (2004). Sistema de apoyo para manejo de inventarios en presencia de donaciones. Trabajo especial de grado. Venezuela: Universidad Metropolitana.

Gardner, M., Scott, J., & Horan, B. (2008). MiRTLE. *EDUCAUSE Review*, 43(5). Retrieved January 28, 2009 from http://connect.educause.edu/Library/EDUCAUSE+Review/MiRTLE/47238?time=122340170.

Garrett, J. J. (2005). Ajax: A New Approach to Web Applications. Retrieved Oct 14, 2008, from http://www.adaptivepath.com/publications/essays/archives/000385.php.

Gavriilidis, A. (2009). Greek riots 2008: A mobile Tiananmen . In Economides, S., & Monastiriotis, V. (Eds.), The return of street politics? Essays on the December riots in Greece (pp. 15–19). London: LSE Reprographics Department.

GDACS. (2008). Global Disaster Alert and Coordination System GDACS. Retrieved June 30, 2009, from http://www.gdacs.org

Gee, J. P. (2003). What video games have to teach us about learning and literacy. New York: Palgrave Macmillan. Retrieved May 20, 2009 from http://www.amazon.com/Video-Games-Learning-Literacy Second/dp/1403984530/ref=sr_1_1?ie=UTF8&s=books&qid=1245466968&sr=8-1.

Geser, H. (1997). The System of Public Media in Transition. Some contemporary "Megatrends" and their Implications for Social Theory and Research. Retrieved October, 10, 2009, from http://socio.ch/intcom/t_hgeser05.htm

Geser, H. (2002). Towards a (Meta-)Sociology of the Digital Sphere. Retrieved April 6, 2009, from http://socio.ch/intcom/t_hgeser13.htm

Geurts, P., Fillet, M., Seny, D., Meuwis, M., Malaise, M., Merville, M., & Wehenkel, L. (2005). Proteomic mass spectra classification using decision tree based ensemble methods. *Bioinformatics (Oxford, England)*, 21(15), 3138–3145. doi:10.1093/bioinformatics/bti494

Ginsberg, J., Mohebbi, M. H., Patel, R. D., Brammer, L., Smolinski, M. S., & Brilliant, L. (2009). Detecting influenza epidemics using search engine query data. Nature, 457(7232), 1012–1014. doi:10.1038/nature07634

Gitlin, T. (1980). The whole world is watching. Mass media in the making and unmaking of the new left. Berkeley, CA: University of California Press.

Glanz, M. H. (2004). Usable Science 8: Early Warning Systems: Do's and Don'ts. National Center for Atmospheric Research Report of workshop, 20–23 October 2003, Shanghai, China. Boulder, CO.

Glossary of Risk Management. (2009). Retrieved June, 20, 2009, from www.merrea.org

Godet, M. (1990). Integration of scenarios and strategic management: Using relevant, consistent and likely scenarios. Futures, 22(7), 730–739. doi:10.1016/0016-3287(90)90029-H

Goldsztajn, M. (2005). Sistema de apoyo para la redistribución de vehículos en caso de desbordamiento de quebradas. Trabajo especial de grado. Venezuela: Universidad Metropolitana.

Gombay, E., & Serban, D. (2005). An adaptation of Pages CUSUM test for change detection. Periodica Mathematica Hungarica, 50, 135–147. doi:10.1007/s10998-005-0007-7

Gómez, M., & Zapata, E. (2001). Catástrofes naturales: Plan de acción para evaluación y asistencia. Trabajo especial de grado. Venezuela: Universidad Metropolitana.

Government of Japan. (2006). Japan's Natural Disaster Early Warning Systems and International Cooperative Efforts. Tokyo, Japan: Technical Report - Early Warning Sub-Committee of the Inter-Ministerial Committee on International Cooperation for Disaster Reduction.

Granot, H. (1997). Emergency inter-organizational relationship. Disaster Prevention and Management, 6(5), 305–310. doi:10.1108/09653569710193736

Graves, R. J. (2004, May). Key technologies for emergency response. Paper presented at the International Community on Information Systems for Crisis Response (ICSCRAM2004) conference, Brussels.

Greco, S., Matarazzo, B., & Slowinski, R. (2001). Rough sets theory for multicriteria decision analysis. European Journal of Operational Research, 129(1), 1–47. doi:10.1016/S0377-2217(00)00167-3

Green, C., & Penning-Rowsell, E. (2007). More or less than words? Vulnerability as discourse . In McFadden, L., Nicholls, R. J., & Penning-Rowsell, E. (Eds.), Managing Coastal Vulnerability. Amsterdam: Elsevier.

Grishman, R., Huttunen, S., & Yangarber, R. (2003). Information Extraction for Enhanced Access to Disease Outbreak Reports. Journal of Biomedical Informatics, 35(4), 236–246. doi:10.1016/S1532-0464(03)00013-3

Guidotti, T. L. (2004). Why do public health practitioners hesitate? Journal of Homeland Security and Emergency Management, 1(4), 403. doi:10.2202/1547-7355.1069

Gundel, S. (2005). Towards a new typology of crises. Journal of Contingencies and Crisis Management, 13(3), 106–115. doi:10.1111/j.1468-5973.2005.00465.x

Güngör, Z., & Arikan, F. (2000). Application of fuzzy decision making in part-machine grouping. International Journal of Production Economics, 63(2), 181–193. doi:10.1016/S0925-5273(99)00010-9

Guo, Z., Fang, F., & Whinston, A. B. (2006). Supply chain information sharing in a macro prediction market. Decision Support Systems, 42(3), 1944–1958. doi:10.1016/j.dss.2006.05.003

Gustavsson, P. M., Wemmergård, J., Garcia, J. J., & Larsson, M. N. (2006). (CML). In Simulation Interoperability Workshop 2006. Stockholm, Sweden: Expanding the Management Language Smorgasbord Towards Standardization of Crisis Management Language.

Häkkinen, M. T., & Sullivan, H. T. (2007). Effective Communication of Warnings and Critical Information: Application of Accessible Design Methods to Auditory Warnings. In B. Van de Walle, P. Burghardt & C. Nieuwenhuis (Ed.), Proceedings of the 4th International Conference on Information Systems for Crisis Response and Management ISCRAM2007 (pp. 167-171). Newark, NJ.

Harrald, J. R. (2002). Web enabled disaster and crisis response: What have we learned from the September 11th. In 15th Bled eCommerce Conference Proceedings "eReality: Constructing the eEconomy (17-19th June 2002). Retrieved October 4, 2008, from http://domino. fov.uni-mb.si/proceedings.nsf/Proceedings/D3A6817C 6CC6C4B5C1256E9F003BB2BD/$File/Harrald.pdf

Harrald, J. R. (2009). Achieving agility in disaster management. International Journal of Information Systems for Crisis Response Management, 1(1), 1–11.

Hatjefthymiades, S., Marias, G. F., & Priggouris, I. (2006). SCIER: Sensor and Computing Infrastructure for Environmental Risks. Retrieved October 7, 2009 from http://62.103.200.120/projects/scier2/site/Assets/ D01_Project_Presentation_v01.pdf

Heath, R. (1995). The Kobe earthquake: some realities of strategic management of crises and disasters. Disaster Prevention and Management, 4(5), 11–24. doi:10.1108/09653569510100965

Heckerman, D., Mamdani, A., & Wellman, M. (1995). Real-world applications of Bayesian networks. Communications of the ACM, 38(3), 24–26. doi:10.1145/203330.203334

Hefeeda, M., & Bagheri, M. (2007). Wireless Sensor Networks for Early Detection of Forest Fires. Pisa, Italy. In Proceedings of the 1st International Workshop on Mobile Ad hoc and Sensor Systems for Global and Homeland Security (MASS-GHS), Pisa, Italy.

Heinrichs, W., Youngblood, P., Harter, P., & Deve, P. (2008). Simulation for team training and assessment: Case studies of online training in virtual worlds. World Journal of Surgery, 32, 161–170. doi:10.1007/s00268-007-9354-2

Helbig, N., Löwe, H., & Lehning, M. (2009). Radiosity approach for the shortwave surface radiation balance in complex terrain. Journal of the Atmospheric Sciences. .doi:10.1175/2009JAS2940.1

Held, V. (2001). Technological Options for an Early Alert of the Population, Zivilschutzforschung [Bonn, Germany.]. Zentralstelle für Zivilschutz BBK, 45, 64–130.

Hellas Grid. (n.d.). Retrieved September 10, 2008, from www.hellasgrid.gr.

Hellstrom, T. (2006). Critical infrastructure and systemic vulnerability: Towards a planning framework. Safety Science, 45(33), 415–430.

Helmer, M., & Hilhorst, D. (2006). Natural disasters and climate change. Disasters, 30(1), 1–4. doi:10.1111/ j.1467-9523.2006.00302.x

Hernández, J. G., & García, M. J. (2006). The Importance of the Procurement Function in Logistics. Proceedings ICIL'2006, Kaunas University of Technology, Lithuania 149-157.

Hernández, J. G., & García, M. J. (2007). Investigación de operaciones y turismo en Revista de Matemática . Teoría y Aplicaciones, 14(2), 221–238.

Hernández, J. G., García, M. J., & Hernández, G. J. (2009). Influence of Location Management in supply chain Management. Global Business And Technology Association (GBATA 2009). Reading Book; Prague, Czech Republic (pp. 500-507).

Hernández, J. G., García, M. J., & Nieto, C. (2005). Modelos matemáticos y su ayuda en caso de catástrofes. Universidades y Riesgo. Una vitrina desde la UCV; Facultad de Arquitectura Universidad Central de Venezuela; Caracas, Venezuela; Publicado en CD del evento Hábitat y Riesgo el rol de las Universidades.

Hernández, L., & Rodríguez, V. (2005). Herramienta de apoyo para la ayuda del control poblacional y preservación del núcleo familiar ante una catástrofe. Trabajo especial de grado. Venezuela: Universidad Metropolitana.

Herold, S., Sawada, M., & Wellar, B. (2005, June). Integrating geographic information systems, spatial databases and the internet: a framework for disaster management. In Proceedings of the 98th Annual Canadian Institute of Geomatics Conference, Ottawa, Canada.

Heugens, P., & van Oosterhout, J. (2001). To boldly go where no man has gone before: integrating cognitive and physical features in scenario studies. Futures, 33(10), 861–872. doi:10.1016/S0016-3287(01)00023-4

Hewitt, A., Spencer, S., Mirliss, D., & Twal, R. (2009). Preparing the graduate student for virtual world simulations: Exploring the potential of an emerging technology. *Innovate Journal of Online Education, 5*(6). Retrieved September 20, 2009 from http://innovateonline.info/index.php?view=article&id=690.

Hewitt, A., Spencer, S., Ramloll, R., & Trotta, H. (2008). Expanding CERC beyond public health: Sharing best practices with healthcare managers via virtual learning. *Health Promotion Practice, 9*(4), (Special Supplement, Pandemic), 83-87.

Hewitt, K. (1998). Excluded perspectives in the social construction of disaster . In Quarantelli, E. L. (Ed.), *What is a disaster? Perspectives on the question* (pp. 75–91). London: Routledge.

Hill, A. (2005). Information Technology & GIS Employment During an Emergency Event. *ESRI Professional Papers*. Retrieved November 2005, from http://gis.esri.com/library/userconf/ proc05/papers/pap1024.pdf

Hodgkin, J., Belton, V., & Koulouri, A. (2005). Supporting the intelligent MCDA user: A case study in multi-person multi-criteria decision support. *European Journal of Operational Research, 160*(1), 172–189. doi:10.1016/j.ejor.2004.03.007

Hofstede, G. (1980). Motivation, leadership, and organization: do American theories apply abroad? *Organizational Dynamics*, 42–63. doi:10.1016/0090-2616(80)90013-3

Hohenhasu, S., Hohenhas, J., Saunders, M., Vadergrift, J., Kohler, T., & Manikowskie, M. (2008). Emergency response: lessons learned during a community hospital's in situ fire simulation. *Journal of Emergency Nursing: JEN, 34*(4), 352–354. doi:10.1016/j.jen.2008.04.025

Holly, W. (2008). Tabloidisation of political communication in the public sphere. In R, Wo & R. Ko (Eds.), *Handbook of Communication in the Public Sphere* (pp. 317-344). Berlin, New York: Mouton de Gruyter.

Horrigan, J. B. (2007). A Typology of Information and Communication Technology Users. *PEW/INTERNET & AMERICAN LIFE PROJECT*. Washington, DC: PEW/INTERNET.

Hospital Business Week. (2008). UFS Burn Center tests virtual reality disaster training: Gaming culture offers endless options for training. Retrieved September 20, 2009 from http://www.homeland1.com/ homeland-security-products/cpr-training-simulation-equipment/articles/429022-ufs-burn-center-tests-virtual-reality-disaster-training/.

Huffman, E., & Prentice, S. (2008, March). Social Media's New Role in Emergency Management. *Emergency Management and Robotics for Hazardous Environments*, Albuquerque, New Mexico.

Hughes, A. L., Palen, L., Sutton, J., & Vieweg, S. (2008). "Site-Seeing" in Disaster: An examination of on-line social convergence. In F. Fiedrich & B. Van de Walle, (Eds.), *Proceedings of the 5th International ISCRAM Conference*. Washington, DC, USA, May 2008.

IFRC (International Federation of Red Cross and Red Crescent Societies). (2005). *World Disasters Report 2005: Focus on Information in Disasters*. Bloomfield, CT: Kumarian Press Inc.

Ignizio, J. (1976). *Goal Programming and extensions*. Massachusetts: Lexington Books.

Indian Government Release.

INFORMS. (2009). OR/MS Resource Collection. Retrieved October 1, 2009, from www.informs.org/Resources/Computer_Programs/.

Intergovernmental Panel on Climate Change (IPCC). (2001). *Climate Change 2001: The Scientific basis (Contribution of working group I to the Third Assessment Report of the IPCC)*. Cambridge: Cambridge University press.

International Risk Governance Council. (2005). *Risk governance: Towards and integrative approach*. White paper no.1. Retrieved August 5, 2009, from http://www.irgc.org/IMG/pdf/IRGC_WP_No_1_Risk_Governance__reprinted_version_.pdf

International Strategy for Disaster Reduction (ISDR). (2004). Living with risk: A global review of disaster reduction initiatives. Geneva, Switzerland: United Nations Publications.

Ishak. R. (2005). Special Report: Disaster Planning and Management. NCD Malaysia 2004, 3(2).

ITU. (2005) International Telecommunication Union Web Site. Retrieved May 2009, from http://www.itu.int/ITU-R/information/emergency/index.asp

Iztúriz, A., Barrientos, Y., Ruíz, S., & Vierma de Bezada, M. (2007 b). Structured instructional game SALTARIESGO to promote safety culture at Vargas State, Venezuela. In S. Wang et al., (Eds.), Strategy and Implementation of Integrated Risk Management. International Disaster Reduction Conference. (pp. 472-475). Harbin, China: Qunyan Press.

Iztúriz, A., Barrientos, Y., Ruíz, S., & Vierma de Bezada, M. (2008). Juego instruccional estructurado: Ludograma para la prevención de riesgos en el Estado Vargas, Venezuela. Research in Geographic Education, 10(1), 33–47.

Iztúriz, A., Barrientos, Y., Ruíz, S., & Vierma de Bezada, M. (2008). Un ludograma para la prevención de riesgos en el Estado Vargas. Paper presented at the meeting XIII Congreso Nacional de tecnologías de la Información Geográfica. Universidad de Las Palmas de Gran Canaria y el Grupo de Tecnologías de la Información Geográfica de la Asociación de Geógrafos Españoles. Las Palmas de Gran Canaria, España.

Iztúriz, A., Tineo, A., Barrientos, Y., Pinzón, R., Ruíz, S., & Montilla, J. (2007a). El Juego instruccional como estrategia de aprendizaje sobre riesgos socio-naturales. Revista Educere, 11(36), 103–112.

Jackson, P. (2005). Hard task of draining New Orleans. British Broadcasting Corporation NEWS. Retrieved September 8, 2009, from http://news.bbc.co.uk/go/pr/fr/-/2/hi/americas/4209394.stm

Jacobsen, M. (2005). ASGARD – System Description 1.1, EU Joint Research Centre. Retrieved June 30, 2009, from http://dma.jrc.it/new_site/documents/AsgardSystemDescription.pdf

Jacquet-Lagreze, E., & Siskos, J. (1982). Assessing a set of additive utility functions for multicriteria decision-making, the UTA method. European Journal of Operational Research, 10(2), 151–164. doi:10.1016/0377-2217(82)90155-2

Jain, S. & McLean, C.R. (2006). An integrating framework for modeling and simulation for incident management. Journal of Homeland Security and Emergency Management, 3(1), Article 9.

Jakobson, G., Buford, J. & Lewis, L. (2008). Models of feedback and adaptation in multi-agent systems for disaster situation management (6943). Orlando, USA.

Jaksch, S., Pfennigschmidt, S., Sandkuhl, K., & Thiel, C. (2003). Information Logistic applications for information-on-demand scenarios: concepts and experiences from the WIND project. In Proceedings of the 29th Conference on EUROMICRO (pp. 41-147). Belek, Turkey.

James, A. M., & Rashed, T. (2006). In their own words: Utilizing weblogs in quick response research . In Guibert, G. (Ed.), Learning from Catastrophe: Quick Response Research in the Wake of Hurricane Katrina (pp. 57–84). Boulder: Natural Hazards Center Press.

James, E. (2006). Clean or corrupt: tsunami aid in Aceh. Canberra, Australia: The Australian National University, Asia Pacific School of Economics and Government. Retrieved from http://apseg.anu.edu.au

Jarventaus, J. (2007). Virtual threat, real sweat. Training & Development, 61(5), 72–78.

Jayawardane, A. K. W. (2006). Disaster mitigation initiatives in Sri Lanka. In Proceedings of the International Symposium on Management System for Disaster Prevention, March 9-11, 2006, Kochi, Japan.

Jensen, E. (1999). Teaching with the brain and mind. Alexandria, CA: ASCD.

Jensen, E. (2000). Moving with the brain in mind. Educational Leadership, 58(3), 34–37.

Jessop, A. (2004). Sensitivity and robustness in selection problems. Computers & Operations Research, 31(4), 607–622. doi:10.1016/S0305-0548(03)00017-0

Jiménez, A., Ríos-Insua, S., & Mateos, A. (2006). A generic multi-attribute analysis system. Computers & Operations Research, 33(4), 1081–1101. doi:10.1016/j.cor.2004.09.003

JMA. (Japan Meteorological Agency (JMA). (2008). What is an earthquake early warning? Retrieved September, 30, 2009 from http://www.jma.go.jp/jma/en/Activities/eew1.html

Joint Commission International. (2008). Emergency Management in Healthcare: An All Hazards Approach. Chicago, IL: JCAHO.

Jones, A. L. (2008). Seismic eruption. Retrieved September, 30, 2009 from http://www.smate.wwu.edu/ slibrary/SeismicEruption/main.html

Jones, G., & Hicks, J. (2004). 3D online learning environments for emergency preparedness and homeland security training. Retrieved September 20, 2009 from http://courseweb.unt.edu/gjones/pdf/Jones_ elearn04.pdf.

Joomla. (2009). Retrieved May 1, 2009, from http://www.joomla.org/.

Jotshi, A., Gong, Q., & Batta, R. (2009). Dispatching and routing of emergency vehicles in disaster mitigation using data fusion. Socio-Economic Planning Sciences, 43(1), 1–24. doi:10.1016/j.seps.2008.02.005

Joyce, M. (2008). Campaign: Digital tools and the Greek riots. Retrieved August 1, 2009, from http://www.digiactive.org/2008/12/22/digital-tools-and-the-greek_riots

Kahneman, D., Slovic, P., & Tversky, A. (1982). Judgment under Uncertainty: Heuristics and biases. Cambridge: Cambridge University Press.

Kamp, U., Growley, B. J., Khattak, G. A., & Owen, L. A. (2008). GIS-based landslide susceptibility mapping for the 2005 Kashmir earthquake region. Geomorphology, 101, 631–642. doi:10.1016/j.geomorph.2008.03.003

Kaufman, D. Q., Stasson, M. F., & Hart, J. W. (1999). Are the tabloids always wrong or it that just what we think? Need for cognition and perceptions of articles in print media . Journal of Applied Social Psychology, 29(9), 1984–1997. doi:10.1111/j.1559-1816.1999.tb00160.x

Kazusa, S. (2006). Disaster management of Japan. In Proceedings of the International Symposium on Management System for Disaster Prevention, March 9-11, 2006, Kochi, Japan.

Keller, C., Siegrist, M., & Gutscher, H. (2006). The Role of the Affect and Availability Heuristics in Risk Communication. Risk Analysis, 26(3), 631–639. doi:10.1111/j.1539-6924.2006.00773.x

Kelvin, S., Rodolfo, S., & Fernando, S. (2006). Global sea-level rise is recognized, but flooding from anthropogenic land subsidence is ignored around northern Manila Bay, Philippines. Disasters, 30(1), 118–139. doi:10.1111/j.1467-9523.2006.00310.x

Kettlewell, J. (2005a, January 6). Early warning technology – is it enough? British Broadcasting Corporation NEWS. Retrieved January 6, 2005, from http://news.bbc.co.uk/go/pr/fr/-/2/hi/science/nature/4149201.stm

Kettlewell, J. (2005b, March 25). Tsunami alert technology – the iron link. British Broadcasting Corporation NEWS. Retrieved March 25, 2005, from http://news.bbc.co.uk/go/pr/fr/-/2/hi/science/nature/4373333.stm

Kicinger, R., Arciszewski, T., & De Jong, K. (2005). Evolutionary Computation and Structural Design: A Survey of the State-of-the-art. Computers & Structures, 83, 1943–1973. doi:10.1016/j.compstruc.2005.03.002

King, B. J. (2008). Web 2.0 and education: Literature review of tools & technologies to enhance education. Retrieved from http://www.scribd.com/doc/9010043/Web-20-Tools-for-Higher-Education-Literature-Review-by-Brian-J-King.

Kirkpatrick, D. (1975). "Techniques for evaluating programs." Parts 1, 2, 3 and 4. Evaluating Training Programs. ASTD.

Kirkwood, C. (1992). An Overview of Methods for Applied Decision Analysis. Interfaces, 22(6), 28–39. doi:10.1287/inte.22.6.28

Kittur, A., Suh, B., & Chi, E. H. (2008). Can you ever trust a wiki?: impacting perceived trustworthiness in wikipedia. In Proceedings of the ACM 2008 Conference on Computer Supported Cooperative Work (San Diego, CA, USA, November 08 - 12, 2008). CSCW '08. ACM, New York, NY, (pp.477-480). Retrieved August 7, 2009, from http://doi.acm.org/10.1145/1460563.1460639

Klein, R. J. T., Nicholls, R. J., & Thomalla, F. (2003). Resilience to natural hazards: how useful is this concept? Environmental Hazards, 5(1–2), 35–45.

Kojadinovic, I. (2005). An axiomatic approach to the measurement of the amounts of interaction among criteria or players. Fuzzy Sets and Systems, 152(3), 417–435. doi:10.1016/j.fss.2004.11.006

Kosucu, B., Irgan, K., Kucuk, G., & Baydere, S. (2009). FireSenseTB: A Wireless Sensor Networks TestBed for Forest Fire Detection. In the 5th ACM International Wireless Communications and Mobile Computing Conference (IWCMC'09), Leipzig, Germany.

Kosuda, S., Ichihara, K., Watanabe, M., Kobayashi, H., & Kusano, S. (2000). Decision-Tree sensitivity analysis for cost-effectiveness of Chest 2-Fluoro-2-D-[18 F] Fluorodeoxyglucose positron emission tomography in patients with pulmonary nodules (Non-small cell lung carcinoma) in Japan. Chest, 117, 346–353. doi:10.1378/chest.117.2.346

Kouzmin, A., Jarman, A. M. G., & Rosenthal, U. (1995). Inter-organizational policy processes in disaster management. Disaster Prevention and Management, 4(2), 20–37. doi:10.1108/09653569510082669

Kreimer, A., & Arnold, M. (Eds.). (2000). Managing disaster risk in emerging economies. Washington, DC: World Bank. doi:10.1596/0-8213-4726-8

Krimsky, S. (2007). Risk communication in the internet age: The rise of disorganized skepticism. Environmental Hazards, 7, 157–164. doi:10.1016/j.envhaz.2007.05.006

Kucuk, G., Kosucu, B., Yavas, A., & Baydere, S. (2008). FireSense: Forest Fire Prediction and Detection System using Wireless Sensor Networks. In the 4th IEEE/ACM International Conference on Distributed Computing in Sensor Systems (DCOSS'08), Santorini Island, Greece.

Kumar, S. V., & Bhagavanulu, D. V. S. (2007). Flood simulation and inundation mapping of Adyar river: A case study using GIS. Disaster and Development: Journal of the National Institute of Disaster Management, 1(2), 155–168.

Kuppuswamy, S., et. al. (2009). Women, information technology and disaster management:

Kuutti, K. (1996). Activity Theory as a Potential Framework for Human-Computer Interaction Research . In Nardi, B. (Ed.), Context and Consciousness: Activity Theory and Human-Computer Interaction. Cambridge, MA: MIT Press.

Lagadec, P. (2005). Crisis management in the 21st century: "Unthinkable" events in "inconceivable" contexts. Ecole Polytechnique - Centre National de la Recherche Scientifique, Cahier No 2005-003. Retrieved August 5, 2009, from http://hal.archives-ouvertes.fr/docs/00/24/29/62/PDF/2005-03-14-219.pdf

Laituri, M., & Kodrich, K. (2008). On line disaster response community: People as sensors of high magnitude disasters using internet GIS. Sensors (Basel, Switzerland), 8(5), 3037–3055. doi:10.3390/s8053037

Lave, J., & Wenger, E. (1990). Situated Learning: Legitimate Peripheral Participation. Cambridge, UK: Cambridge University Press.

Learning Light. (2007). Second Life®and Virtual worlds.

Lecuna, M., & Roa, L. (2005). Logística para reestablecer el ambiente de infraestructura después de una catástrofe de origen natural. Trabajo especial de grado. Venezuela: Universidad Metropolitana.

Lee, E., & Wong, K. W. (2008). A review of using virtual reality for learning. In (Z. Pan et al. Eds.) Transactions on Edutainment 1, LINCS 5080, 231-241.

Lee, S. (2001). On solving unreliable planar location problems. Computers & Operations Research, 28(4), 329–344. doi:10.1016/S0305-0548(99)00120-3

Lehning, M., & Wilhelm, C. (2005). Integral Risk Management and Physical Modelling for Mountainous Natural Hazards . In Albeverio, S., Jentsch, V., & Kantz, H. (Eds.), Extreme Events in Nature and Society (p. 349). Springer.

Lehning, M., Bartelt, P. B., Brown, R. L., Fierz, C., & Satyawali, P. (2002a). A physical SNOWPACK model for the Swiss Avalanche Warning Services. Part II: Snow Microstructure. Cold Regions Science and Technology, 35(3), 147–167. doi:10.1016/S0165-232X(02)00073-3

Lehning, M., Bartelt, P. B., Brown, R. L., Fierz, C., & Satyawali, P. (2002b). A physical SNOWPACK model for the Swiss Avalanche Warning Services. Part III: Meteorological Boundary Conditions, Thin Layer Formation and Evaluation. Cold Regions Science and Technology, 35(3), 169–184. doi:10.1016/S0165-232X(02)00072-1

Lehning, M., Löwe, H., Ryser, M., & Raderschall, N. (2008). Inhomogeneous precipitation distribution and snow transport in steep terrain. Water Resources Research, 44, W07404. .doi:10.1029/2007WR006545

Lehning, M., Völksch, I., Gustafsson, D., Nguyen, T. A., Stähli, M., & Zappa, M. (2006). ALPINE3D: A detailed model of mountain surface processes and its application to snow hydrology. Hydrological Processes, 20, 2111–2128. doi:10.1002/hyp.6204

Lem, A. (2008). Letter from Athens: Greek riots and the news media in the age of twitter. Retrieved August 3, 2009, from http://www.alternet.org/media/113389/letter_from_athens:_greek_riots_and_the_news_media_in_the_age_of_twitter

Lempert, R., Groves, D., Popper, S., & Bankes, S. (2006). A General, Analytic Method for Generating Robust Strategies and Narrative Scenarios. Management Science, 52(4), 514–528. doi:10.1287/mnsc.1050.0472

Lenhart, A., Kahne, J., Middaugh, E., Macqill, A., Evans, C., & Vitak, J. (2008). Teens, video games, and civics: Teens gaming experiences are diverse and include significant social interaction and civic engagement. Pew Internet & American Life Project. Retrieved May 12, 2009 from http://www.pewinternet.org/Presentations/2009/17-Teens-and-Social-Media-An-Overview.aspx

Li, Y., & Liao, X. (2007). Decision support for risk analysis on dynamic alliance. Decision Support Systems, 42(4), 2043–2059. doi:10.1016/j.dss.2004.11.008

Liebes, T. (2005). Viewing and Reviewing the Audience: Fashions in Communication Research . In Cu, J., & Gu, M. (Eds.), Mass Media and Society (4th ed., pp. 356–374). London: Edward Arnold.

Lieckens, K., & Vandaele, N. (2007). Reverse logistics network design with stochastic lead times. Computers & Operations Research, 34(2), 395–416. doi:10.1016/j.cor.2005.03.006

Lin, H. W., Nagalingam, S. V., & Lin, G. C. I. (2009). An interactive meta-goal programming-based decision analysis methodology to support collaborative manufacturing. Robotics and Computer-integrated Manufacturing, 25(1), 135–154. doi:10.1016/j.rcim.2007.10.005

Linder, R. (2006). Wikis, webs, and networks: Creating connections for conflict-prone settings. Washington: Council for Strategic and International Studies. Retrieved August 5, 2009, from http://www.csis.org/component/option,com_csis_pubs/task,view/id,3542/type,1/

Linge, J. P., Steinberger, R., Weber, T. P., Yangarber, R., van der Goot, E., Al Khudhairy, D. H., & Stilianakis, N. I. (2009). Internet surveillance systems for early alerting of health threats. Euro Surveillance : European Communicable Disease Bulletin, 14, 1–2.

Liu, S., Palen, L., Sutton, J., Hughes, A., & Vieweg, S. (2008) In Search of the Bigger Picture: The Emergent Role of On-Line Photo Sharing in Times of Disaster. In Proceedings of the 2008 ISCRAM Conference, Washington, DC.

López, A., & Sánchez, C. (2005). Empresas de riesgo mayor: Recuperación de su logística. Trabajo especial de grado. Venezuela: Universidad Metropolitana.

López, L., & Pérez, M. (2006). Deslaves e inundaciones: Sistema de emergencia para el municipio Baruta. Trabajo especial de grado. Venezuela: Universidad Metropolitana.

Loschiavo (2009). The future of emergency preparedness training. BAM INTEL: Leadership through thinking. Retrieved on September 20, 2009 from http://bamintel.blogspot.com/2009/06/future-of-emergency-preparedness.html

Loudni, S., Boizumault, P., & David, P. (2006). On-line resources allocation for ATM networks with rerouting. Computers & Operations Research, 33(10), 2891–2917. doi:10.1016/j.cor.2005.01.016

Louka, M., & Balducelli, C. (2001). Virtual reality tools for emergency operations support and training. In Proceedings of TIEMS 2001 (The International Emergency Management Society), Oslo, June 2001. Retrieved on September 20, 2009 from http://www2.hrp.no/vr/publications/tiems2001.pdf

Lowenstein, G. A. (1996). Out of control: Visceral influences on behaviour. Organizational Behavior and Human Decision Processes, 65(3), 272–292. doi:10.1006/obhd.1996.0028

Lu, D., Chou, Y., & Yuan, H. (2005). Paradigm in the institutional arrangement of protected areas management in Taiwan-a case study of Wu-Wei-Kang Waterfowl Wildlife Refuge in Ilan, Taiwan. Environmental Science & Policy, 8(4), 418–430. doi:10.1016/j.envsci.2005.03.013

Luo, J. (2007). Buyer-vendor inventory coordination with credit period incentives. International Journal of Production Economics, 108(1-2), 143–152. doi:10.1016/j.ijpe.2006.12.007

Madden, M., & Fox, S. (2006). Riding the Waves of "Web 2.0." Pew Internet Project, October 5, 2006.

Madoff, L. C. (2004). ProMED-mail: an early warning system for emerging diseases. Clinical Infectious Diseases, 39(2), 227–232. doi:10.1086/422003

Mahmoud, M., Liu, Y., Hartmann, H., Stewart, S., Wagener, T., & Semmens, D. (2009). A formal framework for scenario development in support of environmental decision-making. Environmental Modelling & Software, 24(7), 798–808. doi:10.1016/j.envsoft.2008.11.010

Malmold, A., & Balmaceda, M. (2005). ¡Alerta SISMO! Prevención sísmica en las escuelas. EIRD Informa, (10).

Malmold, A., & Balmaceda, M. (2005). La universidad y la formación para la construcción de planes de contingencia. Paper presented at the meeting I Encuentro Internacional y 2do Encuentro Nacional Educación Superior y Riesgos. Hábitat y Riesgo el Rol de las Universidades Universidad Central de Venezuela. Caracas, Venezuela.

Malmold, A., & Balmaceda, M. (2006). La multimedia ¡Alerta SISMO II! Plan de emergencia familiar. EIRD Informa, (13), 71.

Manatakis, D. V., Manolakos, E. S., Xanthopoulos, G., Roussos, A., & Viegas, D. (2010). In-silico estimation of the temperature field induced by a moving fire – Predictive modeling and validation using prescribed burn data. In the VI International Conf. on Forest Fire Research, Coimbra, Portugal.

Mandel, J., Beezley, J., Bennethum, L., Chakraborty, S., Coen, J., & Douglas, C. L, Hatcher, J., Kim, M., & Vodacek, A. (2007). A Dynamic Data Driven Wildland Fire Model. Computational Science - ICCS 2007 (pp. 1042-1049).

Manoj, B. S., & Baker, A. H. (2007). Communication Challenges in Emergency Response. Communications of the ACM, 50(3), 51–53. doi:10.1145/1226736.1226765

Manolakos, E. S., Manatakis, D., & Xanthopoulos, G. (2008). Temperature Field Modeling And Simulation of Wireless Sensor Network Behavior During a Spreading Wildfire, Lausanne – Switzerland. In Proceedings of the European Signal Processing Conference, EUSIPCO.

Mansourian, A., Rajabifard, A., Valadan Zoej, M. J., & Williamson, I. (2004). Facilitating Disaster Management Using SDI. Journal of Geospatial Engineering, 6(1), 39–44.

Mansourian, A., Rajabifard, A., Zoeja, M. J. V., & Williamson, I. (2006). Using SDI and web-based system to facilitate disaster management. Journal of Computers & Geosciences, 32, 303–315. doi:10.1016/j.cageo.2005.06.017

Manyena, S. B. (2006). The concept of resilience revisited. Disasters, 30(4), 433–450.

Marincioni, F. (2007). Information technologies and the sharing of disaster knowledge: The critical role of professional culture. Disasters, 31(4), 459–476. doi:10.1111/j.1467-7717.2007.01019.x

Martínez, M. (1994). La investigación cualitativa etnográfica en educación. México: Editorial Trillas.

Martins, L., & Daltrini, B. M. (1999). Activity Theory: a Framework to Software a JAIIO –Requirements Elicitation. WER'99 - Workshop en Requerimentos, 28 Jornadas Argentinas de Informática e Investigación Operativa. SADIO – IFIP.

Marty, C., Phillips, M., Lehning, M., Wilhelm, C., & Bauder, A. (2009). Climate change and natural hazards in the Grisons. Schweizerischen Zeitschrift für Forstwesen, 160, 201–209. doi:10.3188/szf.2009.0201

Mathur, A. R., Ventura-Traveset, J., Montefusco, C., Toran, F., Plag, H.-P., & Ruiz, L. (2006). Provision of emergency communciation messages through SBAS: the ESA ALIVE concept . In ION GNSS 2005 (pp. 2969–2975). Long Beach, California: Proceedings.

Mayer, H. (2003). Strengthening our homeland security: A collaborative planning approach to readiness enhancement. Journal of City and State Public Affairs, 10, 18–23.

Mazzarella, C., & Ríos, P. (2006). Desarrollo y validación de un sistema computarizado para el aprendizaje de un contenido de genética. Investigación y Postgrado, 21(2), 11–42.

McAllister, I. (1993). Sustaining relief with development: Strategic issues for the Red Cross and Red Crescent. Boston, MA: Marinus Nijhoff Publishers.

McEntire, D. A., & Fuller, C. (2002). The need for a holistic theoretical approach: an examination from El Niño disasters in Peru. Disaster Prevention and Management, 11(2), 128–140. doi:10.1108/09653560210426812

McEntire, D. E. (2002). Coordinating Multi-organizational Responses to Disasters: Lessons from the March 28, 2000, Forth, Worth Tornado. Disaster Prevention and Management, 11(5), 369–379. doi:10.1108/09653560210453416

McGinley, M., Turk, A., & Benet, D. Design (2006). Criteria for Public Emergency Warning Systems. In B. Van de Walle, P. Burghardt and C. Nieuwenhuis (Ed.), Proceedings of the 3rd International Conference on Information Systems for Crisis Response and Management ISCRAM2006 (pp. 154-164). Newark, NJ.

McLaughlin, P., & Dietz, T. (2008). Structure, agency and environment: toward an integrated perspective on vulnerability. Global Environmental Change, 18(1), 99–111. doi:10.1016/j.gloenvcha.2007.05.003

McLaughlin, S. (2003). Emergency Management Handbook. American Society for Healthcare Engineering of the American Hospital Association. Chicago: ASHE.

McMahon, M. (1997, December). Social Constructivism and the World Wide Web- A Paradigm for Learning. Paper presented at the meeting of the ASCILITE, Perth Australia.

McManus, J. H. (1994). Market-driven journalism: Let the citizen beware?London: Sage.

McQuail, D. (1998). Commercialisation and Beyond. In D. McQ & K. Si (Eds.), Media Policy. Convergence, Concentration and Commerce (pp. 107-127) London: Sage.

McQuail, D. (2005). McQuail's Mass Communication Theory (5th ed.). Thousands Oaks, CA: Sage.

Meier, P. (2008, Mar 26). Upgrading the Role of ICT in Conflict Early Warning/Response. Paper presented at the annual meeting of the ISA's 49th Annual Convention, Bridging Multiple Divides, Hilton San Francisco, San Francisco, CA, USA Online from http://www.al-lacademic.com/meta/p254277_index.html

Meissen, U., & Voisard, A. (2007). Situation-based Alerting Strategies in Early Warning Systems. In Proceedings of the International Conference Wireless Applications and Computing IADIS 2007, Lisbon, Portugal.

Meissen, U., & Voisard, A. (2008). Increasing the effectiveness of early warning via context-aware alerting. In F. Friedrich (Ed.), ISCRAM 2008, 5th International Conference on Information Systems for Crisis Response and Management. Proceedings. CD-ROM: May 4-7, 2008 (pp.431-440). The George Washington University. Washington, DC.

Meissen, U., Auge, J., & Fengler, M. (2007). SAFE - Sensor-Actuator-based Early-Warning System for Hazard Protection in Extreme Weather Conditions. In Proceedings of the 1st Conference on Research for Climate Protection and Protection from Climate Impacts klimazwei, (pp. 56-57). Berlin, Germany.

Meissner, A., Luckenbach, T., Risse, T., Kirste, T., & Kirchner, H. (2002). Design challenges for an integrated disaster management communication and information system. In DIREN 2002. The First IEEE Workshop on Disaster Recovery Networks. New York, June 24, 2002, IEEE Computer Society Press, Los Alamitos.

Mepham, M. (2005). Soduko (Traducido por Javier García Sanz). Caracas, Venezuela: Editorial Melvin.

Merz, M., Hiete, M., & Bertsch, V. (2009). Multicriteria decision support for business continuity planning in the event of critical infrastructure disruptions. International. Journal of Critical Infrastructures, 5(1/2), 156–174. doi:10.1504/IJCIS.2009.022854

Miceli, R., Sotgiu, I., & Settanni, M. (2008). Disaster preparedness and perception of flood risk: A study in an alpine valley in Italy. Journal of Environmental Psychology, 28(2), 164–173. doi:10.1016/j.jenvp.2007.10.006

Michalowski, W., Kersten, G., Koperczak, Z., & Szpakowicz, S. (1991). Disaster management with NEGO-PLA. Expert Systems with Applications, 2, 107–120. doi:10.1016/0957-4174(91)90108-Q

Michlmayr, G., & Lehning, M.Holzmann. H., Koboltschnig, G., Mott, R., Schöner, W., & Zappa. M. (2008). Application of Alpine3D for glacier mass balance and runoff studies at Goldbergkees, Austria. Hydrological Processes. doi:.doi:10.1002/hyp.7102

Midkiff, S. F., & Bostian, C. (2002). Rapidly-Deployable Broadband Wireless Networks for Disaster and Emergency Response. In Proceedings of the First IEEE Workshop on Disaster Recovery Networks. New York City, USA.

Mileti, D. S., Darlington, J. D., Passarini, E., Forest, B. C., & Myers, M. F. (1995). Toward an integration of natural hazards and sustainability. Environment and Progress, 17(2), 117–126.

Mingers, J., & Rosenhead, J. (2004). Problem structuring methods in action. European Journal of Operational Research, 152(3), 530–554. doi:10.1016/S0377-2217(03)00056-0

Moleiro, A., & Rojas, K. (2006). Sistema de apoyo para el manejo de afectados en una catástrofe. Trabajo especial de grado. Venezuela: Universidad Metropolitana.

Monteiro, R., & Rodrigues, G. (2006). A system of integrated indicators for socio-environmental assessment and eco-certification in agriculture-ambitec-agro. Journal of Technology Management & Innovation, 1(3), 47–59.

Montells, L., Montero, S., Díaz, P., & Aedo, I. (2006). SIGAME: Web-based system for resources management on emergencies. In B. Van de Walle & M. Turoff (Eds.), Proceedings of the 3rd International ISCRAM Conference (B.), Newark, NJ (USA), May 2006 (pp. 1-5).

Moore, L. K. (2006). Emergency Communications: The Emergency Alert System (EAS) and All-Hazard Warnings. Congressional Research Service Report. Washington, DC: Library of Congress.

Morris, M., & Ogan, Ch. (1996). The Internet as Mass Medium. The Journal of Communication, 46(1), 39–50. doi:10.1111/j.1460-2466.1996.tb01460.x

Moser, C. (1998). The asset vulnerability framework: Re-assessing ultra-poverty reduction strategies. World Development, 26(1), 1–19. doi:10.1016/S0305-750X(97)10015-8

Moss, M. L., & Townsend, A. M. (2006). Disaster forensics: Leveraging crisis information systems for social science. In Van de Walle, B. and Turoff, M. (Eds.) Proceedings of the 3rd International ISCRAM Conference, Newark, NJ (USA), May 2006.

Mott, R., Faure, F., Lehning, M., Löwe, H., Hynek, B., & Michlmayr, G. (2008). Simulation of seasonal snow cover distribution for glacierized sites (Sonnblick, Austrian Alps). Annals of Glaciology, 49, 155–160. doi:10.3189/172756408787814924

Murai, S. (2006). Monitoring of disasters using remote sensing GIS and GPS. In Proceedings of the International Symposium on Management System for Disaster Prevention, March 9-11, 2006, Kochi, Japan.

Murphy, T. (2006). Knowledge Management, Emergency Response, and Hurricane Katrina. International Journal of Intelligent Control and Systems, 11(4), 199–208.

Mustajoki, J., Hämäläinen, R. P., & Sinkko, K. (2007). Interactive computer support in decision conferencing: Two cases on off-site nuclear emergency management. Decision Support Systems, 42(4), 2247–2260. doi:10.1016/j.dss.2006.07.003

Mykhalovskiy, E., & Weir, L. (2006). The Global Public Health Intelligence Network and early warning outbreak detection: a Canadian contribution to global public health. Canadian Journal of Public Health, 97(1), 42–44.

Nakamura, K., Iwai, S., & Sawaragi, T. (1982). Decision Support Using Causation Knowledge Base. Systems, Man and Cybernetics . IEEE Transactions on, 12(6), 765–777.

Nakamura, Y., & Saita, J. (2007). UrEDAS, the earthquake early warning system: Today and tomorrow . In Gasparini, P., Manfredi, G., & Zschau, J. (Eds.), Earthquake early warning systems (pp. 249–281). Berlin: Springer-Verlag. doi:10.1007/978-3-540-72241-0_13

NDT Resource Center. (2009). Teaching with the Constructionivist Learning Theory. Retrieved May 12, 2009 from http://www.ndt-ed.org/TeachingResources/ClassroomTips/Constructivist%20_Learning.htm.

Neuman, R. W. (2002). The futur of the Mass Audience. In D. McQ (Ed.), McQuail's Reader of Mass Communication Theory (pp. 364-374). London: Sage

Nguyen, T. A. (2004). An object-oriented model for adaptive high-performance computing on the computational GRID. PhD Thesis no 3079, EPFL, Switzerland.

Nguyen, T. A., & Kuonen, P. (2007). Programming the Grid with POP-C++. [FGCS]. Future Generation Computer Systems, 23(1), 23–30. doi:10.1016/j.future.2006.04.012

Nguyen, T. A., Pasin, M., & Kuonen, P. (2007). POP-C++, Parallel Object programming C++: User and Installation Manual. University of Applied Sciences of Western Switzerland, Fribourg.

Nielsen, P. J. (1995). Ten Usability Heuristics. Retrieved Oct 14, 2008, from http://www.useit.com/papers/heuristicsheuristic_list.html.

Noji, E. K. (Ed.). (1997). The public health consequences of disasters. New York: Oxford University Press, Inc.

Nudell, M., & Antokol, N. (1988). The handbook of effective emergency and crisis management. Lexington, MA: Lexington Books.

Nunamaker, J. (1997). Future research in group support systems: needs, some questions and possible directions. International Journal of Human-Computer Studies, 47(3), 357–385. doi:10.1006/ijhc.1997.0142

O'Brien, F. (2004). Scenario planning--lessons for practice from teaching and learning. European Journal of Operational Research, 152(3), 709–722. doi:10.1016/S0377-2217(03)00068-7

O'Hagan, A., & Oakley, J. (2004). Probability is perfect, but we can't elicit it perfectly. Reliability Engineering & System Safety, 85(1-3), 239–248. doi:10.1016/j.ress.2004.03.014

Oberoi, S. V., & Thakur, N. K. (2005). Disaster Preparedness in The Hills: Natural Hazard Modeling Using GIS and Remote Sensing. In Proceedings of the 2005 ASCE International Conference on Computing in Civil Engineering, Cancun, Mexico.

Oblinger, D. 2004. The next generation of educational engagement. Journal of Interactive Media in Education, 8. Retrieved May 6, 2009 from http://jime.open.ac.uk/2004/8/oblinger-2004080duspaper.html.

Ogalde, I., & González, M. (2008). Nuevas tecnologías y educación. Diseño desarrollo, uso y evaluación de materiales didácticos. Cuidad de México, México: Editorial Trillas.

Ohmura, A. (2001). Physical basis for the temperature-based melt-index method. Journal of Applied Meteorology, 40, 753–761. doi:10.1175/1520-0450(2001)040<0753:PBFTTB>2.0.CO;2

Ornager, S. (n.d.). Media & Communication: How can partnership with the media support ESD? UNESCO Communication & Information, Bangkok.

OSGi Alliance. (2009). http://www.osgi.org.

Otten, J., Heijningen, B., & Lafortune, J. F. (2004, May). The virtual crisis management centre. An ICT implementation to canalise information. Paper presented at the International Community on Information Systems for Crisis Response Management (ISCRAM2004), Brussels.

Owen, R., & Bannerman, L. (2009). Italy in desperate race to save the buried after the earthquake. The Times. Retrieved April 7, 2009, from http://www.timesonline.co.uk/tol/news/world/europe/article6047691.ece

Pack, D., & Coleman, C. (2008). Assessing interoperability in emergency management standards . In Southeastcon, 2008 (pp. 334–339). Huntsville, AL: IEEE.

Page, E. S. (1954). Continuous Inspection Schemes. Biometrika, 41, 100–115.

Palen, L. (2008). On line social media in crisis events. Educase Quartely, 3, 76–78.

Palen, L., & Liu, S. B. (2007). Citizen communications in crisis: Anticipating the future of ITC-supported public participation. In CHI 2007 Proceedings (pp. 727-735), San Jose, CA, USA.

Palen, L., Hiltz, S. R., & Liu, S. B. (2007). On line forums supporting grassroots participation in emergency preparedness and response. Communications of the ACM, 50(3), 54–58. doi:10.1145/1226736.1226766

Palen, L., Vieweg, S., Sutton, J., Liu, S. B., & Hughes, A. (2007). Crisis informatics: Studying crisis in a networked world. Paper presented at the Third International Conference on e-Social Science (e-SS). Retrieved August 5, 2009, from http://www.cs.colorado.edu/~palen/Papers/iscram08/CollectiveIntelligenceISCRAM08.pdf

Palmer, J. (2008, May 3). Emergency 2.0 is coming to a website near you. New Scientist, 24–25. doi:10.1016/S0262-4079(08)61097-0

Pan American Health Organization (PAHO). (1985). Disaster Chronicles No. 3-Earthquake in Mexico September 19 and 20, 1985. Program of Emergency Preparedness and Disaster Relief Coordination. Retrieved June 28, 2009, from http://www.helid.desastres.net/

Paprika, Z., & Kiss, I. (1985). Interactions in decision support systems: Division of labor in DSSs. Engineering Costs and Production Economics, 8(4), 281–289. doi:10.1016/0167-188X(85)90053-9

Paquet, C., Coulombier, D., Kaiser, R., & Ciotti, M. (2006). Epidemic intelligence: a new framework for strengthening disease surveillance in Europe. Euro Surveillance : European Communicable Disease Bulletin, 11, 665. Retrieved from http://www.eurosurveillance.org/ViewArticle.aspx?ArticleId=665.

Peneva, V., & Popchev, I. (2003). Properties of the aggregation operators related with fuzzy relations. Fuzzy Sets and Systems, 139(3), 615–633. doi:10.1016/S0165-0114(03)00141-6

Pérez, C., Priol, T., & Ribes, A. (2003). A Parallel CORBA Component Model for Numerical Code Coupling. [IJHPCA]. International Journal of High Performance Computing Applications, 17(4), 417–429. doi:10.1177/10943420030174006

Perrow, C. (1984). Normal Accidents: Living with High-Risk Technologies. New York: Basic Books.

Perry, R. W., & Quarantelli, E. L. (Eds.). (2005). What is a disaster? New answers to old questions. Philadelphia: XLibris.

Persona, A., Regattieri, A., Pham, H., & Battini, D. (2007). Remote control and maintenance networks and its applications in supply chain management. Journal of Operations Management, 25(6), 1275–1291. doi:10.1016/j.jom.2007.01.018

Peters, H. P. (2009). Natural disaster and the media. International Strategy for disaster Reduction ISDR Retrieved April 13, 2009, from http://www.chmi.cz/katastrofy/peters.html

PHYSorg.com. (2006, March 3). Ask your car radio! Retrieved September, 30, 2009 from www.physorg.com/news11399.html

Picard, R. G. (2005). Money, Media and the Public Interests. In O. Ge. & K. H. Ja (Eds.), The Institutions of Democracy: The Press (pp. 337-350). New York: Oxford University Press.

Polgreen, P. M., Chen, Y., Pennock, D. M., & Nelson, F. D. (2008). Using internet searches for influenza surveillance. Clinical Infectious Diseases, 47(11), 1443–1448. doi:10.1086/593098

Polsky, C., Neff, R., & Yarnal, B. (2007). Building comparable global change vulnerability assessments: the vulnerability scoping diagram. Global Environmental Change, 17(3-4), 472–485. doi:10.1016/j.gloenvcha.2007.01.005

Popp, R., Armour, T., Senator, T., & Numrych, K. (2004). Countering Terrorism Through Information Technology. Communications of the ACM, 47(3), 36–43. doi:10.1145/971617.971642

Pouliquen, B., Kimler, M., Steinberger, R., Ignat, C., Oellinger, T., Blackler, K., et al. (2006). Geocoding multilingual texts: Recognition, Disambiguation and Visualisation. In Proceedings of the 5th International Conference on Language Resources and Evaluation LREC2006 (pp. 53-58). Genoa, Italy, 24-26 May 2006.

Prasad, B. (1995). JIT quality matrices for strategic planning and implementation. International Journal of Operations & Production Management, 15(9), 116–142. doi:10.1108/01443579510099706

Puras, J. C., & Iglesias, C. A. (2009). Disasters2.0. Application of Web2.0 technologies in emergency situations. In J. Landgren & S. Jul (Eds.), Proceedings of the 6th International ISCRAM Conference – Gothenburg, Sweden, May 2009.

Quarantelli, E. L. (1998). What is a disaster? Perspectives on the question. New York: Routledge.

Quarantelli, E. L. (2007). Problematical aspects of the information/ communication revolution for disaster planning and research: ten non-technical issues and questions. Disaster Prevention and Management, 6(2), 94–106. doi:10.1108/09653569710164053

Rabinovich, E., Dresner, M. E., & Evers, P. T. (2003). Assessing the effects of operational processes and information systems on inventory performance. Journal of Operations Management, 21(1), 63–80. doi:10.1016/S0272-6963(02)00041-4

Raderschall, N., Lehning, M., & Schär, C. (2008). Fine scale modelling of the boundary layer wind field over steep topography. Water Resources Research, 44, W09425. .doi:10.1029/2007WR006544

Raheja, N., Ojha, R., & Mallik, S. R. (1999). Role of internet-based GIS in Effective Natural Disaster Management. Retrieved June, 2009, from http://www.gisdevelopment.net/technology/gis/techgi0030.htm

Rajabifard, A., Mansourian, A., Williamson, I., & Valadan Zoej, M. J. (2004). Developing Spatial Data Infrastructure to Facilitate Disaster Management. In Proceedings GEOMATICS 83 Conference, Tehran, Iran.

Rajbhandari, R., & Subedi, J. (2005). Disaster informatics: Issues and future of information management. Proceedings of International Conference on Disaster Management: Achievements and Challenges (pp. 146-150). Nepal Engineering College, Kathmandu

Raman, M., Terry, R., & Lorne, O. (2006). Knowledge Management System for Emergency Preparedness: An Action Research Study. In Proceedings of the 39th Hawaii International Conference on System Sciences (pp. 1-10).

Ramloll, R. (2008, July). Using Web 2.0 & Table Top exercises in Safety and Health Profession Topics. Webinar presented for American Society of Safety Engineers.

Raskob, W., Gering, F., & Bertsch, V. (2009). Approaches to visualisation of uncertainties to decision makers in an operational Decision Support System. In Proceedings of the 6th International Conference on Information Systems for Crisis Response and Management (IS-CRAM 2009).

Ravichandran, N. (2007). A finite horizon inventory model: An operational framework. International Journal of Production Economics, 108(1-2), 406–415. doi:10.1016/j.ijpe.2006.12.057

Raymond, M. (2004). Personalized Emergency Alerting System . In Extreme Mark-Up Languages 2004. Montreal, Quebec: WIRLED PEAS-World Information Resources, Localized Environment Distribution.

Razavi, A. R., Gill, H., Ahlfeldt, H., & Shahsavar, N. (2008). Non-compliance with a postmastectomy radiotherapy guideline: Decision tree and cause analysis. BMC Medical Informatics and Decision Making, 8(41). doi:.doi:10.1186/1472-6947-8-41

Reese, S. D., Grani, A., & Danielan, L. H. (1994). The structure of news sources on television: A network analysis of 'CBS News, ' 'Nightline', 'MacNeil/Lehrer,' and 'this Week with David Brinkley.' . The Journal of Communication, 44(2), 84–107. doi:10.1111/j.1460-2466.1994.tb00678.x

Renn, O. (1992). The social Amplification of Risk. Theoretical Foundations and Empirical Applications. The Journal of Social Issues, 48(4), 137–160. doi:10.1111/j.1540-4560.1992.tb01949.x

Renn, O., & Walker, K. (Eds.). (2008). Global Risk Governance: Concept and Practice Using the IRGC Framework. Dordrecht, The Netherlands: Springer. doi:10.1007/978-1-4020-6799-0

Report, C. S. T. B. Computer Science and Telecommunications Board (1999). Summary of a Workshop on Information Technology Research for Crisis Management. National Research Council, Committee on Computing and Communications research to Enable Better Use of Information Technology in Government. Washington, DC: National Academy Press. Retrieved November 2005, from http://books.nap.edu/html/itr_crisis_mgmt

República Bolivariana de Venezuela. (2009). Ley de gestión Integral de Riesgos Socionaturales y Tecnológicos. Gaceta Oficial N° 39.095 del 9 de enero de 2009. Caracas, Venezuela: Asamblea Nacional.

Rettew, J. (2009). Crisis communication and social media. Paper presented at the AIM2009 Conference. Retrieved August 4, 2009, from http://www.slideshare.net/AIM_Conference/crisis-communications-and-social-media-jim-rettew-the-red-cross-2009-aim-conference

Revkin, A. C. (2004). With No Alert System, Indian Ocean Nations Were Vulnerable. The New York Times, Retrieved October 30, 2009, from http://www.nytimes.com/2004/12/27/science /27science.html

Ríos-Insua, D., & French, S. (1991). A framework for sensitivity analysis in discrete multi-objective decision-making. European Journal of Operational Research, 54(2), 176–190. doi:10.1016/0377-2217(91)90296-8

Rodríguez, D., & Zabala, J. (2002). Trasporte para el traslado de heridos y afectados desde refugios a centros de asistencia y albergues. Trabajo especial de grado. Venezuela: Universidad Metropolitana.

Rogers, E. M. (2003). Diffusion of News of the September 11 Terrorist Attacks . In No, M. A. (Ed.), Crisis Communications: Lessons from September 11 (pp. 17–30). Lanhalm, MA: Rowman & Littlefield Publishing Group.

Rohweder, L., & Virtanen, A. (2008). Learning for a sustainable future. Innovative solutions from the Baltic Sea region. Uppsala, Swedish: The Baltic University Press.

Rommelfanger, H. (2007). A general concept for solving linear multicriteria programming problems with crisp, fuzzy or stochastic values. Fuzzy Sets and Systems, 158(17), 1892–1904. doi:10.1016/j.fss.2007.04.005

Rose-Pehrsson, S. L., & Hart, S, J., Street, T. T., Williams, F. W., Hammond, M. H., Gottuk, D. T., Wright, M. T., & Wong, J. T. (2003). Early warning fire detection system using a probabilistic neural network. Fire Technology, 39(2), 147–171. doi:10.1023/A:1024260130050

Rose-Pehrsson, S. L., Shaffer, R. E., Hart, S. J., Williams, F. W., Gottuk, D. T., Strehlen, B. D., & Hill, S. A. (2000). Multi-criteria Fire Detection Systems Using a Probabilistic Neural network. Sensors and Actuators. B, Chemical, 69(3), 325–335. doi:10.1016/S0925-4005(00)00481-0

Roubelat, F. (2000). Scenario Planning as a Networking Process. Technological Forecasting and Social Change, 65(1), 99–112. doi:10.1016/S0040-1625(99)00125-0

Roy, B., & Vincke, P. (1981). Multicriteria analysis: survey and new directions. European Journal of Operational Research, 8(3), 207–218. doi:10.1016/0377-2217(81)90168-5

Ruhrmann, G., & Göbbel, R. (2007). Veränderungen der Nachrichtenfaktoren und Auswirkungen auf die journalistische Praxis in Deutschland. Retrieved April 2, 2009, from http://www.netzwerkrecherche.de/docs/ruhrmann-goebbel-veraenderung-der-nachrichtenfaktoren.pdf

Russell, S., & Norvig, P. (2003). Artificial intelligence. Upper Saddle River: Prentice Hall.

Sagun, A., Anumba, C. J., & Bouchlaghem, D. (2006a). Coping with Extreme Events in the Built Environment: ICT for Disaster Mitigation and Collaboration. In Proceedings of APSEC 2006 Asia Pacific Structural Engineering and Construction Conference, Challenges Toward Sustainable Construction, Kuala Lumpur, Malaysia.

Sagun, A., Bouchlaghem, D., & Anumba, C. J. (2006b). Improving Safety and Security Through the Deployment of ICT in Disaster Management and Mitigation. In INCITE-ITCSED 2006 World Conference on IT in Design and Construction, New Delhi, India.

Sagun, A., Bouchlaghem, D., & Anumba, C. J. (2009). A Scenario-Based Study on Information Flow and Collaboration Patterns in Disaster Management. Disasters: The Journal of Disaster Studies . Policy and Management, 33(2), 214–238.

Sakarovitch, M. (1979). Techniques mathématiques de la recherche opérationnelle III – Optimisation dans les réseaux. France: Université Scientifique et Médicale Institut National Polytechnique de Grenoble.

Samuelson, P. A., & Nordhaus, W. D. (1986). Economía Duodécima edición. España: McGraw-Hill. (English, 1985, Economics, USA: McGraw-Hill).

Santos-Reyes, J. (2007). Early warning coordination centres: a systemic view. In S. Hernandez, & C.A. Brebbia (Eds.), Engineering Nature-2007: First International Conference on the Art of Resisting Extreme Natural Forces (pp.111-120). Sussex, England, UK: WIT PRESS.

Santos-Reyes, J., & Beard, A. N. (2001). A systemic approach to fire safety management. Fire Safety Journal, 36, 359–390. doi:10.1016/S0379-7112(00)00059-X

Santos-Reyes, J., & Beard, A. N. (2009). Analysis of Tabasco's flooding by applying the SDMS model. In Proceedings of the 2nd International Conference on Risk Analysis and Crisis Response, 19-21 October, 2009, Beijing, China.

Sarimveis, H., Patrinos, P., Tarantilis, Ch. D., & Kiranoudis, Ch. T. (2007). Dynamic modelling and control of supply chain systems: A review. Computers & Operations Research, 35(11), 3530–3561. doi:10.1016/j.cor.2007.01.017

Scalem, M., Sincar, A. K., Bandyopadhyay, S., & Sinha, S. (2005). A Decentralized Disaster Management Information Network (DDMIN) for Coordinated Relief Operations. In 9th World Multiconference on Systemics, Cybernetics and Informatics (WMSCI 2005), Orlando, USA.

Schafer, W., Carroll, J. M., Haynes, S. R., & Abrams, S. (2008). Emergency management planning as collaborative work. Journal of Homeland Security and Emergency Management, 5(1), 10. doi:10.2202/1547-7355.1396

Schaller, D. R., & Allison-Bunnell, S. (2003). Practicing what we teach: How learning theory can guide development of online educational activities. Paper presented at Museums and the Web 2003, Charlotte, NC, March. Retrieved May 6, 209 from http://www.archimuse.com/msw2003papers/schaller/ schaller.html.

Schlauch, B. (2008). Game theory 2.0: Second life for disaster training. Retrieved September 20, 2009 from http://www.homeland1.com/homeland-security-products/technology/articles/427835-game-theory-2-0-second-life-for-disaster-training/

Schnaars, S. (1987). How to develop and use scenarios. Long Range Planning, 20(1), 105–114. doi:10.1016/0024-6301(87)90038-0

Schnaars, S., & Ziamou, P. (2001). The essentials of scenario writing. Business Horizons, 44(4), 25–31. doi:10.1016/S0007-6813(01)80044-6

Schoeder, R. (2002). Social Interaction in Virtual Environments: Key Issues, Common Themes, and a Framework for Research . In Schoeder, R. (Ed.), The Social Life of Avatars: Presence and Interaction in Shared Virtual Environments (pp. 1–18). London: Springer-Verlag.

Schoemaker, P. (1993). Multiple Scenario Development: Its Conceptual and Behavioral Foundation. Strategic Management Journal, 14(3), 193–213. doi:10.1002/smj.4250140304

Scholz, R., & Tietje, O. (2002). Embedded Case Study Methods. Integrating Quantitative and Qualitative Knowledge. Thousand Oaks: Sage Publications.

Scott, M. A. (2002). Modeling and analysis of multi-commodity network flows via goal programming. Master of Science in Operations Research dissertation, Air Force Institute of Technology.

Seddigh, N., Nandy, B., & Lambardis, J. (2006). An Internet Public Alerting System: A Canadian Experience. In B. Van de Walle, P. Burghardt & C. Nieuwenhuis (Eds.), Proceedings of the 3rd International Conference on Information Systems for Crisis Response and Management ISCRAM2006 (pp. 141-146). Newark, NJ.

Seeger, M. W. (2008). Disasters and Communication. In W. Do (Ed.), The International Encyclopedia of Communication. Oxford: Blackwell Publishing. Retrieved March 24, 2009, from http://www.communicationencyclopedia.com/subscriber/tocnode?id=g9781405131995_chunk_g97814051319959_ss43-1

September 20, 2009 from http://www.submityourarticle.com/articles/Peter-Yellowlees-5256/virtual-reality-67144.php.

Shafer, G. (1976). A mathematical theory of evidence. Princeton, NJ: Princeton Univ. Press.

Shankar, K. (2008). Wind, Water, and Wi-Fi: New Trends in Community Informatics and Disaster Management. The Information Society, 24(2), 116–120. doi:10.1080/01972240701883963

Sharma, A., & Gupta, M. (1998). Reducing urban risk through community participation (TDR Project Progress Report). Delhi, India: Sustainable Environment and Ecological Development Society (SEEDS).

Shaw, M. (1996). Civil society and media in global crisis: Representing distant violence. London: Pinter.

Shen, S., & O'Hare, G. M. P. (2004). Agent-Based resource selection for grid computing. Workshop 4: International Workshop on Agents and Autonomic Computing and Grid Enabled Virtual Organizations (pp. 658-672).

Shim, J., Warkentin, M., Courtney, J., Power, D., Sharda, R., & Carlsson, C. (2002). Past, present, and future of decision support technology. Decision Support Systems, 33(2), 111–126. doi:10.1016/S0167-9236(01)00139-7

Siegrist, M., & Gutscher, H. (2006). Flooding risks: A comparison of lay people's perception and expert's assessments in Switzerland. Risk Analysis, 26(4), 971–979. doi:10.1111/j.1539-6924.2006.00792.x

Siegrist, M., & Gutscher, H. (2008). Natural hazards and motivation for mitigation behaviour: People cannot predict the affect evoked by a severe flood. Risk Analysis, 28(3), 771–778. doi:10.1111/j.1539-6924.2008.01049.x

Sillem, S., & Wiersma, E. (2006). Comparing Cell Broadcast and Text Messaging for Citizen Warning. In B. Van de Walle, P. Burghardt & C. Nieuwenhuis (Ed.), Proceedings of the 3rd International Conference on Information Systems for Crisis Response and Management ISCRAM2006 (pp. 147-153). Newark, NJ.

Singer, E., & Endreny, P. M. (1987). Reporting hazards: Their benefits and costs. The Journal of Communication, 37(3), 10–26. doi:10.1111/j.1460-2466.1987.tb00991.x

Singhal, K., Singhal, J., & Starr, M. K. (2007). The domain of production and operations management and the role of Elwood Buffa in its delineation. Journal of Operations Management, 25(2), 310–327. doi:10.1016/j.jom.2006.06.004

Sivathanu, C., & Tseng, L. K. (1996). Fire Detection Using Near-IR Radiation and Source Temperature Discrimination. National Institute of Standards and Technology. Annual Conference on Fire Research.

Skouri, K., Konstantaras, I., Papachristos, S., & Ganas, I. (2009). Inventory models with ramp type demand rate, partial backlogging and Weibull deterioration rate. European Journal of Operational Research, 192(1), 79–92. doi:10.1016/j.ejor.2007.09.003

Slattery, C., Syvertson, R., & Krill, S. Jr. (2009). The eight step training model: Improving disaster management leadership. Journal of Homeland Security and Emergency Management, 6(1), Article 8.

Sontag, M. (2009). A learning theory for 21st –century students. Innovate, 5(4). Retrieved May 6, 2009 from http://www.innovateonline.info/index.php?view=article&id=524&action=article.

Sorensen, J. H., Sorensen, B. V., Smith, A., & Williams, Z. (2009). Results of An Investigation of the Effectiveness of Using Reverse Telephone Emergency Warning Systems in the October 2007 San Diego Wildfires. Report prepared for the U.S. Department of Homeland Security. Retrieved August 4, 2009, from http://galain-solutions.com/resources/San$2520DiegoWildfires$2520Report.pdf

Sprague, R., & Watson, H. (1986). Decision Support Systems. USA: Prentice Hall.

Srivastava, S. K., Hedge, V. S., & Jayaraman, V. (2007). Integrating technological interventions and a community-centric approach for disaster-risk reduction. Disaster and Development: Journal of the National Institute of Disaster Management, 1(2), 111–118.

Stahl, M. T. (2005). Open-source software: not quite endsville. Drug Discovery Today, 10(3), 219–222. doi:10.1016/S1359-6446(04)03364-1

Steinberger, R., Fuart, F., van der Goot, E., Best, C., von Etter, P., & Yangarber, R. (2008). Text Mining from the Web for Medical Intelligence. In F.-S. Françoise, D. Perrotta, J.Piskorski & R. Steinberger (Eds.), Mining Massive Data Sets for Security (pp. 295-310). Amsterdam: IOS Press. Retrieved from http://langtech.jrc.it/Documents/2009_MMDSS_Medical-Intelligence.pdf

Steinberger, R., Pouliquen, B., & van der Goot, E. (2009). An Introduction to the Europe Media Monitor Family of Applications. In Proceedings of the Workshop 'Information Access in a Multilingual World: Transitioning from Research to Real-World Applications SIGIR-CLIR2009. Boston, USA. 19-23 July 2009.

Steinle, E., & Vögtle, T. (2001). Automated Extraction and Reconstruction of Buildings in Laserscanning Data for Disaster Management. In E. P. Baltsavias, A. Gruen, & L. Van Gool (Eds.), Proceedings of the Workshop Automatic Extraction of Man-Made Objects from Aerial and Space Images (III), Swets & Zeitlinger, Lisse, The Netherlands.

Stenchion, P. (1997). Development and disaster management. *Australian Journal of Emergency Management*, 12(3), 40–44.

Stephenson, R., & Anderson, P. S. (1997). Disasters and the Information Technology Revolution. *Disasters*, 21(4), 305–334. doi:10.1111/1467-7717.00065

Stephenson, W. D., & Bonabeau, E. (2007). Expecting the unexpected: The need for a networked terrorism and disaster response strategy. *Homeland Security Affairs*, III(1). Retrieved August 5, 2009, from http://www.hsaj.org/?article=3.1.3

Stewart, S. I., Radeloff, V. C., Hammer, R. B., & Hawbaker, T. J. (2007). Defining the Wildland Urban Interface. *Journal of Forestry*, 105, 201–207.

Stewart, T. (2005). Dealing with uncertainties in MCDA. In Multiple Criteria Decision Analysis: State of the Art Surveys, International Series in Operations Research & Management Science (pp. 445-470). Springer.

Stewart, T. J., Janssen, R., & Herwijnen, M. (2004). A genetic algorithm approach to multiobjective land use planning. *Computers & Operations Research*, 31(14), 2293–2313. doi:10.1016/S0305-0548(03)00188-6

Stössel, F., Guala, M., Fierz, C., Manes, C., & Lehning, M. (Manuscript submitted for publication). On the micrometeorology of surface hoar on mountain snow covers. *Water Resources Research*.

Sutton, J., Palen, L., & Shklovski, I. (2008). Backchannels on the front lines: Emergent uses of social media in the 2007 Southern California Wildfires. In F. Fiedrich & B. Van de Walle (Eds.), Proceedings of the 5th International ISCRAM Conference, Washington, DC, USA, May 2008.

Systems, E. M. (1997). EMS (pp. 1–44). Project Evaluation Matrix.

Szczurek, M. (1989). Simulaciones y juegos instruccionales. Un hipertexto. Trabajo de Ascenso. Universidad Pedagógica Experimental Libertador. Instituto Pedagógico de Caracas. Trabajo Inédito. Caracas, Venezuela: Autor.

Taber, N. (2008). Emergency response: E-learning for paramedics and firefighters. *Simulation & Gaming*, 39(4), 515–527. doi:10.1177/1046878107306669

Tavana, M. (2004). A subjective assessment of alternative mission architectures for the human exploration of Mars at NASA using multicriteria decision making. *Computers & Operations Research*, 31(7), 1147–1164. doi:10.1016/S0305-0548(03)00074-1

Telecommunications Regulatory Commission of Sri Lanka. (1998). Pilot Study on the Use of Telecommunications in Disaster and Emergency Situations in Sri Lanka. In association with United Nations Office for the Coordination of Humanitarian Affairs, Working Group on Emergency Telecommunications and ICO Global Communications Interim Report. Retrieved November, 2009, from http://www.reliefweb.int/telecoms/tampere/slcs.html#EXECUTIVE%20SUMMARY

TETRA. (2008). Terrestrial Trunked Radio TETRA. Retrieved June 30, 2009, from http://www.tetra-association.com

The Risk Communicator. (2008). Social media and your emergency communication efforts. *The Risk Communicator*, 1, 3-6. Retrieved June 15, 2009, from http://emergency.cdc.gov

Tian, J., Wang, Y., Li, H., Li, L., & Wang, K. (2007). DSS development and applications in China. *Decision Support Systems*, 42(4), 2060–2077. doi:10.1016/j.dss.2004.11.009

Tietje, O. (2005). Identification of a small reliable and efficient set of consistent scenarios. *European Journal of Operational Research*, 162(2), 418–432. doi:10.1016/j.ejor.2003.08.054

Tinker, T., & Fouse, D. (Eds.). (2009). Expert round table on social media and risk communication during times of crisis: Strategic challenges and opportunities. Special report. Retrieved August 6, 2009, from http://www.boozallen.com/media/file/Risk_Communications_Times_of_Crisis.pdf

Townsend, A. M., & Moss, M. L. (2005). Telecommunications Infrastructure in Disasters: Preparing Cities for Crisis Communications. Graduate School of Public Service, New York University.

Trendberth, K. (2005). Uncertainty in hurricanes and global warming. Science, 308, 1753–1754. doi:10.1126/science.1112551

Tress, B., & Tress, G. (2003). Scenario visualisation for participatory landscape planning—a study from Denmark. Landscape and Urban Planning, 65(3), 161–178. doi:10.1016/S0169-2046(02)00219-0

Troy, D. A., Carson, A., Vanderbeek, J., & Hutton, A. (2008). Enhancing community-based disaster preparedness with information technology. Disasters, 32(1), 149–165. doi:10.1111/j.1467-7717.2007.01032.x

Tsamboulas, D. A., & Mikroudis, G. K. (2006). TRANSPOL: A mediator between transportation model and decision makers' policies. Decision Support Systems, 42(2), 879–897. doi:10.1016/j.dss.2005.07.010

tsunami affected districts of Tamil Nadu. International Journal of Innovation and Sustainable Development, 4.

Turban, E. (2001). Decision Support Systems and Intelligent Systems. New York: Prentice Hall.

Tziros, T. (2008). The riots in new media. Article in Newspaper "Makedonia" on December 15, 2008. [In Greek]. Retrieved August 5, 2009, from http://www.makthes.gr/index.php?name=News&file=article&sid=30225

UN DHA. (1992). Internationally Agreed Glossary of Basic Terms Related to Disaster Management. UN DHA (United Nations Department of Humanitarian Affairs) (1992, December). Retrieved from.

UN/ISDR. (2006). Developing Early Warning Systems: A Checklist. Third International Conference on Early Warning EWC III. Bonn, Germany.

UNISDR (UN International Strategy for Disaster Reduction). (2005). Building the resilience of nations and communities to disasters. In Proceedings of the World Conference on Disaster Reduction (pp. 18-22). January 2005, Kobe, Hyogo, Japan. United Nations Inter-Agency Secretariat of the International Strategy for Disaster Reduction, Geneva.

UN-ISDR. (2007). Words into action: A guide for implementing the Hyogo Framework. Geneve: UN- ISRD.

UN-ISRD. (2006). Let our children teach us!India: Books for Change.

United Nations Development Programme (UNDP). (2002). Deepening democracy in a fragmented world (Human Development Report 2002). New York, USA. Retrieved June 28, 2009, from http://hdr.undp.org/en/reports/global/hdr2002/

United Nations Development Programme (UNDP). (2004). Reducing Disaster Risk: A Challenge for Development (p. 146). New York: United Nations Development Programme, Bureau for Crisis Prevention and Recovery.

United Nations Development Programme (UNDP). (2005). Survivors of the tsunami: One year later. Retrieved June 28, 2009, from http://www.iotws.org/ev_en.php?ID=1685_201&ID2=DO_TOPIC

United Nations Human Settlements Programme (UN-HABITAT). (2006). Retrieved August 25, 2006, from http://www.unhabitat.org/

United Nations International Strategy for Disaster Reduction (UNISDR). (2004). Living With Risk: A Global Review of Disaster Reduction Initiatives.

United Nations. (2005). Hyogo Framework for Action 2005-2015: Building the Resilience of Nations and Communities to Disasters, Geneva: International Strategy for Disaster Reduction [ISDR] World Conference on Disaster Reduction (A/CONF.206/6). Consulta: 2007. Marzo 18. Disponible en: www.unisdr.org.wcdr

United Nations. (2005). UN Millenium Development Goals Report 2005. New York: United Nations. Retrieved May 14, 2006, from: http://unstats.un.org/unsd/mi/pdf/MDG%20Book.pdf

United Nations. (2006). Global Survey of Early Warning Systems. Bonn, Germany: ISDR Platform for the Promotion of Early Warning PPEW.

unpan1.un.org/intradoc/groups/public/.../UN-PAN025913.pdf

Urbina-Medina, H. (2006, Enero 14). Los niños y los desastres. Sociedad de Puericultura y Pediatría. Encartado El Nacional (pp. 12-13).

US Department of Homeland Security (USDHS). (2003). Fact Sheet: National Emergency Training Center. Retrieved May11, 2009 from http://www.dhs.gov/xnews/releases/press_release_0192.shtm.

US Department of Homeland Security (USDHS). (n.d.). About HSEEP. Retrieved March 15, 2009 from https://hssep.dhs.gov/pages/1001_About.aspx.

US Fed News Service. Including US State News. (2008, July 14). Using Web 2.0 & Table Top Exercises in Safety and Health Professional Topics of Upcoming American Society of Safety Engineers' Webinars. Washington, DC.

USGS (United States Geological Survey). (2009). Prompt Assessment of Global Earthquake for Response (PAGER). Retrieved on June 20, 2009, from http://earthquake.usgs.gov/eqcenter/pager/background.php

Vakis, R. (2006). Complementing natural disasters management: the role of social protection. Social Protection Advisory Service, Washington, DC: World Bank. Retrieved June 26, 2009, from http://www.preventionweb.net/english/professional/publications/v.php?id=2491

van der Heijden, K. (2007). Scenarios: The Art of Strategic Conversation. Chichester: Wiley.

Van Nieuwpoort, R. V., Maassen, J., Hofman, R., Kielmann, K., & Bal, H. E. (2002). Ibis: an efficient Java-based grid programming environment. In Proceedings of joint ACM-ISCOPE conference on Java Grande table of contents, Seattle, Washington, USA (pp.18- 27).

Vanneschi, M. (2002). The programming model of AS-SIST, an environment for parallel and distributed portable applications. Parallel Computing, 28(12), 1709–1732. doi:10.1016/S0167-8191(02)00188-6

Varaganti, R. (2009). Personalized Forest Fire Evacuation Grid System. Unpublished MSc Thesis, University of Bedfordshire, UK.

Vasant, P. M., Barsoum, N. N., & Bhattacharya, A. (2008). Possibilistic optimization in planning decision of construction industry. International Journal of Production Economics, 111(2), 664–675. doi:10.1016/j.ijpe.2007.03.006

Vieweg, S. (2008). Social networking sites: Reinterpretation in crisis situations. Workshop on Social Networking in Organizations CSCW08, San Diego USA, November 9. Retrieved August 5, 2009, from http://research.ihost.com/cscw08-socialnetworkinginorgs/papers/vieweg_cscw08_workshop.pdf

Vieweg, S., Palen, L., Liu, S. B., Hughes, A. L., & Sutton, J. (2008). Collective intelligence in disaster: Examination of the phenomenon in the aftermath of the 2007 Virginia Tech shooting. In F. Fiedrich & B. Van de Walle (Eds.), Proceedings of the 5th International IS-CRAM Conference, Washington, DC, USA, May 2008.

Vincent, D., Sherstyul, A., Burgess, L., & Connolly, K. (2008). Teaching mass casualty triage skills using immersive three dimensional virtual reality. Academic Emergency Medicine, 15(11), 1160–1171. doi:10.1111/j.1553-2712.2008.00191.x

Viviroli, D., Zappa, M., Gurtz, J., & Weingartner, R. (2009). An introduction to the hydrological modelling system PREVAH and its pre- and post-processing-tools. Environmental Modelling & Software, 24(10), 1209–1222. .doi:10.1016/j.envsoft.2009.04.001

von Lubitz, D. K. J. E., Beakley, J. E., & Patricelli, F. (2008). All hazards approach' to disaster management: the role of information and knowledge management, Boyd's OODA Loop, and network-centricity. Disasters, 32(4), 561–585.

von Lubitz, D. K. J. E., Beakley, J. E., & Patricelli, F. (2008). Disaster management: The structure, function, and significance of network-centric operations. Journal of Homeland Security and Emergency Management, 5(1). Retrieved August 5, 2009, from http://www.bepress.com/jhsem/vol5/iss1/42/

von Winterfeldt, D., & Edwards, W. (1986). Decision Analysis and Behavioral Research. Cambridge: Cambridge University Press.

von Winterfeldt, D., & Fasolo, B. (2009). Structuring decision problems: A case study and reflections for practitioners. European Journal of Operational Research, 199(3), 857–866. doi:10.1016/j.ejor.2009.01.063

Vroom, V. H., & Yetton, P. W. (1973). Leadership and decision making. Pittsburgh, PA: University of Pittsburgh Press.

Vygotsky, L. S. (1978). Mind and Society. Cambridge, MA: Harvard University Press.

Wagner, M. (2009). Second Life® helps save, improve lives. InformationWeek Healthcare. Retrieved September 20, 2009 from http://www.informationweek.com/news/healthcare/patient/showArticle. jhtml?articleID=22.

Wang, W., Rivera, D. E., & Kempf, K. G. (2007). Model predictive control strategies for supply chain management in semiconductor manufacturing. International Journal of Production Economics, 107(1), 56–77. doi:10.1016/j.ijpe.2006.05.013

WAPMERR (World Agency of Planetary Monitoring and Earthquake Risk Reduction). 2009. Retrieved on June 20, 2009, from http://www.wapmerr.org/qlarm.asp

Ward, P. (2000). Effective Disaster Warnings. Washington, DC: Report of the Working Group on Natural Disaster Information Systems – Subcomitee on Natural Disaster Risk Reduction.

Watthayu, W., & Peng, Y. (2004). A Bayesian network based framework for multi-criteria decision making. In Proceedings of the 17th International Conference on Multiple Criteria Decision Analysis (pp. 6-11).

Weber, Ch. A., Current, J. R., & Benton, W. C. (1991). Vendor selection criteria and methods. European Journal of Operational Research, 50(1), 2–18. doi:10.1016/0377-2217(91)90033-R

Weber, M. (1987). Decision making with incomplete information. European Journal of Operational Research, 28(1), 44–57. doi:10.1016/0377-2217(87)90168-8

Weistroffer, R., Smith, C., & Subhash, N. (2005). Multiple Criteria Decision Support Software . In Figueira, J., Greco, S., & Ehrgott, M. (Eds.), Multiple Criteria Decision Analysis: State of the Art Surveys (pp. 989–1018). New York: Springer.

Wever, N., Lehning, M., Clifton, A., Rüedi, J.-D., Nishimura, K., & Yamaguchi, S. (2009). Verification of moisture budgets during drifting snow conditions in a cold wind tunnel. Water Resources Research, 45. .doi:10.1029/2008WR007522

White, C., Plotnick, L., Aadams-Moring, R., Turoff, M., & Hiltz, S. R. (2008). Leveraging a Wiki to Enhance Virtual Collaboration in the Emergency Domain. In Proceedings of the 41st HICSS.

Willitts-King, B., & Harvey, P. (2005). Managing the risks of corruption in humanitarian relief operations. Overseas Development Institute (ODI). Retrieved June 28, 2009, from http://www.odi.org.uk/publications/index.html

Wilson, H. C. (2000). Emergency response preparedness: small group training. Part I: Training and learning styles. Disaster Prevention and Management, 9(2), 105–116. doi:10.1108/09653560010326987

Wilson, J. M., Polyak, M. G., Blake, J. W., & Collmann, J. (2008). A heuristic indication and warning staging model for detection and assessment of biological events. Journal of the American Medical Informatics Association, 15(2), 158–171. doi:10.1197/jamia.M2558

Winerman, L. (2009). Crisis Communication. Nature, 457(7228), 376–378. doi:10.1038/457376a

World Health Organization, Division of Mental Health. (1992). Psychological Consequences of Disaster: Prevention and Management. Geneva: World Health Organization.

Wyss, M. (2005). Earthquake loss estimates applied in real time and to megacity risk assessment. In B. Van de Walle and B. Carlé (Eds.) Proceedings of the 2nd International ISCRAM Conference, Brussels, Belgium, April 2005 (pp. 297-299).

Yager, R. R. (1987). On the Dempster-Shafer Framework and New Combination Rules. Information Sciences, 41, 93–137. doi:10.1016/0020-0255(87)90007-7

Yellowlees, P. (2009). Virtual reality in medicine – many evolving uses and advantages. Retrieved.

Zappa, M., Pos, F., Strasser, U., Warmerdam, P., & Gurtz, J. (2003). Seasonal water balance of an Alpine catchment as evaluated by different methods for spatially distributed snowmelt modeling. Nordic Hydrology, 34, 179–202.

Zemp, H., & Bonfadelli, H. (2008). Hochwasserereignisse im Spiegel der Presse. In G.R. Be & C. He (Eds.), Ereignisanalyse Hochwasser 2005, Teil 2 – Analyse von Prozessen, Massnahmen und Gefahrengrundlagen (pp. 347-362). Bern, CH: Bundesamt für Umwelt BAFU, Eidgenössische Forschungsanstalt WSL, Umwelt-Wissen, 2508.

Zemp, H., & Bonfadelli, H. (2008). Hochwasserereignisse im Spiegel der Presse. In G.R. Be & C. He (Eds.), Ereignisanalyse Hochwasser 2005, Teil 2 – Analyse von Prozessen, Massnahmen und Gefahrengrundlagen (pp. 347-362). Bern, CH: Bundesamt für Umwelt BAFU, Eidgenössische Forschungsanstalt WSL, Umwelt-Wissen, 2508.

Zervas, E., Mpimpoudis, A., Anagnostopoulos, C., Sekkas, O., & Hadjiefthymiades, S. (2009). Multisensor Data Fusion for Fire Detection and Monitoring. International Journal on Multi-Sensor, Multi-Source Information Fusion.

Zervas, E., Sekkas, O., Hadjiefthymiades, S., & Anagnostopoulos, C. (2007). Fire Detection in the Urban Rural Interface through Fusion techniques, Pisa, Italy. In Proceedings of the 1st International Workshop on Mobile Ad hoc and Sensor Systems for Global and Homeland Security (MASS-GHS), Pisa, Italy.

Zevallos, O. (1996). Ocupacion de laderas: Incremento del riesgo por degradacion ambiental urbana en Quito, Ecuador. In Fernandez, M. A. (Ed.), Ciudades en riesgo: Degradacion ambiental, riesgos urbanos y desastres en America Latina (pp. 165–178). Lima: ITDG Publishing.

Zhanqing, L., Khananian, A., Fraser, R., & Cihlar, J. (2001). Automatic Detection of Fire Smoke using Artificial Neural Networks and Threshold Approaches applied to AVHRR Imagery. IEEE Trans. On Geoscience and Remote Sensing.

Zhao, B., Taucer, F., & Rossetto, T. (2009). Field investigation on the performance of building structures during the 12 May 2008 Wenchuan earthquake in China. Engineering Structures, 31, 1707–1723. doi:10.1016/j.engstruct.2009.02.039

Zimmermann, H. (2005). Recent developments in emergency telecommunications. In B. Van de Walle & B. Carle (Eds.), Proceedings of the 2nd International ISCRAM Conference Brussels, Belgium (pp. 327-334).

Zisman, A. (2000). An Overview of XML. Computing & Control Engineering Journal, 11(4), 165–167. doi:10.1049/cce:20000405

About the Contributors

Eleana Asimakopoulou has a first degree (University of Luton, UK) and an MA (University of Westminster, UK) in Architecture and a PhD (Loughborough University, UK) in managing natural disasters using Grid technology. She is currently a visiting lecturer at the Department of Computer Science and Technology at the University of Bedfordshire. Eleana is a reviewer in several international conferences and journals and her research interests include emergency management, response and planning for disasters, business continuity, construction and risk management, and also advanced ICT methods (such as grid and other forms of applicable collaborative technologies) for disaster management.

Nik Bessis is currently a principal lecturer (associate professor) in the Department of Computer Science and Technology at University of Bedfordshire (UK). He obtained a BA from the TEI of Athens and completed his MA and PhD at De Montfort University (Leicester, UK). His research interest is the analysis, research, and delivery of user-led developments with regard to data integration, annotation, and data push methods and services in distributed environments with a particular focus on next generation and grid technologies for the benefit of various virtual organizational settings, including disaster management. He is involved in and leading a number of funded research and commercial projects in these areas. Dr. Bessis has published numerous papers and articles in international conferences and journals, he serves as a conference organizer and he is the editor of two books and, the Editor-in-Chief of the International Journal of Distributed Systems and Technologies (IJDST).

* * *

Yolanda Barrientos is a teacher of Biology and General Sciences. Universidad Pedagógica Experimental Libertador–Instituto Pedagógico de Caracas (UPEL-IPC). M.Sc. in Biology. Ottawa University. Canada Ph.D in Biology. Saint Andrews University, Scotland. Associated lecturer at UPEl-IPC.

Mathias Bavay, age 33, holds a "Diplôme d'ingénieur" from the French Ecole Supérieure d'éléctricité (SUPELEC 1999). He received his PhD in plasma physics from Paris University in 2002, simulating and designing experiments of electrically driven magnetic flux compression on Z (Sandia National Laboratories), for the "Délégation Générale pour l'Armement". He then led the power generation experimental program for CRT Holdings (Columbus, GA), before joining Ivanhoe Technologies (Albuquerque, NM) in 2004 to work on pulse power projects. He joined the Swiss Federal Institute for Snow and Avalanche Research (Davos) in 2006 and now does snow cover numerical modeling.

Alan N Beard is a reader in Fire Safety Engineering at Heriot-Watt University, Edinburgh, UK. His research is in the broad area of modelling in relation to fire safety, including deterministic and probabilistic modelling, as well as qualitative research, in particular applying the concepts of systems to safety and natural disasters management. More generally, he is helping to develop a framework for the acceptable use of fire models in fire safety decision-making. Dr Beard is Co-Editor of The Handbook of Tunnel Fire Safety, which was published in 2005. He studied physics at Leicester University and in 1972 was awarded a PhD in theoretical physics from Durham University.

Evgenia Belyaeva received her Master's Degree in Linguistics/Translations from the Vyatka State University of Humanities in 2005 (graduated with honours). She has worked as a free-lance translator and a teacher from 2003 to 2005. In 2006 she joined the Optima team at the JRC of the EU as an external consultant, where she sets up multilingual categories, works on evaluations, multilingual resources, helps EMM with her linguistic knowledge and is responsible for the quality of the database for persons and organizations. Shortly she will start a PhD program at the University of Lugano in the field of discourse analysis and online news communication across languages.

Matilde V. de Bezada is a teacher of Biology and General Sciences. Universidad Pedagógica Experimental Libertador–Instituto Pedagógico de Caracas. M.Sc. in Biology. Ottawa University. Canada. Professor at UPEL-IPC.

Stefano Bucci has worked as a freelance developer since 2000. He received his degree as Informatics Engineer (Artificial Intelligence and Robotics) from Politecnico di Milano, Italy, in 2006. He has ten years experience in web applications development mostly using Java and open source tools. He is an expert in Java, J2EE and open source operating systems. In 2008 he joined the EMM/Optima team at the Joint Research Centre working on blog monitoring.

Rui Chen is an assistant professor of Information Systems in Miller College of Business at Ball State University. His research interests are in the areas of information assurance, emergency management, coordination and collaboration, and information technology outsourcing. He has published in Journal of the AIS, Decision Support Systems, Communications of the ACM, Information Systems and e-Business Management, and Journal of Information Privacy and Security. He has presented his research work at conferences such as ICIS and AMCIS, among others. He is a Microsoft Certified System Engineer and Microsoft Certified Database Administrator.

David Chester is a graduate of the Universities of Durham and Aberdeen and has spent most of his academic life at the University of Liverpool. He is currently Reader (Associate Professor) in the Department of Geography, and has worked on volcanoes and volcanic hazards for more than thirty years. A priest in the Anglican (Episcopalian) Church, David Chester has more recently become interested in religious responses to natural disasters.

Tina Comes studied Mathematics, literature and philosophy at the Universität Trier, Germany, Université Lille I, France and the the Friedrich-Alexander Universität Erlangen-Nürnberg, Germany. She holds a diploma in Mathematics from the Friedrich-Alexander Universität Erlangen-Nürnberg. At present, she works as a researcher in the interdisciplinary research unit on 'Technique Assessment and

Risk Management' at the Institute for Industrial Production (IIP) at Karlsruhe Institute of Technology (KIT). Her major research areas are risk management, multi-criteria decision analysis, scenario analysis and intelligent distributed reasoning systems.

Miranda Dandoulaki is an expert in risk management with considerable field experience. She has studied civil engineering (NTUA 1981) and owns a MSc in regional development and a PhD in spatial planning. She has worked for the Greek Earthquake Planning and Protection Organization (1994-2002) and the Joint Research Centre of the European Commission (2004-2008) and has served as Vice Director of the European Centre for the Prevention and Forecasting of Earthquakes. She is presently appointed by the Greek National Centre for Public Administration and Local Government as a scientific and research officer in the Institute of Training. Dr.Dandoulaki has significant research experience and has published in books and scientific journals. Her current research interests lay in multidisciplinarity of disaster management and in spatial aspects of risk management.

Hamdan Al Ghasyah Dhanhani is a Captain in the Dubai Police Force. He has a Masters Degree in Police Studies from the Police Academy, Egypt. He is currently reading for a PhD degree at the University of Bedfordshire on the evaluation of the response capability of the United Arab Emirates to the impact of natural hazards.

Angus Duncan is a graduate of the University of Durham and University College London. He is currently Head of the Research Graduate School at the University of Bedfordshire. He is a co-author (with David Chester and others) of Mount Etna Anatomy of a volcano – 1985 and Volcanoes of Southern Italy – 2003. More recently he has worked in the Azores and Costa Rica on the response of human communities to natural hazards.

Flavio Fuart has worked as Scientific/Technical officer at the Joint Research Centre of the European Commission since 2005, with a one year break when he worked as free-lance consultant for the UN. He develops user interface components, analytical tools for media monitoring and domain-specific derivations of the EMM system. He shows particular interest in MedISys as he had previously worked for ten years as Project Manager, Software Architect and Developer in the R&D department of a leading Slovenian solution provider of information systems for the healthcare sector. Flavio holds a University Degree in Computer Engineering from the University of Ljubljana.

María J. García had combined her managerial experience, mainly from Minimax Consultores, C. A., and educational, mostly from the Universidad Metropolitana of Caracas, Venezuela, to increase his investigations, already above one hundred and sixty, mainly in the areas of Evaluation and Management of Projects, Managerial and Social Decision making, Logistics, Risk Management and Operations research, have been presented or published in different countries, having publications and offering their reports, chats or conferences in: Germany, Italy, Czech Republic, Iceland, Lithuania, Spain, France, Portugal, Panama, Paraguay, Uruguay, Brazil, Cuba, Mexico, Argentina and Chile besides attending as guest speaker, in reiterated occasions, in lectures to relevant events in Colombia, Peru, Costa Rica, Brazil and their own country, Venezuela.

Monica Gemo is working as a scientific technician at the JRC. She got her MSc in electric engineering from Politecnico di Milano, Italy, in 2001. From 2002 to 2007, she has worked as a research engineer at the Telecommunications Laboratory of the Catholic University of Louvain, Belgium, specialising in biomedical applications for emergency health care units. In 2008 she became a member of the OPTIMA team developing tools and services for real-time alerts and world press reviews from internet sources (collation, multilingual information extraction, semi-automatic edition, trend/impact follow-up).

María Angélica González Lara is a teacher of Science Computing. Universidad Pedagógica Experimental Libertador–Instituto Pedagógico de Caracas. Teacher of Computing Laboratory at San Antonio de Papua II School.

Erik van der Goot is one of the original developers of MedISys/EMM and started work on open source monitoring in 2002. He has developed many of the EMM components, is responsible for the overall system design and is still actively developing software. He currently leads a team of 20 developers and researchers working on open source text information mining and analysis. His original background is in Mechanical Engineering, he holds an MSc in Computer Science and has been developing software since 1978. Erik is Dutch and has worked in the Netherlands, the UK and for the last 14 years at the JRC in Italy.

Stathes Hadjiefthymiades received his B.Sc., M.Sc. and Ph.D. degrees in Informatics from the Dept. of Informatics and Telecommunications, University of Athens (UoA). He also received a Joint Engineering-Economics M.Sc. from the National Technical University of Athens. In 1992 he joined the Greek consulting firm Advanced Services Group. In 1995 he joined the Communication Networks Laboratory (CNL) of UoA. During the period 2001–2002, he served as a visiting assistant professor at the University of Aegean, Dept. of Information and Communication Systems Engineering. On the summer of 2002 he joined the faculty of the Hellenic Open University, Patras, Greece, as an assistant professor. Since December 2003, he is in the faculty of the Dept. of Informatics and Telecommunications, University of Athens, where he is presently an assistant professor. He has participated in numerous EU & National projects. His research interests are in the areas of web engineering, mobile/pervasive computing and networked multimedia. He has contributed to over 150 publications in these areas. Since 2004 he co-ordinates the Pervasive Computing Research Group of CNL.

Matina Halkia is an architect engineer and human computer interaction expert. Her core research interests lie in augmented reality/ICT applications in physical/architectural space. She is currently working on emergency response map validation and urban analysis for remote sensing applications in human security contexts. She has studied History of Art and Architecture at Tufts University under a Fulbright Scholarship and Media Arts and Sciences at the Massachusetts Institute of Technology. She has worked as an interaction designer in Boston (1997-1998), and as principal investigator at Starlab Laboratories, a technology think-tank, in Brussels (1998-2001). Since 2001 she works for the Joint Research Centre of the European Commission, currently at the Global Security and Crisis Management Unit, as a scientific officer.

José Hernández had combined his managerial experience, mainly from Minimax Consultores, C. A., and educational, mostly from the Universidad Metropolitana of Caracas, Venezuela, to increase his investigations, already above one hundred and sixty, mainly in the areas of Evaluation and Management of Projects, Managerial and Social Decision making, Logistics, Risk Management and Operations research, have been presented or published in different countries, having publications and offering their reports, chats or conferences in: Germany, Italy, Czech Republic, Iceland, Lithuania, Spain, France, Portugal, Panama, Paraguay, Uruguay, Brazil, Cuba, Mexico, Argentina and Chile besides attending as guest speaker, in reiterated occasions, in lectures to relevant events in Colombia, Peru, Costa Rica, Brazil and their own country, Venezuela.

Anne M. Hewitt (PhD) is an Associate Professor in the Master of Healthcare Administration program at Seton Hall University where she is also Director of the Seton Center for Community Health. She was selected as a national Community-Campus Partnership for Health Fellow and has received grants from both federal agencies and non-profit foundations. She has numerous publications and presents at both national conferences and virtual events. Her research interests include public health policy, nonprofit agency evaluations and innovations in health professions teaching.

Michael Hiete holds a diploma and PhD in Geoecology, both from the Technical University Carolo-Wilhelmina at Brunswick, Germany. He is a Scientific Research Assistant and the Head of the inter-disciplinary research team on 'Technique Assessment and Risk Management' at the IIP, the University of Karlsruhe (TH), Germany. His major research areas include the techno-economic assessment of emission abatement options, the modelling and analysis of critical infrastructure, risk management and multi-criteria decision analysis.

Ana T. Iztúriz Moreau is a teacher of Earth's Sciences. Universidad Pedagógica Experimental Libertador – Instituto Pedagógico de Caracas (UPEL-IPC). M.Sc. and Dr. in Geological Sciences. Universidad Central de Venezuela. Facultad de Ingeniería (UCV-FI). Head of the Research Center Estudio del Medio Físico venezolano. Department of Ciencias de la Tierra (UPEL-IPC). Professor at UPEL-IPC.

Delilah Al Khudhairy graduated with a first class honours BSc (Eng.) and PhD degrees in Materials Science Engineering from London University. She was employed at British Gas, UK, before joining the Joint Research Centre, European Commission, in 1990. Her research areas have covered x-ray and electron scattering (scanning and transmission), x-ray diffraction, numerical modelling, and remote sensing. She has been the head of the Global Security and Crisis Management, GLOBESEC, Unit at the Joint Research Centre since 2004. Today, the GLOBESEC Unit specialises in open source text information mining and analysis, geo-spatial information extraction and analysis, statistical data mining and analysis, and collaborative crisis management technologies. On the basis of this expertise, the GLOBESEC Unit provides scientific and technical support to EU policies concerned with crisis management.

Pierre Kuonen obtained a Master degree in electrical engineering from the Swiss Federal Institute of Technology (EPFL) in 1982. After six year of experience in industry he joined the Computer Science Theory Laboratory at EPFL in 1988 and received his Ph.D degree in 1993. From that time, he steadily worked in the field of parallel and distributed computing. First at EPFL where he founded and managed the GRIP (Parallel Computing Research Group) then at the University of Applied Science of Valais. Since 2003 he is full professor at the University of Applied Science of Fribourg in the Information and Communication technologies department where he is leading the GRID & Ubiquitous Computing Group.

Sunitha Kuppuswamy is a Lecturer of Media Sciences in Anna University Chennai. She had been serving as one of the content providers for Anna FM – India's First Campus Community Radio. She is also a Consultant for the United Nations Development Programme (UNDP). She's certified by the Centre for Disaster Management, LBSNAA as an Incident Manager after the completion of 'Certified Incident Managers' Programme' in 2007. She has been trained by ADRC, Japan & NDMA, India on Community Based Hazard Map Development in 2007. Her areas of academic interest include Science Communication, Community Based Disaster Risk Management, Development Communication, Environmental Communication, and Media & Disaster Management. She has published two books Tally (TNCDW 2006), This is how we communicate: ICT initiatives by NGOs in tsunami affected areas of Tamilnadu (TNTRC 2007). She has published research papers and articles in peer reviewed journals. She has presented papers at several National and International Conferences. Ms. Sunitha has undertaken an ICT Study, for the United Nations Development Programme (UNDP), in 2007-2008. The study focuses on Communication Needs for Coastal Communities in tsunami affected districts of Tamilnadu. She is a member of the Indian Broadcasting Society & Indian Science Congress Association.

JinKyu Lee is Assistant Professor of Management Science and Information Systems in Spears School of Business, Oklahoma State University. His current research interests include adoption of information & communication technologies for public/commercial services, information security and privacy, and inter-organizational information sharing. Prof. Lee has published in various journals and conference proceedings such as Decision Support Systems (DSS), Communications of the ACM (CACM), Information Systems Frontiers, IEEE Transactions on Systems, Man and Cybernetics (SMC), International Conference on Information Systems (ICIS), and Hawaii International Conference on System Sciences (HICSS).

Michael Lehning is the Head of the "Snow and Permafrost" Research Unit at SLF Davos and an Atmospheric Physicist. His major research areas include: Interaction Snow - Atmosphere; Natural Hazards Forecasting; Turbulence and Boundary Layer Flows and Mountain Hydrology. He is currently responsible for the development of the "Swiss Experiment," an ETH CCES activity to install a large and common platform for environmental science observations and modeling. He has started and lead the developed the well-known model systems SNOWPACK and Alpine3D. Other work includes the investigation of (snow) precipitation and its distribution in Alpine terrain, particle transport at the earth's surface and climate change scenario calculations.

Jens P. Linge is a scientific officer at the Joint Research Centre of the European Commission focusing on open-source media monitoring in the area of public health. Before joining the European Commission, he has worked in the structural bioinformatics unit at the Pasteur Institute in Paris and for a renowned patent law firm in Munich. From 1997 to 2000, he has been a Ph.D. student at the European Molecular Biology Laboratory (EMBL) in Heidelberg. Jens holds a master's degree in physics and a master's degree in biochemistry, both obtained at the Ruhr-University Bochum in Germany.

Dimitrios V. Manatakis was born in Athens, Greece. He is currently, a Ph.D. candidate at the National and Kapodistrian University of Athens, Department of Informatics & Telecommunications. He has obtained with honors the M.Sc. degree in Signal Processing for Communications and Multimedia from the same Department (2007), and the BEng degree from the Department of Electronics, Technological Educational Institution (TEI) of Lamia (2004). In 2004 Dimitrios was awarded a prize from the Greek State Scholarships Foundation after ranking among the first three students in his class. He has participated as researcher in the EC founded FP-6 research project SCIER (2006-2008). His research interests are mainly in Statistical and Distributed Signal Processing, Wireless Sensor Networks, Data Fusion, Machine Learning, Estimation Theory and their applications to Environmental Risk Management.

Elias S. Manolakos is currently an Assoc. Professor with the University of Athens, Dept. of Informatics and Telecommunications. He has received the Dipl. Ing. degree from NTU Athens, the MSc in Electrical Engineering from U. of Michigan, Ann Arbor and the PhD from U. of Southern California, Los Angeles (1989). Before returning to Greece (2004) he has been with the faculty of the ECE Dept. Northeastern University, Boston (Assoc. Professor with tenure). His research in the US has been supported by NSF, DARPA, EPA, and recently by EC (FP6, FP7) and GSRT Greece. His research interests are in the areas of signal/image processing systems (parallel and distributed DSP, embedded DSP, sensor networks), machine learning, and their applications in multi-disciplinary domains, such as computational biology, ecological modeling and environmental risk management. Prof. Elias S. Manolakos has authored, or coauthored with his students, more than 100 refereed publications in journals and conference proceedings. He is a Senior Member of the IEEE and has organized several IEEE Conferences and Workshops. Prof. Manolakos has served in the Editorial Board of the IEEE Trans. on Signal Processing, IEEE Signal Processing Letters, Journal of Signal Processing Systems, Springer Publ. etc. E-mail: eliasm at di.uoa.gr.

Ulrich Meissen is a research scientist at Fraunhofer ISST where he heads the research group on Early Warning Systems. He received his Master degree in industrial engineering from the University of Karlsruhe and he spent one year at Trinity College Dublin during his studies. His main subjects during his studies were operations research, artificial intelligence, and environmental-engineering. After his studies he worked as an IT-Manager for the Versicherungskammer Bayern, managing development projects in the area of applications for damage estimation. Since 2000 he has been working as a research scientist at the Fraunhofer ISST, where his main research fields are Information Logistics, Location-based Services and Early Warning Systems. His ongoing Ph.D. work is about effective alerting in Early Warning Systems for better disaster mitigation. In the recent years he was responsible for several applied research projects in the field of Early Warning and Disaster Alert Systems, one of which evolved to the largest commercial weather warning system in Europe with currently 420.000 subscribed users in Germany, Austria, Switzerland, Poland and Hungary. Currently he is responsible for the central systems engineering and integration of the German-Indonesian Tsunami Early Warning System.

Danielle Mirliss an Associate Director at Seton Hall University's Teaching, Leaning and Technology Center. She is responsible for overseeing the Instructional Design and Training team in a variety of faculty and student support initiatives. She has been part of the Instructional Design team at Seton Hall for six years and has a broad range of experience in mobile computing, pen-based computing, course redesign, online learning, virtual worlds and assessment. Before coming to Seton Hall, she was an Instructional Designer at The Distance Learning Project, an online division of Columbia University. Mirliss is a candidate for the Ed.D in Instructional Technology and Media from Teachers College, Columbia University. She also received her Ed.M from the same program and her M.A. in Industrial and Organizational Psychology from New York University.

Peter Noringhton received his PhD from the University of Bedfordshire in 2009 with a thesis in cognitive authentication techniques, where he currently works as e-PDP Development Officer. He has worked in the education, hospitality and journalism sectors. His research interests center around cooperative and collaborative systems and usability.

Kumaresh Rajan has an MBA from SUNY Buffalo with concentrations in Finance and Information Assurance. He also possesses a Bachelor's of Science Degree in Computer Science from the University of Oklahoma. Professionally he has worked as a software engineer in the consulting and financial sector. His interests include developing enterprise J2EE web applications and utilizing Web 2.0 technologies to create rich internet applications. He has independently created a web application, www.kanakku. com, utilizing Ajax and JSF.

Hejamadi Raghav Rao is a Professor of MIS at SUNY Buffalo. He graduated from Krannert Graduate School of Management, Purdue University with a PhD. he received an MBA from Delhi University and a BTech in Chemical Engineering from IIT Kanpur.

Larry Rivas is a teacher of Earth's Sciences. Universidad Pedagógica Experimental Libertador – Instituto Pedagógico de Caracas (UPEL-IPC). M.Sc. in Physical Geography (Candidate). Laboratory assistant at Department of Ciencias de la Tierra (UPEL-IPC).

Simón Ruíz B is a teacher in Chemistry and General Sciences. Universidad Pedagógica Experimental Libertador–Instituto Pedagógico de Caracas (UPEL-IPC). M.Sc. in Chemistry. Instituto Venezolano de Investigaciones Científicas (IVIC). Dr. in Education. (UPEL-IPC). Associated lecturer at UPEl-IPC.

Aysu Sagun is the Pathway Leader in Interior Design in the Cambridge School of Arts at Anglia Ruskin University. She received her PhD, MFA and BFA degrees from the Department of Interior Architecture and Environmental Design at Bilkent University. She taught design studio, drawing and CAD in architecture and interior architecture departments. Her research interests include crowd dynamics, disaster management, ICT, virtual environments, web-based design education and collaborative design studios. She has publications in proceedings of international conferences and The Journal of Disaster Studies, Policy and Management, International Journal of Design and Technology Education, and Journal of Art and Design Education.

Jaime Santos-Reyes is a lecturer at Instituto Politecnico Nacional, Mexico, whose main research interests are safety management systems, accident and risk analysis, reliability engineering and natural disasters. He obtained a PhD from Heriot-Watt University, UK, in 2001. Since then he has conducted research on safety management on offshore installations, on the UK railway network, in tunnels and on natural disasters. Jaime has a degree in Mechanical Engineering from the IPN, Mexico and an MSc in Thermal Power and Fluid Engineering from UMIST, UK. He has spent 3 years working as a RA at Heriot-Watt University. He also spent some years working in the oil and gas industry.

Frank Schultmann is Professor at the Karlsruhe Institute of Technology (KIT) and Director of the Institute for Industrial Production (IIP) and the French-German Institute for Environmental Research (DFIU). In addition, he is Adjunct Professor at the University of Adelaide, Australia. He studied Business

Engineering at the Universität Karlsruhe (TH) and received a Ph.D. in Economics from the Faculty of Economics and Business Engineering of the Universität Karlsruhe (TH). Previous to his present positions he was Professor at the Department of Computer Science at the University of Koblenz-Landau and holder of the Chair of Business Administration, Construction Management and Economics at the University of Siegen.

Odysseas Sekkas received his B.Sc. in Informatics from the Department of Informatics & Telecommunications at the National and Kapodistrian University of Athens, Greece in 2003. Afterwards, he attended the Postgraduate Program of the same Department and received his M.Sc. in Communication Systems and Data Networks in 2005. Nowadays he is a Ph.D. candidate in the above Department in the research area of Pervasive Computing and especially in Context-Aware Data Management. He is also member of the Pervasive Computing Research Group (p-comp) and the Communication Networks Laboratory (CNL) of the National and Kapodistrian University of Athens. He has participated in several projects founded by European and National founded research projects. His research interests are mainly focused in the areas of pervasive and mobile computing, wireless sensor networks and data fusion.

Susan Spencer is the Program Director for the online Master of Healthcare Administration degree program with the College of Arts and Sciences at Seton Hall University in South Orange, New Jersey. Susan has a Ph.D. in Education Leadership, Management and Policy from Seton Hall University and a Masters degree in Counseling from the University of South Florida. Before coming to Seton Hall University, she held faculty and counseling positions as well as a Student Services Directorship position. Currently, Susan has developed online courseware and also teaches research methods and statistics online for the MHA program.

Ralf Steinberger is a Research Scientist in the field of Language Technology at the European Commission's Joint Research Centre in Ispra, Italy. He received his M.A. degree (1991) in Theoretical Linguistics at LMU University in Munich, Germany, and his Ph.D. (1994) in Computational Linguistics at the University of Manchester Institute of Science and Technology (UMIST), United Kingdom. Before joining the European Commission in 1998, he worked at the Institute for Applied Information Sciences IAI in Germany, at UMIST, at the Kyushu Institute of Technology in Japan, and at Sharp Laboratories of Europe in Oxford, UK. His research interests include multilingual and cross-lingual text analysis applications, with a focus on methods to provide cross-lingual information access and to help users digest large multilingual document collections.

Jishnu Subedi Ph.D. in Civil Engineering and MBA, is currently working as Chief of Research and Training Unit of Department of Civil Engineering at Institute of Engineering, Tribhuvan University. Before joining the institute, he worked as Researcher in the Disaster Management and Planning Unit of United Nations Centre for Regional Development. He also served as Vice-Principal (2006-2007) and Coordinator for Masters in Disaster Risk Management (2004-2007) at Nepal Engineering College. He has visited Network of Humanitarian Assistance Universities in Europe and Ehime University in Japan as Research Scholar in the year 2007 and 2009 respectively.

Riad Twal is an Instructional Designer with the Teaching, Learning, and Technology Center at Seton Hall University. He supports faculty members with the design, integration and assessment of technology resources for courses and programs. While completing his graduate work, Riad was an associate instructor at Indiana University, and taught various technology courses in the teacher education program. Riad is a graduate Fairfield University, where he earned an M.A. in Media and Educational Technology, and Indiana University, where he earned an M.S. and Ed.S. in Instructional Systems Technology.

Ravikanth Varaganti graduated in Computer Science and Engineering in 2007. In 2009, he received his MSc in Computer Science from University of Bedfordshire. Currently, he is a PhD student at the University of Bedfordshire. His research interests include Grid Computing, Virtual Organizations and Distributed-Parallel computing. He is also working as a part-time lab demonstrator at University of Bedfordshire teaching Java and UML to undergraduate students.

Agnès Voisard heads the "Location-based Services" Department at Fraunhofer ISST and is an adjunct faculty at the Free University of Berlin, Department of Computer Science. She received her Ph.D. degree in computer science from the University of Paris at Orsay (Paris XI) and INRIA (French Institut National Recherche en Informatique et en Automatique). After a post-doctoral year at the University of Munich (LMU) she was appointed Assistant Professor of Computer Science at the Free University of Berlin where she obtained her 'habilitation'. She then worked at Kivera (a startup on navigation systems), Oakland, California, as its system architect. In February 2003 she joined Fraunhofer ISST. Her areas of expertise include spatial databases, event notification systems, location-based services, and interoperability in information systems. She has published many articles on these topics at international conferences and has participated in numerous program committees. She is a member of various editorial boards including GeoInformatica, was General Chair of the 5th International Symposium on Spatial Databases (SSD'97), and Program co-Chair of the 10th ACM GIS conference. She co-authored the book "Spatial Database – with application to GIS", Morgan Kaufmann/Elsevier, 2001 and co-edited the book "Location-based Services", Morgan Kaufmann/Elsevier, 2004.

Niek Wijngaards received his PhD in 1999 on the topic of self-modifying agent systems using a re-design process. Since 1998 he worked at the University of Canada as a Postdoctoral-Fellow and at the VUA, where he was an assistant professor from 2000 to 2004 at the Intelligent Interactive Distributed Systems group. Since October 2004 he works for Thales Research & Technology Netherlands as senior researcher and program manager. He is fully employed at D-CIS Lab. Wijngaards is involved in research on actor-agent teams as well as their practical applications at e.g. the Dutch Police organisation and the Dutch Railroads.

Roman Yangarber obtained his MS and PhD (2000) from New York University (NYU), USA, in Computer Science, concentrating in Computational Linguistics. Prior to moving to Finland in 2004, he was Assistant Research Professor at the Courant Institute of Mathematical Sciences, NYU, where he specialized in natural language processing, and machine learning for automatic acquisition of semantic knowledge from plain text, in particular, from news streams. He has been an organizer, editorial board member and program committee member for a number of scientific events, conferences, organizations and journals, and has served on evaluation panels for funding agencies. Since coming to work at the University of Helsinki, he has held the post of Acting Professor, and currently leads and participates in several research projects, in text mining and linguistic analysis.

Helena Zemp is currently a PhD. candidate and research assistant at the Institute of Mass Communication and Media Research (IPMZ) at the University of Zürich in Switzerland. Her research interests and current studies focus on media coverage of natural disasters and risk communication. The research is interdisciplinary, combining media content analysis with audience research and the historical context; thus, providing an under-examined perspective on natural hazard communications. The research addresses the implications of effective communication with regard to natural hazards in a fast changing media environment and new ICTs for risk awareness and crisis management. In this capacity Ms. Zemp has contributed to a number of interdisciplinary forums focussed on both studying disaster communications and developing frameworks for crisis management. The research project is co-funded by the Federal Office for the Environment (FOEN), the University of Zürich and a private environmental foundation.

Index

A

Activity Theory 164, 165, 169, 170, 171, 175, 177, 178
Alerting Technologies (AT) 108, 121
Alpine3D 237, 240, 241, 242, 243, 245, 258, 260
avalanche 241

B

blog 137, 138, 143, 144, 145, 149, 150, 151, 152, 153, 154, 161, 162, 167, 168, 171, 172, 174

C

C++ 237, 240, 244, 245, 248, 249, 250, 251, 252, 253, 254, 255, 256, 257, 258, 260
calibration 238, 239
catastrophe 36, 150, 201, 202, 203, 209, 210, 211, 212, 214, 215
chemical, biological, radiological and nuclear (CBRN) 131
chemical refineries 222
Citizen Owned Sensors (COS) 264
civic services 66
Civil works 212
climate change 80
cognitive learning 35, 38
collaborative intelligence systems 143, 144
Common Object Request Broker Architecture (CORBA) 245, 246, 260
communication infrastructure 96, 102
communication system 96, 98
Computing Subsystem (CS) 267, 268, 273
crisis communication 179, 180, 181, 183, 191, 192, 195

crisis communication skill 179, 183, 192
crisis management 143, 144, 145, 146, 147, 148, 149, 152, 153, 154, 155, 156, 157, 159
cyclone 22, 23, 24, 25, 30, 77, 116

D

Danish Emergency Management Agency (DEMA) 223, 232
Data Distribution System (DDS) 30
Data Management System (DMS) 30
decision making 221, 222, 223, 224, 228, 229, 231, 232, 235
Decision Support System (DSS) 30, 201, 202, 203, 208, 209, 210, 211, 212, 213, 215, 220, 286, 287, 335
decision tree 211, 217
Decreasing digits (Dd) 202, 203, 209, 210
Department of Homeland Security (DHS) 99
developed world 237
digital divide 27, 28
Digital Subscriber Line (DSL) 28, 118
digital technology 48, 182
disaster aware 26
disaster communication 83
disaster informatics 80, 82, 91, 93
disaster management 1, 2, 3, 4, 6, 7, 8, 10, 12, 13, 16, 19, 20, 26, 27, 29, 30, 33, 34, 43, 48, 49, 50, 51, 58, 59, 60, 70, 75, 76, 81, 82, 83, 85, 87, 90, 91, 92, 95, 96, 97, 98, 99, 100, 101, 102, 103, 104, 105, 106, 112, 122, 144, 158, 160, 161, 164, 166, 168, 170, 171, 172, 173, 175, 176, 177, 178, 179, 180, 181, 182, 184, 185, 186, 194, 195, 198, 282, 285, 330, 332, 337